Wastewater Quality Monitoring and Treatment

Water Quality Measurements Series

Series Editor

Philippe Quevauviller
European Commission, Brussels, Belgium

Published Titles in the Water Quality Measurements Series

Hydrological and Limnological Aspects of Lake Monitoring
Edited by Pertti Heinonen, Giuliano Ziglio and André van der Beken

Quality Assurance for Water Analysis
Authored by Philippe Quevauviller

Detection Methods for Algae, Protozoa and Helminths in Fresh and Drinking Water
Edited by André van der Beken, Giuliano Ziglio and Franca Palumbo

Analytical Methods for DrinkingWater: Advances in Sampling and Analysis
Edited by Philippe Quevauviller and K. Clive Thompson

Biological Monitoring of Rivers: Applications and Perspectives
Edited by Giuliano Ziglio, Maurizio Siligardi and Giovanna Flaim

Wastewater Quality Monitoring and Treatment
Edited by Philippe Quevauviller, Olivier Thomas and André van der Beken

Forthcoming Titles in the Water Quality Measurements Series

Rapid Chemical and Biological Techniques for Water Monitoring
Edited by Philippe Quevauviller, Catherine Gonzalez and Richard Greenwood

Wastewater Quality Monitoring and Treatment

PHILIPPE QUEVAUVILLER
European Commission, DG Environment, Brussels, Belgium

OLIVIER THOMAS
Environment and Sustainable Development Institute, University of Sherbrooke, Quebec, Canada

ANDRÉ VAN DER BEKEN
Department of Hydrology and Hydraulic Engineering, Free University of Brussels, Brussels, Belgium

John Wiley & Sons, Ltd

Telephone (+44) 1243 779777

Email (for orders and customer service enquiries): cs-books@wiley.co.uk
Visit our Home Page on www.wileyeurope.com or www.wiley.com

Other Wiley Editorial Offices

John Wiley & Sons Inc., 111 River Street, Hoboken, NJ 07030, USA

Jossey-Bass, 989 Market Street, San Francisco, CA 94103-1741, USA

Wiley-VCH Verlag GmbH, Boschstr. 12, D-69469 Weinheim, Germany

John Wiley & Sons Australia Ltd, 42 McDougall Street, Milton, Queensland 4064, Australia

John Wiley & Sons (Asia) Pte Ltd, 2 Clementi Loop #02-01, Jin Xing Distripark, Singapore 129809

John Wiley & Sons Canada Ltd, 6045 Freemont Blvd, Mississauga, Ontario, L5R 4J3, Canada

Wiley also publishes its books in a variety of electronic formats. Some content that appears in print may not be available in electronic books.

Library of Congress Cataloging-in-Publication Data

Wastewater quality monitoring and treatment / [edited by] Philippe Quevauviller, Olivier Thomas, André van der Beken.
p. cm.
Includes bibliographical references and index.
ISBN-13: 978-0-471-49929-9
ISBN-10: 0-471-49929-3
1. Sewage – Purification – Quality control. 2. Water quality – Measurement.
3. Water quality management. I. Quevauviller, Ph. II. Thomas, Olivier. III. Beken, André van der.

TD745.W345 2006
628.1′68 – dc22 2006018012

British Library Cataloguing in Publication Data

A catalogue record for this book is available from the British Library

ISBN-10: 0-471-49929-3
ISBN-13: 978-0-471-49929-9

Typeset in 10.5/12.5pt Times New Roman by TechBooks, New Delhi, India
Printed and bound in Great Britain by TJ International Ltd, Padstow, Cornwall
This book is printed on acid-free paper responsibly manufactured from sustainable forestry in which at least two trees are planted for each one used for paper production.

Contents

Wastewater Quality Monitoring and Treatment Edited by P. Quevauviller, O. Thomas and A. van der Beken

Series Preface

Water is a fundamental constituent of life and is essential to a wide range of economic activities. It is also a limited resource, as we are frequently reminded by the tragic effects of drought in certain parts of the world. Even in areas with high precipitation, and in major river basins, over-use and mismanagement of water have created severe constraints on availability. Such problems are widespread and will be made more acute by the accelerating demand on freshwater arising from trends in economic development.

Despite the fact that water-resource management is essentially a local, river-basin-based activity, there are a number of areas of action that are relevant to all or significant parts of the European Union and for which it is advisable to pool efforts for the purpose of understanding relevant phenomena (e.g. pollution, geochemical studies), developing technical solutions and/or defining management procedures. One of the keys for successful cooperation aimed at studying hydrology, water monitoring, biological activities, etc., is to achieve and ensure good water quality measurements.

Quality measurements are essential to demonstrate the comparability of data obtained worldwide and they form the basis for correct decisions related to management of water resources, monitoring issues, biological quality, etc. Besides the necessary quality control tools developed for various types of physical, chemical and biological measurements, there is a strong need for education and training related to water quality measurements. This need has been recognized by the European Commission which has funded a series of training courses on this topic, covering aspects such as monitoring and measurements of lake recipients, measurements of heavy metals and organic compounds in drinking and surface water, use of biotic indexes, and methods to analyse algae, protozoa and helminths. In addition, series of research and development projects have been or are being developed.

This book series will ensure a wide coverage of issues related to water quality measurements, including the topics of the above-mentioned courses and the outcome of recent scientific advances. In addition, other aspects related to quality control tools (e.g. certified reference materials for the quality control of water analysis) and

Wastewater Quality Monitoring and Treatment Edited by P. Quevauviller, O. Thomas and A. van der Beken

monitoring of various types of waters (river, wastewater, groundwater) will also be considered.

This book *Wastewater Quality Monitoring and Treatment* is the sixth one of the series; it has been written by experts in wastewater policy, treatment and analytical science and offers the reader an overview of existing knowledge and trends in wastewater monitoring features.

The Series Editor – Philippe Quevauviller

Preface

The European Community decided in 1991 to obligate all the Member States to be equipped with wastewater treatment plants for all the cities whose wastewater organic loads are greater than 15 000 equivalent-inhabitants, before the 31st December 2000, and 2000 equivalent-inhabitants before the 31st December 2005. In this context, the quality of the treated wastewater must be better than reference values for some variables such as BOD (biological oxygen demand), COD (chemical oxygen demand), TSS (total suspended solids), global nitrogen and total phosphorus. These obligations generate a huge range of activities within the European Union, including research and technological developments, and similar trends can be observed, e.g. in the USA and Canada.

Unfortunately, wastewater monitoring procedures are prone to many drawbacks because of difficulties to accurately and frequently measure the necessary variables, which essentially rely on 'classical' monitoring approaches involving sampling, storage and laboratory analysis. The only way to make progress in wastewater treatment (and hence to comply with related regulations) is to ensure that the plants are able to work with unqualified reliability which implies that reliable monitoring of the wastewater quality and quantity and of the treatment efficiency should be performed for the characterisation of raw and treated wastewaters and for the control of the plant itself.

This book reflects this awareness by summarising different views on wastewater treatment-related monitoring and control. The book is composed of six different parts. The first part provides an overview of EU and US wastewater policies, standard methodologies, reference materials and discusses sampling assistance, biosensors and alternative methods. Sewer quality control is examined in the second part, including considerations on sewage characterisation and evolution, flow measurements and monitoring in rural areas. This is followed, in the third part, by chapters concerning urban wastewater treatment plant control and, in the fourth, by industrial wastewater treatment plant control. Part 5 discusses monitoring in the context of discharges and receiving medium, including water quality modelling. Finally,

Wastewater Quality Monitoring and Treatment Edited by P. Quevauviller, O. Thomas and A van der Beken

socio-economic aspects are considered in the sixth part, with a focus on data collection and merging, as well as training.

This book has been written by experts in the field of wastewater treatment policy, control and monitoring. It provides an overview of the existing knowledge in wastewater monitoring and identifies emerging needs, which will be of direct interest to policy makers, water scientists and industries, and analytical control laboratories.

Philippe Quevauviller, Olivier Thomas and André van der Beken

List of Contributors

Av Aharoni	Mekorot Ltd, 9 Lincoln St, PO Box 20128, Tel-Aviv 61201, Israel
Gianni Andreottola	Department of Civil and Environmental Engineering, University of Trento, Via Mesiano 77, 38050 Trento, Italy
Juan Azcárate	Ayuntamiento de Madrid, Departamento de Aguas y Saneamiento, C/Barceló, 6 1°, 28004 Madrid, Spain
Olivier Bernard	INRIA - Projet COMORE, BP93, 06902 Sophia-Antipolis Cedex, France
Davide Bixio	Aquafin NV, Dijkstraat 8, 2630 Aartselaar, Belgium
Carmen Cámara	Departamento de Química Analítica, Facultad de Ciencias Químicas, Universidad Complutense de Madrid, Ciudad Universitaria s/n, 28040 Madrid, Spain
Jean-Luc Cécile	Institut de Régulation et d'Automation, 23 Chemin des Moines, 13200 Arles, France
Victor Cerdà	Department of Chemistry, University of Balearic Islands, Carretera de Valldemossa, km 7.5, 07122 Palma de Mallorca, Spain
Benoît Chachuat	Ecole Polytechnique Fédérale de Lausanne, Laboratoire, d'Automatique, Station 9, 1015 Lausanne, Switzerland
Haim Cikurel	Mekorot Ltd, 9 Lincoln St, PO Box 20128, Tel-Aviv 61201, Israel

Wastewater Quality Monitoring and Treatment Edited by P. Quevauviller, O. Thomas and A. van der Beken

Martijn Devisscher	Dijkstraat 8, 2630 Aartselaar, Belgium
Christian Dietz	Departamento de Química Analítica, Facultad de Ciencias Químicas, Universidad Complutense de Madrid, Ciudad Universitaria s/n, 28040 Madrid, Spain
Estelle Dupuit	24 rue de Gerofosse, 91150 Etampes, France
José M. Estela	Department of Chemistry, University of Balearic Islands, Carretera de Valldemossa, km 7.5, 07122 Palma de Mallorca, Spain
Paola Foladori	Department of Civil and Environmental Engineering, University of Trento, Via Mesiano 77, 38050 Trento, Italy
Ann van Griensven	BIOMATH, Ghent University, Coupure Links 653, 9000 Ghent, Belgium
Greet De Gueldre	Aquafin NV, Dijkstraat 8, 2630 Aartselaar, Belgium
Jaap De Koning	Delft University of Technology, Stevinweg 1, PO Box 5048, GA Delft 2600, The Netherlands
Claudia Lasagna	AMGA S.p.A., via Piacenza 54, 16138, Genova, Italy
Yolanda Madrid	Departamento de Química Analítica, Facultad de Ciencias Químicas, Universidad Complutense de Madrid, Ciudad Universitaria s/n, 28040 Madrid, Spain
Geneviève Marcoux	Civil Engineering Department, Sherbrooke University, 2500 bld de l'Université, Sherbrooke, Quebec, J1K 2R1, Canada
Charles S. Melching	Department of Civil and Environmental Engineering, Marquette University, PO Box 1881, Milwaukee, WI 53201-1881, USA
Thomas Melin	RWTH Aachen University, Turmstrasse 46, Aachen 52056, Germany
Marie-Noëlle Pons	Laboratoire des Sciences du Genie Chimique, LSGC - CNRS-Groupe ENSIC, 1 rue Grandville, BP 20451, 54001 Nancy Cedex, France

Olivier Potier Laboratoire des Sciences du Genie Chimique, LSGC - CNRS-Groupe ENSIC, 1 rue Grandville, BP 20451, 54001 Nancy Cedex, France

Marie-Florence Pouet Environment and Sustainable Development Institute, Sherbrooke University, 2500 bld de l'Université, Local A6-1021, Sherbrooke, Quebec, J1K 2R1, Canada

Philippe Quevauviller European Commission, DG Environment (BU9 3/142), Rue de la Loi 200, 1049, Brussels, Belgium

Aldo Ravazzini Delft University of Technology, Stevinweg 1, PO Box 5048, GA Delft 2600, The Netherlands

Carmen Rebollo Ayuntamiento de Madrid, Departamento de Aguas y Saneamiento, C/Barceló, 6 1°, 28004 Madrid, Spain

Michael J. Scott Process Measurement Technology, 27 West Green, Barrington, Cambridge CB2 SRZ, UK

Boudewijn Van De Steene Aquafin NV, Dijkstraat 8, 2630 Aartselaar, Belgium

Jean-Philippe Steyer INRA, URO50, Laboratoire de Biotechnologie de l'Environnement, Avenue des etangs, 11100 Narbonne, France

Chris Thoeye Aquafin NV, Dijkstraat 8, 2630 Aartselaar, Belgium

Olivier Thomas Environment and Sustainable Development Institute, Sherbrooke University, 2500 bld de l'Université, Local A6-1021, Sherbrooke, Quebec, J1K 2R1, Canada

Evelyne Touraud Ecole des Mines d'Alès, 6 Avenue de clavières, 30319 Alès Cèdex, France

Véronique Vandenberghe BIOMATH, Ghent University, Coupure Links 653, 9000 Ghent, Belgium

Peter Vanrolleghem BIOMATH, Ghent University, Coupure Links 653, 9000 Ghent, Belgium

Marco Vian Department of Civil and Environmental Engineering, University of Trento, Via Mesiano 77, 38050 Trento, Italy

Violeta Vincevicienne European Commission, DE Environment (BU93/168), Rue de la Loi 200, 1049 Brussels, Belgium

Thomas Wintgens RWTH Aachen University, Turmstrasse 46, Aachen 52056, Germany

Giuliano Ziglio Department of Civil and Environmental Engineering, University of Trento, Via Mesiano 77, 38050 Trento, Italy

1.1
Wastewater Regulation

Violeta Vinceviciene

Disclaimer. The views expressed herein are those of the author and do not necessarily represent the views of the European Commission.

Wastewater Quality Monitoring and Treatment Edited by P. Quevauviller, O. Thomas and A. van der Beken

1.1.1 URBAN WASTEWATER TREATMENT REGULATION IN THE EUROPEAN UNION

1.1.1.1 Urban Wastewater Treatment Directive in the Context of European Union Water Legislation

European water policy began in the 1970s with the adoption of so-called Community Environmental Action Programmes (EAPs) and legally binding legislation. The first EAP covered the period 1973–1976, and the latest – the sixth EAP – covers the period 2001–2010 and has four priority areas: climate change; nature and biodiversity; environment and health; and management of natural resources and waste. One out of the eight actions set up in the EAP relates to sustainable use and quality of water, where the measure to improve application of water legislation is underlined (European Commission, 2002).

Parallel to the political programmes, three waves of the European Union (EU) water legislation can be distinguished. The first wave of legislation used water quality-oriented approach; the second covered review and update of regulations from the first wave and addressed new legislation related to the emission-control approach. The Urban Wastewater Treatment Directive (UWWTD) 91/271/EEC (European Commission, 1991) and the Integrated Pollution Prevention and Control Directive (IPPCD) 96/91/EC adopted during the second wave are mainly dealing with urban and industrial wastewater (European Commission, 1996).

The third wave started with using integrated approach of those two, mutually reinforcing each other. The integrated approach takes into account two aspects: limiting pollution at the source by setting emission limit values (or emission standards); and establishing water quality objectives (or quality standards) for water bodies. This approach is in accordance with the principles established in the EU Treaty: i.e. the precautionary principle, high level of environmental protection, principle of preventive action and rectification of pollution at the source, polluter pays principle and integration of environmental protection into other Community Policies. The new EU policy area started with the adoption of the Water Framework Directive (WFD) 2000/60/EC (European Commission, 2000) with which the UWWTD is closely linked. The implementation of the UWWTD forms the cornerstone part of the programme of measures of WFD to be included in river basin management plans with the objective to achieve good ecological status of surface waters by 2015. However, the UWWTD sets up only minimum requirements to achieve this objective. More stringent measures than those prescribed in the UWWTD for urban wastewater treatment may be required in some specific cases when having sensitive water bodies or water bodies being at risk of becoming sensitive. One of the main problems of surface water bodies is eutrophication, thus specific treatment requirements for nitrogen and phosphorus removal shall be set up for urban wastewater discharges to receiving waters.

1.1.1.2 Main Aspects of Wastewater Treatment Directives

Related Community legislation

The UWWTD 91/271/EEC is the main piece of EU environmental legislation dealing with urban wastewater. IPPCD 96/91/EC sets up the provisions for wastewater from certain large industrial sectors. Dangerous Substances Directive (DSD) 76/464/EEC and daughter directives control discharges of wastewater containing certain dangerous substances (European Commission, 1976).

Community measures on emissions including water discharges on sector-oriented approach included in the IPPCD 96/61/EC set up emission limit values for large industrial installations of specific industrial sectors. However, water contamination depends on the quantity of discharges which may also stem from smaller plants. The Directive requires fixing emission limit values in the individual permits for installations that come under this Directive. Referring to the Directive the pollutants covered are those 'likely to be emitted from installation in significant quantities, having regards to their nature and their potential to transfer pollution from one medium to the other (water, air and land)'. Very important to fix emission limit values for 12 'main polluting substances' listed in the Annex III of the Directive.

The Directive 76/464/EEC fixed the framework conditions for discharges of dangerous substances into waters. The Directive establishes two lists of substances and groups of substances to be addressed: list I contains substances considered toxic, persistent or bioaccumulative; list II other polluting substances, which have a deterious effect on the aquatic environment and which depend on the characteristics and location of the water into which they are discharged. The Directive requires all discharges containing list I or list II substances to be authorized. The authorizations had laid down emission limit values for these substances. Under the Directive emission limit values and quality objectives were fixed for 17 substances, 'taking into account the best technical means available'. For the list II substances Member States have to establish programmes in order to reduce water pollution, and set timetables for their implementation. The programmes had to include quality objectives for water, and individual authorizations had to be issued in such a way that these quality objectives shall be respected.

Principles and requirements of Urban Wastewater Treatment Directive

The UWWTD **concerns** collection, treatment and discharge of urban wastewater from agglomerations and **aims** to protect the environment from being adversely affected by the disposal of insufficiently treated urban wastewater and discharges of wastewater from food-processing industries.

The Directive **applies**: (a) to all agglomerations having the organic load of more than 2000 population equivalent (p.e.);[1] (b) to agglomerations with less than 2000 p.e. having collecting systems in place; and (c) to food-processing industries having the load of more than 4000 p.e. and discharging treated wastewater directly to receiving waters.

The requirements for treatment level are defined depending on the **agglomeration size** and type of receiving water body where treated wastewater is discharged.

Types of wastewater covered by the Directive are urban, domestic, and industrial wastewater:

- Urban wastewater means domestic wastewater or the mixture of domestic wastewater with industrial wastewater and/or run-off rain water.
- Domestic wastewater means wastewater from residential settlements and services which originates predominantly from the human metabolism and from household activities.
- Industrial wastewater means any wastewater, which is discharged from premises used for carrying on any trade or industry, other than domestic wastewater and run-off rainwater.

Four main principal obligations are laid down in the directive: **planning; regulation; monitoring; and information and reporting.**

The planning aspect requires:

- To designate sensitive areas (sensitive water bodies) in accordance with three specific criteria, and to review their designation every 4 years; to identify relevant hydrologic catchment area of this sensitive area, and to ensure that all discharges from agglomerations with more than 10 000 p.e. located in sensitive area and their catchment shall have more stringent treatment (containing nutrients removal) in place.

  Surface water body shall be designated as sensitive if it falls into one of the following groups:

  – it is eutrophic or in the nearest future may become eutrophic if protective action is not taken;

  – it is intended for abstraction of drinking water;

  – where further treatment than secondary is necessary to fulfil other Council Directives.
- Less sensitive areas (as an option) can also be specified according to certain criteria.

[1] According to Article 2(6) of the UWWTD, '1 p.e. means the organic biodegradable load having five day biochemical oxygen demand (BOD5) of 60 g of oxygen per day'.

- To establish a technical and financial programme for the implementation of the Directive for construction of sewage networks and wastewater treatment plants addressing treatment objectives within the deadlines set up by the Directive.

Regulation requires ensuring that:

- All urban wastewater generated in agglomerations discharging into urban sewer systems and treatment plants have prior regulation or specific authorization.
- Industrial wastewater discharging into urban sewage networks is based on prior regulation and/or specific authorization; pretreatment requirements, ensuring that: (a) treatment plant operation and sludge treatment will not being impeded; (b) it will be no adverse effect to the environment (including receiving waters); and (c) safe disposal of sewage sludge.
- Food-processing industries have prior regulation and/or specific authorization and permit system.
- All urban wastewater generated in agglomerations with more than 2000 p.e. are supplied with collecting systems, and the capacity of those is such that it ensures to collect all urban wastewater taking into account normal local climatic conditions and seasonal variations.
- National authorities are taking measures in relation to collecting systems to limit pollution of receiving waters from storm water overflows under unusual situations, such as heavy rain.
- Wastewater treatment is provided for all these agglomerations, at the level of treatment specified and within the required deadline:
  - The basic rule for the level of treatment is secondary (i.e. biological) and more stringent treatment in sensitive areas (i.e. with nutrient removal in particular and other pollutant affecting the quality of specific use of the receiving water).
  - For certain discharges in coastal waters treatment might be less stringent, i.e. primary, under specific conditions and subject to agreement of the Commission.
  - For agglomerations with population equivalent of less that 2000 but equipped with collecting system, 'appropriate treatment' has to be provided, i.e. treatment that ensures to meet the relevant quality objectives of the receiving waters.
- Technical requirements on design, construction, operation and maintenance for wastewater treatment plants treating urban wastewater are maintained ensuring adequate capacity of the plant and treatment of urban wastewater generated in agglomeration taking into account normal local climatic conditions and seasonal variations.
- Environment is protected from adverse effects of the discharge of wastewater.

- Environmentally and technically sound reuse or disposal of sewage sludge is subject to general rules, registration or authorization, and requirements of specific inter-linked directives for agricultural reuse (86/278/EEC), incineration (89/429/EEC and 89/369/EEC), and landfill (99/31/EC), are respected. The disposal of sludge to surface waters is banned.

Monitoring requires ensuring that:

- appropriate monitoring capacity of parameters to be monitored;
- proper analysis of samples by using standard methods;
- timely frequency of monitoring for:
  - monitoring of discharges from urban wastewater treatment plants; and
  - monitoring of waters receiving those discharges.

Information and reporting requires ensuring that:

- Adequate cooperation and exchange of information with other member states in cases where discharges of wastewater have a transboundary effect on water quality of shared waters.
- Adequate reporting procedure and databases for the requests from the Commission for information on:
  - transposition of the directive into national legislation, implementation programmes, situation reports on the disposal and reuse of urban wastewater and sewage sludge;
  - status of collecting systems, efficiency of treatment plants (i.e. treatment level and monitoring results) and water quality of receiving waters;
  - status of discharges from food-processing industry to surface waters.
- That the public has access to relevant information and that relevant authorities of member states every 2 years will publish status reports to the public on the status of wastewater collection and treatment and disposal or reuse of sludge.

The Directive is based on a number of **principles** that have been laid down in the Treaty of the European Union, such as precautionary, nondeterioration, sustainable use of water resources, and principle of subsidiary. The implementation of the Directive should not result in deterioration of the current level of environmental protection offered by the member states. Furthermore, the level of protection may be even stricter than the Directive requires in case there is a need to fight deterioration of quality of receiving water bodies and to try to restore waters affected by wastewater discharges.

Parameters and parametric values in the Urban Wastewater Treatment Directive

The Directive **regulates the main conventional pollutants** in treated wastewater discharges from treatment plants. These are: total suspended solids, chemical oxygen demand, biochemical oxygen demand, total nitrogen and total phosphorus. However, other parameters shall also be considered especially when making the assessment of receiving waters to designate sensitive areas and to achieve water quality objectives of water bodies.

The Directive sets up emission limit values for the above-mentioned parameters or by showing treatment efficiency (calculating it through incoming and outgoing pollution load of each regulated parameter). Either the concentration of a pollutant at the discharge point or the reduction rate of pollution load shall apply.

The Directive sets up a general requirement for:

- Treated urban wastewater reuse to ensure that there will be no adverse effect to the environment. However, there are no detailed regulations on treated wastewater quality for the purposes of its reuse for various economic activities.
- The usage of sewage sludge indicating that it shall be re-used whenever appropriate having no adverse affect on the environment.

However, there are no precise provisions in this Directive on setting emission limit values or quality standards to be achieved when having activities of the reuse of these end-products from wastewater treatment process. These two aspects are partly or indirectly regulated by the other EU pieces of legislation.

Sampling and monitoring

The Directive requires establishing a monitoring and inspection programme for compliance assessment of discharges from urban wastewater treatment plants and for assessing the amounts and composition of sludge.

The Directive defines minimum monitoring requirements for treated wastewater by setting sampling frequency, which is dependent on the size of urban wastewater treatment plant. The maximum number of noncomplying samples is also defined in the Directive.

The Directive also defines the standard laboratory methods to be used for the analysis of the samples.

Quality control and assurance

Member States have to ensure laboratory capacities, and laboratories must use the methods specified in the Directive Annex I and to be subject to regular quality control. Accreditation schemes for laboratories are the means of constantly ensuring quality control. Quality control is restricted to the analytical laboratory methods to be used for the analysis of samples. Member States need to have some quality control system in place in the approved laboratories for wastewater analyses.

Taking into account the principle of subsidiary, Member States have a duty to organize and self-control the Directive implementation by setting adequate urban wastewater collection systems and treatment facilities as well as controlling wastewater pollution level through monitoring of urban wastewater against Directive requirements.

The control of annual monitoring results of treated urban wastewater is a duty of Member States to check against the Directive requirements. The data shall be stored in the Member State to analyse trends and impact of discharged wastewater to the environment and to ensure the reporting of the results to the Commission to check the implementation status. Member States have to provide requested data within the deadline of 6 months.

1.1.1.3 Other Related Legislation on Other Types of Wastewater Except Urban

Implementation of the main UWWT Directive is closely linked with other EU legislation, in particular:

- Water Framework Directive 2000/60/EC;
- Nitrates Directive 91/676/EEC;
- Integrated Pollution Prevention and Control Directive 96/91/EC;
- Dangerous Substances Directive 76/464/EEC and its seven daughter directives;
- Sewage Sludge Directive 86/278/EEC;
- Landfill Directive 99/31/EC;
- Incineration Directives 89/429/EEC and 89/369/EEC;
- Environmental Impact Assessment Directive 85/337/EEC;
- Directive on Access to Environmental Information 90/313/EEC;
- Reporting Directive 91/692/EEC and Decision 94/741/EEC;

Particularly relevant issues in these directives concern:

- The provisions under UWWTD are the integral part of basic measures in the programme of measures to be included into river basin management plans under

WFD, without any change of deadlines set under UWWTD for the EU-15 and in line with the transitional periods set up in the Accession Treaty for the EU-10.

- The provision of adequate facilities for either incineration or landfill of sewage sludge.
- The quality requirements for sewage sludge used in agriculture.
- Certain size large installations of food-processing industries covered by UWWTD also fall under requirements of the IPPCD (i.e. for some installations the requirements of both directives overlap). It should be borne in mind that IPPC Directive sets requirements for application of a combined approach (as the WFD does) of emission controls and water quality standards. In each particular case the more stringent approach (setting more stringent treatment standards for wastewater) to reach certain water quality objectives applies.

Integration of EU water legislation does not only imply compliance to the requirements of various related directives but will also involve harmonization and streamlining of monitoring and reporting requirements. Reporting requirements will have to address compliance and the state of, and trends in, the quality of aquatic environment through implementing prevention measures – decrease generation of wastewater at the source, sustainable management of water resources as well as when wastewater is generated – ensuring the adequate required wastewater treatment. The process on harmonization of reporting has been started with the concept and the ambitious goal to have an integrated reporting system in Europe, the so-called Water Information System for Europe (WISE).

1.1.1.4 Conclusions

It should be underlined that the implementation of the UWWTD ranks amongst the most challenging and expensive tasks throughout the range of EU legislation. This is why early and careful consideration of the environmental and technical aspects of the Directive is of significant importance.

Much of the environmental legislation in the EU is interrelated, both within a specific sector (e.g. water sector) and between sectors. Consequently, individual Community legal acts cannot be implemented in isolation. For example, implementation of UWWTD 91/271/EEC in advance of another directive can lead to environmental problems, e.g. increased volumes of sewage sludge which then need to be disposed of in accordance with Waste Framework Directive 75/442/EEC. Furthermore, the main measures to implement the UWWTD are one of the main components of the Programme of measures of River basin management plans required by the Water Framework Directive. However, complete implementation of UWWTD is only a minimum requirement to achieve good ecological status of surface waters required by the WFD 2000/60/EC.

The EU regulation on urban wastewater treatment has contributed significantly to the improvement and regulation of wastewater treatment and improvement of quality of discharges of wastewater into receiving waters and has contributed to fighting the pollution at the source and the improvement of quality of surface waters. It is expected that the underlying principles of the UWWTD will be further strengthened and improved when implementing the WFD and by using an integrated approach. Extending the control of wastewater discharges from end-of-pipe to using an integrated approach and meeting water quality standards and objectives to achieve good ecological status of surface waters, boosts the confidence of European citizens concerning the safety and wholesomeness of the use of surface waters for various needs. Close cooperation between the European Commission and Member States is a prerequisite to achieve this target.

1.1.2 URBAN WASTEWATER TREATMENT REGULATION IN THE UNITED STATES

1.1.2.1 Introduction

The main regulatory basis to deal with water pollution control in the United States is the Federal Water Pollution Control Act (FWPCA), known as the Clean Water Act (CWA). It is a comprehensive statute aimed to restore and maintain chemical, physical and biological integrity of the US waters. Enacted originally in 1948, the Act was amended numerous times until it was reorganized and expanded in 1972. The CWA is part of the US main legislation included in the Code of Federal Regulation (CFR) on 18 October 1972 and forms Title 33 of this Code; it continues to be amended almost every year (Deketelaer and Gekiere, 2002).

As authorized by the CWA, the National Pollutant Discharge Elimination System (NPDES) permit programme controls water pollution by regulating point sources that discharge pollutants into waters of the US. Since its introduction in 1972, the NPDES permit programme has been responsible for significant improvements in quality in the water bodies in the US (http://cfpub.epa.gov/npdes/).

1.1.2.2 Development of Urban Wastewater Treatment Regulations

History of the Clean Water Act

The CWA is a law that establishes **environmental programmes**, including the NPDES programme (introduced in 1972) to protect US waters and directs the Environmental Protection Agency (EPA) to issue rules on how to implement this law. The Act does not deal directly with ground water nor with water quantity issues. The statute employs a variety of regulatory and nonregulatory tools to reduce direct pollutant discharges into waterways, finance municipal wastewater treatment facilities, and manage polluted runoff.

For many years after reorganization of the CWA in 1972, the EPA, states, and Indian tribes focused mainly on chemical aspects of water quality. During the last few decades, however, more attention has been given to physical and biological parameters. In the early decades of the Act's implementation, efforts focused on regulating discharges from traditional point sources, such as municipal sewage plants and industrial facilities, with little attention paid to runoff from streets, construction sites, farms, and other 'wet-weather' sources.

Starting from late 1980s, efforts have been concentrated to address and regulate nonpoint sources of pollution and 'wet weather point sources' (like urban storm sewer systems and construction sites).

Evolution of CWA programmes over the last decade has also included a shift from a programme-by-programme, source-by-source, pollutant-by-pollutant approach to watershed-based strategies by using an integrated approach on protecting water bodies.

The Act established **the basic structure for regulating discharges of pollutants into the water bodies** of the US. It gave the EPA the authority to implement pollution control programmes such as setting wastewater standards for industry. The CWA also set up **water quality standards for all contaminants in surface waters**. The Act prohibits discharging any pollutant from a point source into navigable waters without a permit. It also regulates funding of the construction of sewage treatment plants under the construction grants programme.

The current version of the CWA consists of six main titles (International, EC and US Environmental Law; Sands, 2002; Kramer 2003):

- Title I – Research and Related Programmes (Sections 1251–1271 of 33 US CFR, 18 October 1972).
- Title II – Grants for Construction of Treatment Works (Sections 1281–1299).
- Title III – Standards and Enforcement (Sections 1311–1330).
- Title IV – Permits and Licenses (Sections 1341–1345).
- Title V – General Provisions (Sections 1361–1377).
- Title VI – State Water Pollution Control Revolving Funds (Sections 1381–1387).

National Pollutant Discharge Elimination System (NPDES)

The NPDES permit programme **aims** to control water pollution by regulating point source discharges and is **based on** statutory requirements contained in the CWA and regulatory requirements contained in the NPDES regulations.

The NPDES permit system consists of a number of programmes and initiatives and is based on water quality and technology-based permitting regulations.

Industrial, municipal, and other facilities must obtain permits if their discharges go directly to surface waters. Individual homes that are connected to a municipal system, use a septic system, or do not have a surface discharge do not need an NPDES permit. In most cases, the NPDES permit programme is administered by authorized states.

The US EPA Water Permits Division (WPD) of the Office of Wastewater Management (OWM) leads and manages the NPDES permit programme in partnership with 10 EPA regional offices, states, tribes, and other stakeholders.

Total Maximum Daily Loads

Following the integrated approach to the management of water resources, the watershed-based NPDES permitting system is also used in the US. One of the basic elements of this system is a total maximum daily load (TMDL) calculation methodology – a tool for implementing water quality standards, which is based on the relationship between pollution sources and in-stream water quality conditions. The TMDL establishes the allowable loading of pollutants to a water body and provides the basis to establish water quality-based controls. These controls should provide the pollution reduction necessary for a water body to meet water quality standards.

Implementation and control bodies

The primary authority for the implementation and enforcement of the CWA and controlling water pollution in the US is the EPA of OWM together with 10 regional EPAs. Their main responsibility is to promote effective and responsible water use, treatment, disposal and management and to encourage the protection and restoration of the catchments of surface water bodies (http://www.epa.gov/owm/; http://ipl.unm.edu/cwl/fedbook/fwpca.html).

The US EPA OWM WPD in partnership with EPA regional offices, states, tribes, and other stakeholders leads and manages the NPDES permit programme and ensures its effective implementation. It also regulate discharges from point sources (including pipes, ditches, and sanitary or storm sewers) into surface waters such as wetlands, lakes, rivers, estuaries, bays and oceans. The US EPA OWM is also responsible for management of the Clean Water State Revolving Fund, the largest water quality funding source, focused on funding wastewater treatment systems, nonpoint source projects and estuary protection (http://www.epa.gov/owm/).

If changes to the NPDES regulations are needed, then EPA issues (proposed and final) rules related to the NPDES permit programme. When making changes to the NPDES regulations, EPA first develops a proposed rule and provides it in the Federal Register for public review and comment. After receiving public comments, EPA develops a final regulation and publishes it in the Federal Register. Once each year, all final federal rules are compiled into a document called the Code of Federal Regulations (http://www.epa.gov/waterscience/guide/).

1.1.2.3 Highlights of Federal Water Pollution Control Act (Clean Water Act)

The objectives of the CWA are to restore and maintain the chemical, physical and biological integrity of the US waters.

Even prior to the enactment of the 1972 version of the CWA, the Act authorized a number of actions, for example, the action to prepare comprehensive programmes for eliminating and reducing the pollution of interstate waters and tributaries and improving sanitary condition of surface and groundwater (CWA, Sections 1251–1252).

The Act authorizes water quality programmes, requires federal effluent limitations and state water quality standards and permits for the discharge of pollutants into navigable waters, provides enforcement mechanisms, and authorizes funding for wastewater treatment works construction grants and state revolving loan programmes, as well as funding of states' and tribes' water quality programmes. Provisions also address water quality problems in specific regions and specific waterways.

Title II 'Grants for Construction of Treatment Works' of the CWA deals with the **regulation of wastewater treatment management plans and grants**. The Act requires development and implementation of the wastewater treatment management plans and practices using best practicable technology before they discharge pollutants into receiving waters. The confined disposal/discharges of pollution should be so that they will not migrate to cause water and other environmental pollution. It also requires identification of areas with substantial water quality control problems. Furthermore, no NPDES permit may be issued which is in conflict with an approved plan. The Act outlines a programme of grants to state, municipalities or intermunicipal or interstate agencies for the construction of publicly owned treatment works (POTWs).

Title III 'Standards and Enforcement' of the Act deals with **regulation of emission limit values and water quality standards**. The Act prohibits discharge of pollutants except in compliance with emission limit values and other provisions of the Act. Effluent limitations from point sources other than POTWs must be treated using best practicable control technology. Toxic pollutants, defined and otherwise described in the Act, require treatment using the best available technology, which is economically achievable. If it is discharged into POWTs, it must comply with applicable pretreatment requirements. The Act makes it unlawful to discharge any radiological, chemical, or biological warfare agent, any high-level radioactive waste, or any medical waste into navigable waters.

Effluent limitations must be determined for point sources, which are consistent with state water quality standards, including toxic and pretreatment standards. The Act requires establishing procedures to assure water quality standards, developing guidelines to identify and evaluate the extent of nonpoint source pollution, and setting water quality inventory requirements. The Act also requires the EPA to develop national standards of performance for the control of discharge of pollutants from new sources.

When discharges of pollutants from a point source or group of point sources under established emission limit values would have an adverse effect on the receiving water body or on maintenance of water quality necessary to assure protection of public health, public water supplies, agricultural and industrial uses, and the protection and propagation of a balanced population of shellfish, fish and wildlife, and allow recreational activities in and on the water, the EPA must establish emission limit values for the point source or sources which can reasonably be expected to contribute to the attainment or maintenance of water quality (CWA, Sections 1311, 1312, 1314, 1316, 1317).

Continuing public information and education **programmes on recycling and reuse of wastewater (including sludge)** are also required under CWA (Section 1294).

The state must establish a TMDL for those pollutants suitable for maximum daily measurements (CWA, Sections 1311, 1313, 1315). States must identify waters for which controls on thermal discharges are not stringent enough to assure protection and propagation of a balanced indigenous population of shellfish, fish and wildlife. The Act imposes **requirements for storage of monitoring results and reporting requirements, and allows for inspections**. The states must report on their water quality biennially. It also contains extensive provisions on enforcement, with administrative, civil and criminal penalties available for violations (CWA, Sections 1318–1319). The institutional entities which are planning to have discharges of wastewater into water bodies are subject to certification and must obtain federal permits or licenses in order to assure that it will not violate applicable effluent limitations and water quality standards (CWA, Section 1341).

Title IV 'Permits and Licenses' of the Act deals with regulation of permitting systems of wastewater discharges based on the NPDES. According to the NPDES, all industrial sources and publicly owned treatment works must have a permit to discharge pollutants into navigable waters. Discharge must meet the requirements outlined extensively in the CWA and meet federal emission limit values and state water quality standards. The state has to administer its own permit programme in line with the federal programme. The Act also sets up **special provisions on municipal and industrial storm water discharges** (CWA, Section 1342). Discharge permits also comply with the guidelines for determining the degradation level of the waters of the territorial seas, the contiguous zone and the oceans. The guidelines include, for example, the effect of disposal of pollutants on human health or welfare; on marine life, changes in marine ecosystem diversity, productivity and stability, or species and community population changes; the effect of pollutants on aesthetic, recreational and economic values, etc.

The EPA is authorized to prohibit the use of a site for disposal of dredged or fill material in navigable waters if discharges would have an adverse effect on municipal water supplies, shellfish beds, fishery areas, and wildlife or recreational uses. The Act also regulates the disposal and use of sewage sludge through the NPDES system as well. It is required to identify uses of sludge, including disposal, to specify factors to consider in determining measures and practices applicable to each use or disposal, and identify concentrations of pollutants which interfere with each use or disposal (CWA, Section 1345).

Since 1987 the Act regulates nonpoint source pollution by developing management plans and programmes and implementing those (CWA, Section 1329). The Act also regulates development and implementation of management plans for estuaries, establishes a clean lakes programme, regulates thermal discharges, marine sanitation devices, discharges of oil and hazardous substances (CWA, Sections 1321, 1322, 1324, 1326, 1330).

1.1.2.4 Highlights of the National Pollutant Discharge Elimination System within the Clean Water Act

The NPDES programme is legally based on the CWA, NPDES regulations and Federal regulations, and other primary federal laws that also apply to the NPDES permit programme. It is also based on the strategy of permitting for environmental results.

NPDES programme areas

A number of NPDES permit programme areas affect how a municipality handles its sanitary wastewater and storm water runoff. **The major NPDES permit programme areas** where it sets up regulations are: wastewater treatment plants, pretreatment from industrial and commercial facilities to be connected to a publicly owned treatment works, combined sewer overflows, sanitary sewer overflows, storm water sewer overflows, and animal feeding operations.

- **Wastewater treatment plants**. Municipal wastewater must have biological treatment (or so-called secondary treatment). Secondary treatment standards are established by the EPA for POTWs and reflect the performance of secondary wastewater treatment plants. These technology-based regulations apply to all municipal wastewater treatment plants and represent the minimum level of effluent quality attainable by secondary treatment, as reflected in terms of 5-day biochemical oxygen demand (BOD5) and total suspended solids (TSS) removal.

  The secondary treatment standards also provide special considerations regarding combined sewers, industrial wastes, waste stabilization ponds, and less concentrated influent wastewater for combined and separate sewers. In addition, the secondary treatment standards also provide alternative standards established on a case-by-case basis for treatment facilities considered equivalent to secondary treatment (trickling filters and waste stabilization ponds).

- **Pretreatment programme**. POTWs generally are designed to treat domestic sewage only. However, POTWs also receive wastewater from industrial (non-domestic) users. The General Pretreatment Regulations in the Code of Federal Regulations (40 CFR part 403) establishes responsibilities of Federal, State, and local government, industry and the public to implement pretreatment standards and limits, and to control pollutants from the industrial users which may pass through or interfere with POTW treatment processes or which may contaminate sewage sludge. The new version of the Code of Federal Regulations (40 CFR part 403), incorporating the updates/changes on rules made by the EAP in 2005 is scheduled for publication in July 2006.
- **Combined sewer overflows**. During periods of heavy rainfall or snowmelt, the wastewater volume in a combined sewer system can exceed the capacity of the sewer system or treatment plant. For this reason, combined sewer systems are designed to overflow occasionally and discharge excess wastewater directly to water bodies. Combined sewer overflows (CSOs) contain storm water but also untreated human and industrial waste, toxic materials, and debris. The EPA's CSO policy published 19 April 1994 is the national framework for control of CSOs and provides guidance on how communities with combined sewer systems can meet CWA goals in as flexible and cost-effective manner as possible.
- **Storm water overflows**. Most storm water discharges are considered point sources and require coverage by an NPDES permit. The primary method to control storm water discharges is through the use of best management practices. Storm water overflows are regulated by the storm water NPDES permitting programme. In December 2005, the EPA proposed the NPDES Storm Water Multi-Sector General Permit Programme for Industrial Activities, which replaces the MSGP-2000 that expired on 30 October 2005.
- **New policy addressing peak rainfall events discharges**. There so-called 'wet weather discharges' refer to point source discharges that include storm water runoff, CSOs and wet weather sanitary sewer overflows (SSOs). Under the NPDES permit programme, there are three programme areas that address each of the wet weather discharges. To identify and address cross-cutting issues and promote co-ordination, the EPA established the Urban Wet Weather Flows Federal Advisory Committee in 1995 and in 2005 the EPA proposed a new policy for addressing very high or 'peak' flow events at municipal wastewater treatment plants. The policy describes certain management techniques to be used by the operator of a municipal wastewater treatment facility to address very high flows and indicates how the management of peak flows must be documented in NPDES permits. The proposed Peak Wet Weather policy requires that discharges must still meet all the requirements of NPDES permits and that operators demonstrate that all feasible measures are used to minimize wet weather problems. It also prohibits the use of these peak flow management techniques in systems where high peak flows are due to poor system maintenance or a lack of investment in upgrades to improve treatment capacity.

- **Concentrated animal feeding operations.**[2] These are point sources, as defined by the CWA [Section 502(14)], and have the potential of being regulated under the NPDES permitting programmes.

Types of permitting in NPDES

The NPDES programme is mainly based on water-quality- and technology-based permitting.

- **Technology-based permitting**. Effluent limitations are a primary mechanism in NPDES permits to control discharges of pollutants to receiving waters. When developing emission limit values for an NPDES permit, two aspects must be considered: limits based on both the technology available to control the pollutants (i.e. technology-based effluent limits) and limits that are protective of the water quality standards of the receiving water (i.e. water-quality-based effluent limits).

  Technology-based effluent limits in NPDES permits (Chapter 5 of US EPA NPDES Permit Writer's Manual) require a minimum level of treatment of pollutants for point source discharges based on available treatment technologies, while allowing the discharger to use any available control technique to meet the limits. The potential impact of every wastewater discharge on the quality of the receiving water must be considered.

  – For industrial (and other nonmunicipal) facilities, technology-based effluent limits are derived by using:

    (i) national emission limit values guidelines and standards established by the EPA; and/or

    (ii) best professional judgement (BPJ) on a case-by-case basis in the absence of national guidelines and standards.

  – For municipal facilities (POTWs), technology-based effluent limits are derived from national secondary treatment standards.

- **Water-quality-based permitting**. If technology-based effluent limits are not sufficient to ensure that water quality standards will be attained in the receiving water, the CWA and NPDES regulations (Chapter 6 of US EPA NPDES Permit Writer's Manual) require more stringent, water-quality-based effluent limits designed to ensure that water quality standards required by the CWA are maintained.

[2] Agricultural operations where animals are kept and raised in confined situations and where feed is brought to the animals rather than the animals grazing or otherwise seeking feed in pastures.

Water-quality based permits covers the following important issues:

- Watershed-based permits, where the broader context of the watershed in which the discharge is located is considered when setting the limit values.
- Where a watershed is listed as impaired, NPDES permits may need to reflect the results of TMDLs.
- Assessment of whole effluent toxicity (WET) and inclusion of the results into the NPDES, as WET describes the uses of whole effluent toxicity in NPDES permits to protect the receiving water quality from the aggregate toxic effect of a mixture of pollutants in a discharge.
- In cases of more than one discharge point located within a short distance (or closely located) to each other, the option of effluent trading for NPDES permitees is also possible.

Effluent limitations guidelines and standards

The CWA requires the EPA to develop effluent limitation guidelines and standards for different industrial sectors based on the degree of pollutant reduction attainable by an industrial category through the application of pollutant control technologies (http://ecfr.gpoaccess.gov/cgi/t/text/), such as:

- best conventional pollutant control technology for conventional pollutants; best practicable control technology currently available for conventional, toxic and non-conventional pollutants; best available technology (BAT) economically achievable for toxic and nonconventional pollutants; applicable to existing dischargers; and
- new source performance standards for conventional pollutants and applicable to new sources.

To date, the EPA has established guidelines and standards (published in 40 CFR parts 405–499) for more than 50 different industrial sectors (e.g. metal finishing facilities, steam electric power plants, iron and steel manufacturing facilities). Additionally, Section 304(m) of the 1987 Water Quality Act (WQA) requires the EPA to publish a biennial plan for developing new effluent guidelines and a schedule for the annual review and revision of existing promulgated guidelines. All effluent guidelines applicable to an industrial facility are legally binding and must be included in an NPDES permit.

Watershed-based NPDES permitting

This is an integrated approach that synchronizes permits and addresses all stressors within a hydrologically defined drainage basin (EPA 833-B-03-004, published

17 December 2003). Watershed-based permitting can encompass a variety of activities including developing water quality-based effluent limits using a multiple discharger modelling analysis. The ultimate goal of this effort to develop and issue NPDES permits based on water quality trading policy that better protect entire watersheds.

Types of wastewater covered by the CWA and the NPDES

The CWA through the **NPDES permit system regulates all types of industrial wastewater and urban wastewater treated in POTWs**. Discharge must meet the requirements outlined extensively in the CWA and meet federal effluent limitations and state water quality standards. [Regulations and guidance on water quality standards are developed by the EPA and are contained in the 40 CFR part 131 (http://ecfr.gpoaccess.gov/cgi/t/text/).]

Parameters and parametric values

The CWA together with the NPDES sets up the pollutants to be controlled and issues the criteria and standards for the NPDES. (The latest version of criteria as of 10 January 2006 has been issued under 40 CFR part 125.)

- Regulated pollutants
  - Effluent limitations from POTWs are based on secondary treatment where the minimum BOD5, TSS and pH should be controlled and not exceed the numerical values defined by the Secondary Treatment Regulation (40 CFR part 133.102; http://ecfr.gpoaccess.gov/cgi/t/text/). The emission limit values of the conventional pollutants (BOD5, TSS, pH, oil and grease, fecal coli form) shall also be controlled (40 CFR part 401.16).
  - Effluent limitations for industrial wastewater discharges are determined in terms of amounts of constituents and chemical, physical, and biological characteristics of pollutants, the degree of effluent reduction attainable through:

    (i) application of pollutant control technology currently available for classes and categories of industrial point sources;

    (ii) guidelines for pretreatment of pollutants; and

    (iii) individual control strategies for toxic pollutants (CWA, Sections 1313–1314).
  - More stringent limitation of certain pollutants including those necessary to meet water quality standards, treatment standards or schedules of compliance

established according to any State and/or Federal law or regulation shall be also set up and applied if necessary.

- For all toxic pollutants listed in 40 CFR part 401.15 (65 parameters in total), effluent limitations should be implemented as expeditiously as practicable but no later than 3 years after the date of such limitation are promulgated. The effluent limitations shall be reviewed at least every 5 years.
- It shall be unlawful to discharge radiological, chemical, or biological warfare agents, high-level radioactive waste or medical waste.

• Unregulated pollutants. It should be underlined that there could be no emission limit values set up for certain pollutants related to certain activities for which a permit is not required (CWA, Section 1342). Those are:

- agricultural return flows;
- storm water runoff from oil, gas, and mining operations;
- additional pre-treatment of conventional pollutants;
- discharges composed entirely of storm water (however, a permit is required for discharge associated with industrial activity, from municipal separate storm water sewer system serving a population of more than 250 000, and more than 100 000 but less than 250 000 (CWA, Section 1344).

Sampling, compliance monitoring and enforcement

In order to protect human health and the environment and to ensure that environmental laws, regulations and statutory programmes (NPDES) are respected, **compliance monitoring programmes** are used. The actions involve on-site visits of pollution discharge points by qualified inspectors, and a review of the information and monitoring results submitted by NPDES permit holders to EPA or state/tribe regulatory institutions. The EPA also uses compliance incentives and auditing to find and disclose violations. Violations also could be discovered from complaints received by the EPA from the public. Violations discovered may lead to civil or criminal enforcement. Compliance with the environmental laws is the goal, but enforcement is a vital part of encouraging governments, companies and others who are regulated to meet their environmental obligations. The EPA's civil and criminal enforcement programmes are formed to take legal action in both federal and state courts that bring polluters into compliance with federal environmental laws (http://www.epa.gov/compliance/monitoring/).

The monitoring programme requirements include biomonitoring requirements (40 CFR part 125.63), water quality requirements and effluent monitoring programme requirements. Effluents limitation and mass loading will ensure compliance with the requirements of pretreatment, nonindustrial toxics control, and control of

CSOs. The reporting requirements include the results of the listed required monitoring programmes.

Quality control and assurance

In order to ensure quality control, industries and municipalities must use the specific EPA-approved laboratory analytical methods to analyse the chemical and biological components of wastewater, drinking water, sediment, and other environmental samples. These standard methods are required by the CWA, the regulations, rules and guidance prepared by the EPA and are listed in the Electronic Code of Federal Regulations (eCFR) (40 eCFR part 136; http://epa.gov/epacfr40/chapt-I.info/). As the CWA together with the NPDES system regulates all industrial wastewater discharges, there are about 1600 analytical methods officially issued by the EPA, and this process is continuously developing and evolving. In order to classify and to more easily find the right methods for pollutant analysis, in April 2003 the EPA issued a revised edition of the 'Index of EPA Test Methods'. This is an ever changing and evolving product (e.g. companies may ask the EPA to approve a new method or an alternative test procedure to improve performance, decrease hazardous materials in the laboratory, and better protect human health and environment; http://www.epa.gov/ost/methods/).

1.1.2.5 Conclusions

It shall be concluded that the regulations of wastewater collection treatment and control of discharges in the US is regulated in an integrated way. It is based on the CWA's NPDES permitting programme. It enables regulations and rules to be set up to tackle wastewater pollution at the source and to follow emission limit values in order to achieve quality standards of receiving water bodies by using an integrated watershed-based approach. It should be underlined that the TMDL calculation and legally binding implementation schemes provide a comprehensive tool to improve the quality of surface water bodies of the US. It helps to set up a comprehensive permitting system inter-linked with emission trading policy to support the protection and propagation of fish, shellfish, and wildlife and recreation in and on the surface water bodies of the US.

Watershed-based permits are being issued where the pollution concerns relate to regulatory controls involving traditional end-of-pipe or new types of wet-weather concerns. The next step will be to establish a credible watershed-oriented management framework that combines regulatory and other types of more consensus-based control arrangements. The US EPA has established policy guidelines for these type of water quality trading initiatives and such new regional and watershed-based management approaches will aim to achieve a better degree of integration with assessment systems than has been the case over the last 30 years. The anchor of effective management programmes, including wastewater management at the federal level, will be these watershed-oriented permitting and trading systems (Cooter, 2004).

REFERENCES

Cooter, W.S. (2004) *Clean Water Act Assessment Processes in Relation to Changing US Environmental Protection Agency Management Strategies*. Environmental Science and Technology, vol. 38. American Chemical Society, Columbus, OH, pp. 5265–5273.

Deketelaer, K. and Gekiere, J. (Eds) (2002) *International, EC and US Environmental Law: A Comparative Selection of Basic Documents, Volume II*. Kluwer Law International, The Hague, pp. 1523–1733.

European Commission (1976) Council Directive 76/464/EEC of 4 May 1976 on pollution caused by certain dangerous substances discharged into the aquatic environment of the Community. Official Journal No. L 129, 18.5.1976, pp. 23–29.

European Commission (1991) Council Directive 91/271/EEC of 21 May 1991 concerning urban waste water treatment. Official Journal No. L 135, 30.5.1991, pp. 40–52.

European Commission (1996) Council Directive 96/61/EC of 24 September 1996 concerning integrated pollution prevention and control. Official Journal No. L 257, 10.10.1996, pp. 26–40.

European Commission (2000) Directive 2000/60/EC of the European Parliament and of the Council of 23 October 2000 establishing a framework for Community action in the field of water policy. Official Journal No. L 327, 22.12.2000, pp. 1–73.

European Commission (2002) Decision No. 1600/2002/EC of the European Parliament and of the Council of 22 July 2002 laying down the Sixth Community Environment Action Programme. Official Journal No. L 242, 10.9.2002, pp. 1–15.

Kramer, L. (2003) *EC Environmental Law*, 5th Edn. Sweet & Maxwell Ltd., London, pp. 1–28, 244–264.

Sands, Ph. (2002) *Principles of International Environmental Law*, 2nd Edn. Cambridge University Press, Cambridge, pp. 732–734, 776–779.

1.2
Sampling Assistance

Olivier Thomas

1.2.1 WASTEWATER MONITORING CONSTRAINTS

Sampling is a key operation for wastewater monitoring. Whatever the objective, regulation compliance, treatment efficiency, discharge impact, etc., sampling is the first step of the classical analytical procedure before laboratory analysis, and is also carried out for validation of on-site or on-line measurement. Even if the use of

Wastewater Quality Monitoring and Treatment Edited by P. Quevauviller, O. Thomas and A. van der Beken

on-line devices is increasing, the great majority of wastewater quality measurements is carried out in the laboratory, after sampling. Thus, before considering analytical methods for wastewater quality monitoring, based on either standard or alternative procedures, the sampling step must be considered because of its importance as a source of potential errors. With the aim of getting a representative volume of effluent, sampling has to face a lot of specific constraints related to wastewater characteristics. Thus, wastewater sampling is difficult, considering the heterogeneity and variability of effluents, and moreover the evolution of samples during transportation from sampling site to laboratory, related to sample aging.

1.2.1.1 Heterogeneity

As for water, there are several types of wastewater. All types are characterized by their composition heterogeneity. A wastewater is composed of water, carrying a lot of suspended solids and dissolved substances which were not present originally (the pollutants). Wastewater types depend on the nature and concentration of solids and pollutants.

The most frequent type is urban wastewater, mixing municipal wastewater and industrial ones. The composition of municipal wastewater is rather well known and does not vary a lot from one human being to another or one town to another. Typical compositions of urban wastewater have been published (Muttamara, 1996; Metcalf and Eddy, 2003; Degrémont, 2005). The concentration of total suspended solids (TSS) varies from 200 to 600 mg/l, the volatile suspended solids from 200 to 600 mg/l, the biological oxygen demand (BOD) from 100 to 500 mg/l, the chemical oxygen demand (COD) from 200 to 1200 mg/l, the total organic carbon (TOC) from 50 to 300 mg/l, the total nitrogen from 50 to 100 mg/l, and the total phosphorous from 10 to 20 mg/l. These values can be decreased in the case of combined sewer (effect of dilution of rainfall) or increased, depending on the proportion and nature of industrial wastewater collected in the urban area.

Thus, the heterogeneity is related to the diversity of soluble pollutants' nature, and increased when considering emergent pollutants, but also to the nonsoluble fractions distribution: colloids, supra-colloids and settleable suspensions. Table 1.2.1 presents the size distribution of particulates and the coarse chemical composition of the soluble fraction.

The composition of industrial wastewaters is obviously related to the industrial activity (Eckenfleder, 2001; Metcalf and Eddy, 2003; Degrémont, 2005), but above all, to the existence of environmental equipments (e.g. wastewater treatment plant) and investments (e.g. recycling process). Contrary to wastewater of domestic origin, which increases with number of inhabitants, industrial loads are more and more controlled and reduced under regulatory pressure. However, some problems remain for industrial discharges in urban sewers, when the industrial fraction of wastewater is dominant, leading to toxic effect and increasing the heterogeneity.

Table 1.2.1 Dispersion characteristics of the main fractions of wastewater. (Adapted from Sophonsiri and Morgenroth, 2004)

Fraction	Min.	Max.	Mean	Standard deviation	RSD (%)	Results calculated from
Settleable (%) (>100 μm)	7	45	26.3	13.2	50	8 studies
Supra-colloidal (%) (1–100 μm)	12	50	27.4	12.1	44	9 studies
Colloidal (%) (0.1–1 μm)	7	48	15.6	12.6	81	9 studies
Soluble (%) (<0.1 μm)	9	64	37.2	17.4	47	10 studies
Composition						
COD (mg/l)	203	967	496	292	59	7 studies
Protein (% COD)	8	31	19.3	9.1	47	8 studies
Carbohydrate (% COD)	6	18	11.3	4.6	40	9 studies
Lipid (% COD)	7	82	33.2	28.1	87	6 studies
Unidentified (% COD)	8	78	51.4	26.0	51	7 studies

1.2.1.2 Variability

Wastewater variability is due to its composition, changing along the sewer system under the influence of several factors (see Chapter 2.1) and with the mixing of effluents of different origin (municipal and industrial). For an industrial sewer network, the wastewater composition varies from downstream units or workshops to treatment plant, with a decrease in variability under homogenisation effects of mixing and storage tanks. Another variability factor is time, the wastewater production being generally less during the night for domestic activities, or during weekends and holidays for some industries.

For all fractions and chemical compound groups of Table 1.2.1, the variability, expressed as the residual standard deviation (RSD), is around 50 %, except for the colloid fraction and for lipids. It should be noted that, for the soluble fraction, half of the chemical compounds are not actually identified.

The variability can also be estimated from nonparametric measurement like UV absorption spectra, giving qualitative information on the global composition of wastewater (linked to UV absorbing substances). This approach will be explained in Chapter 4.2 on industrial wastewater and discharges.

The heterogeneity and variability of wastewater quality must be taken into account when a monitoring programme is planned.

1.2.1.3 Sampling Ageing

As in sewers, wastewater composition can vary very quickly when sampled. This phenomenon, known as sample ageing, occurs under the influence of at least three factors:

- Firstly, as a heterogeneous medium, agitated in a sewer, suspended solids settle rapidly in the sampling flask modifying the distribution of the fraction size by flocculation, adsorption, etc.
- The second factor is of a chemical nature, with reactions of reduction, complexation, modification of acidic–basic equilibria, etc., occurring when the depletion of dissolved oxygen leads to anaerobic conditions and to variation of redox potential and pH. For example, the adsorption of surfactants on suspended solids, is responsible, in raw or physico-chemically treated wastewater, for colloidal fraction aggregation and, thus, for the increase of suspended solids (Baurès *et al.*, 2004).
- The third factor is probably the most important with the biodegradation effect by microorganisms present in wastewater (coming from domestic waste). The consequence is principally a degradation of organic matter, under aerobic or anaerobic conditions, as it is the case in sewers. This will be explained in Chapter 2.1.

Finally, sample ageing occurs even if the samples are refrigerated (in this case the kinetic of sample evolution is slowed down) and can lead to 20 % variation for some parameters (COD, TSS) in a few hours (Baurès *et al.*, 2004). This implies that samples must be transported to the laboratory for analysis as soon as possible after sampling.

1.2.2 MAIN PROCEDURES FOR WASTEWATER QUALITY MONITORING

1.2.2.1 Sampling

Wastewater sampling is generally performed by one of two methods; grab (manual or spot) sampling or automatic (sequential or composite) sampling. The first method is simple, cheap and largely used, whilst the second is better for monitoring relevance, considering the heterogeneity and variability of wastewater. The choice of a sampling procedure is related to the sampling objective, regulatory requirements, measuring treatment plan efficiency, sewer management, knowledge. Grab sampling is useful for detecting fluctuation in composition, and discharge of pollutants, especially in industrial effluent and storm-sewage investigations (Muttamara, 1996; Metcalf and Eddy, 2003), and automatic sampling is preferred for all other purposes (regulatory, time variation, mass balance, etc.). In any case, the measurement of flow rate during sampling is strongly recommended for pollution loads calculation.

Grab sampling

Grab sampling is like a snapshot, giving instantaneously a volume of wastewater in one point. The reliability of measurement and analysis carried out from a grab sample is thus limited to the composition of wastewater for a given control point at one moment. Nevertheless, grab sampling is extensively used for water and wastewater quality monitoring, and can be very useful for rapid information on a 'slug' discharge, intermittently flows, short term variations checking or analysis or very unstable constituents (phenols, cyanides, volatile organic compounds) (WEF, 1996).

It can be thus complementary to composite sampling. However, even if the grab sampling procedure seems to be simple, several recommendations have to be made, namely the following:

- use of clean and adapted flasks, depending on the analysis to be made;
- choose a sampling site with a homogeneous section preventing wastewater quality variability (as for flow measurement);
- pay attention for sludge, biofilm or sediment on bottom or sides of sampling site;
- be aware to not modify the sample composition just after sampling;
- do not agitate before dissolved oxygen on site measurement or fill up the flask for laboratory measurement;
- use relevant conservation procedure(s) depending on analysis;
- always note the sampling conditions of air temperature and time.

Thus grab sampling is not so easy to do, and cannot be carried out by untrained people.

Automatic sampling

For wastewater quality monitoring, an automatic sample is generally preferred because of the time variability of effluents. Automatic sampling can principally be performed using sequential or integrated mode, depending on time or volume.

- Even if it is the simplest form of automatic sampling, because no other devices are needed other than the automatic sampler, the sequential mode can be carried out several ways. The first one is the full sequential sampling mode with sampling at regular time intervals of a given volume collected in one flask. After one sample, the distributing system moves inside the sampler in order to fill the next flask, i.e. several flasks are placed into the sampler (generally 24 or 12), corresponding to hourly or bi-hourly samples. The composite sequential sampling mode is preferred, when a higher sampling frequency is needed, with the collection of equal volume sub-samples at regular time intervals. A selected volume is sampled with a given frequency (e.g. 200 ml every 15 min) and samples are collected in a same flask of large volume (e.g. 20 l) for a single daily composite sample or in several flasks for hourly or bi-hourly composite samples. In this last case, the collection system of the automatic sampler is constituted of 12 or 24 flasks of 1 or 0.5 l, each corresponding to a period of time of 2 or 1 h, if the sampling period is one full day. This technique is used if the daily variation of effluent characteristics has to be known and is obviously more representative than several grab samples.

- The integrated sampling mode is selected when the knowledge of the daily load has to be known. Instead of sequential samples of fixed volume, taken at regular intervals over a period of 24 h, the volume of each sample is proportional to the mean flow rate of a given time interval. Thus a flow meter, generally a device measuring the height of the water table in a control section where the relation height/flow is known, has to be installed and coupled with the automatic sampler. Samples are collected in a single container in order to have a sample representative of the average of the daily composition of wastewater and the pollution load is calculated as the product of a given parameter by the mean value of flow rate during 24 h. If the evolution of composition and load has to be known, samples are collected, as for hourly or bi-hourly sequential sampling, in 24 or 12 flasks. In this case, the daily load can thus be calculated as the sum of hourly or bi-hourly loads. Sometimes, the volume of samples remains constant, but the time interval is automatically adjusted, inversely proportional to the flow rate (e.g. 200 ml are sampled every 10 m^3). The use of two composite sampling during 24 h, at the inlet and outlet of a wastewater treatment plant, is the most common way to determine the average efficiency of the plant.

In practice

The urban wastewater treatment European Directive (Council Directive of 21 May 1991) indicates in Annex I-D that flow-proportional or time-based 24-h samples shall be collected at the same well-defined point in the outlet and if necessary in the inlet of the treatment plant in order to monitor compliance with the requirements for discharged wastewater laid down in this Directive (see Chapter 1.1). Good international laboratory practices aiming at minimizing the degradation of samples between collection and analysis shall be applied. The minimum annual number of samples shall be determined according to the size of the treatment plant and be collected at regular intervals during the year:

- 2000–9999 p. e.: 12 samples during the first year with four samples in subsequent years, if it can be shown that the water during the first year complies with the provisions of the Directive; if one sample of the four fails, 12 samples must be taken in the year that follows.

- 10 000–49 999 p.e.: 12 samples.
- 50 000 p.e. or over: 24 samples.

International standards provide precise information on sampling design (ISO, 1980), sampling techniques (ISO, 1991) and wastewater sampling (ISO, 1992), which is very close to those of other organizations (APHA, 2005). Among the recommendations it can be noted that automatic composite sampling must be chosen for sewer systems, considering the variability in wastewater composition and the difficulty to have a representative sample in very variable conditions. Some other practical recommendations can be found in technical literature (WEF, 1996; Seldon, 2004).

However, some unstable parameters such as dissolved oxygen, temperature, pH, volatile organic compounds cannot be measured in a composite sample, and a grab one is preferable. The use of grab sampling must be avoided when the objective of sampling is to evaluate the performance of a treatment plant. It can be envisaged for a rapid preliminary diagnosis of a sewer network or assessment impact of treated wastewater discharge in receiving medium. Grab sampling can also be used for the study of combined sewer overflow discharges when an automatic sampler cannot be installed.

When a sampling mode is chosen, the precise sampling location(s) must be selected. In order to have the more representative sample, the sampling site must correspond to a well mixed area of wastewater, preferably in a linear section of a channel, where the flow is sufficient to prevent settling, by keeping wastewater solids in suspension. Sampling points for wastewater treatment plants are proposed in technical literature (WEF, 1996).

If automatic sampling is decided upon, two main techniques can be used. The first one is based on a peristaltic pump (or more rarely piston) the characteristics of which must be sufficient for an isokinetic sampling (aspiration speed close to the velocity of wastewater at the sampling site) and for the hydraulic pressure needed from the sewer up to the sampling system. Another technique based on high vacuum for aspiration gives better results for solids capture but tends to increase the related parameters (total suspended solids and global pollution parameters such as COD).

Moreover, the choice of 12 or 24 sampling flasks is important only for the study of the composition variability of wastewater and the measurement of flow rate or volume during sampling is obligatory for loads calculation.

1.2.2.2 Field Measurement

Field measurement can be carried out on site, either by automatic instruments (online analyser or remote sensors) or by manual systems (handheld instruments or test kits) and is very useful for some monitoring objectives like process control or early warning (Thomas and Pouet, 2005). Field measurement is complementary to the classical procedure, recommended and even required in official texts for regulation monitoring, based on sampling and laboratory analysis. This approach, obligatory for the measurement of temperature and very often for other basic parameters (dissolved oxygen, pH, etc.), is increasingly envisaged in order to obtain rapid information, as is the case for early warning systems (detection of accidental pollution). Unfortunately, the availability of systems for on-site or on-line monitoring is rather limited, if restricted to adapted devices (some instruments, derived from laboratory techniques are too complex and fragile, e.g. chromatographs, to be really useful). However, a relevant control of a treatment process cannot be envisaged without on-line monitoring. Among the commercially available on-line systems, UV analysers [for the rapid estimation of global (TOC, COD, TSS) or specific (nitrate, phenols, anionic surfactants) parameters], specific analysers based on electrochemical analysis (e.g. for nutrients) or other principles (TOC meter, hydrocarbons analyser, etc.), are proposed. Chemical or biological colorimetric test kits are also available for a lot of parameters, either mineral or organic. For all these devices, end-users must be aware of the existence of potential interferences. Thus, waiting for the development of reliable and cheap on-site measurement systems, the classical procedure will be preferred for a lot of specific parameters (metallic compounds, emergent pollutants, etc.).

1.2.2.3 Sample Handling

The aim of this section is to stress sample preservation, between sampling and analysis; this topic is well covered by standards and technical works (WEF, 1996; ISO, 2003; APHA, 2005). The basic principles for good handling and conservation practices are very simple. First of all, the delay of conservation between sampling and analysis must be as short as possible to prevent sample ageing. After sampling, samples must be introduced into wide mouthed polyethylene flasks up to the top. For some parameters, such as hydrocarbons and micro or emergent pollutants, more inert and cleanable material, other plastics or preferably (brown) glass, may be used because of adsorption problems. The volume of flask depends on the analytical process and is précised in the literature (WEF, 1996; ISO, 2003; APHA, 2005).

While filling the flask, the raw sample must be gently agitated before being transferred, in order to ensure that suspended solids are collected and to prevent re-oxygenation during transportation. The flasks are then stored at low temperature (4 °C) until analysis. Obviously all information for traceability (location, date, etc.) must be noted while sampling, and flasks carefully identified. For some parameters, preservatives have to be added to the flask (total metallic compounds, BOD, dissolved oxygen by Winkler titration). For more information on conservation, storage, delay before laboratory analysis, standard recommendations must be considered (ISO, 2003).

1.2.3 INTEREST OF SAMPLING ASSISTANCE

For wastewater, sampling is often a routine operation with a given procedure. For a monitoring programme for treatment plant efficiency, the sampling sites are already located (inlet and outlet of a treatment plant), the duration and the frequency fixed (24 h each month), and parameters identical from one sampling campaign to another (for example: temperature, pH, conductivity, BOD, COD, TOC, TSS, nitrogen forms, total phosphorus). However, for objectives other than process efficiency, the design of a sampling procedure is sometimes not evident. For the impact study of a discharge of treated wastewater in a receiving medium or for the diagnosis of a sewer network, the choice of sampling site is difficult, as well as the other factors (mode, date and duration). This is the reason why sampling assistance has to be envisaged to help the design of specific sampling programmes. Considering that an extensive sampling campaign is not realistic (too complex and too expensive), the first step in sampling assistance is the choice of sampling site and the second one is related to the sampling operations, with adapted on-site complementary measurement for grab or automatic sampling.

1.2.3.1 Choice of Critical Control Points

As for natural water, one key point is the design of the monitoring programme, except in the case of a regulatory survey of a wastewater treatment plant where the location (inlet and outlet) and the time period (24 h) are fixed. The study of a sewer network, for example, or of the impact of a treated wastewater discharge, needs to know where to sample. One way to select the sampling points is to apply the Hazard Assessment and Critical Control Points (HACCP) method.

If a good knowledge of the sampling area, based on experience and detailed geo-referenced maps and leading to the obvious choice of sampling sites, is not possible, the HACCP method will help for the monitoring programme design. Developed and used for risk analysis and mitigation in the agro-food industry (Council Directive of 14 June 1993), the method is based on seven steps, which can be adapted for wastewater monitoring:

(1) Analyse hazards. Identification of potential hazards (biological, chemical, or physical) and monitoring objectives.

(2) Identify control points. From the source to the discharge, identification of control points where potential hazard can be controlled or eliminated (e.g. industrial discharge in sewer, see Chapter 4.2).

(3) Establish critical limits. More than critical limits, the choice of parameters and corresponding sampling constraints should be made.

(4) Monitor critical control points. Procedures might include determining the efforts for the organization of sampling operations (manpower, methods, tools, and management).

(5) Take corrective measures. This point is not crucial for sampling assistance but could be envisaged if sampling sites should be moved (or frequency adjusted) to get more representative information.

(6) Establish verification procedures. Procedures include the appliance of best practices for sampling quality control, including a reliable traceability of the final results of the monitoring.

(7) Set up record-keeping procedures. Record-keeping is essential and would include records of hazards and problems encountered and their control methods, the monitoring of safety requirements, and actions taken to correct potential problems.

Finally, the modified HACCP approach can help in the identification of sampling points and in all sampling operations.

1.2.3.2 Assistance for Grab Sampling

Except in the case of 'historical' surveillance, where the operator knows where, when and how sampling, the full design of a grab sampling programme is not easy. The spatio-temporal variability of wastewater composition is a constraint, contrary to sampling locations, very often related to the inlet and outlet of a treatment plant and to the discharge stream of treated wastewater. The main objective being the relevance of the information expected from sample analysis (representativity of sample), the location and the procedure (date and method) should be well defined. Once the critical control points are identified (see above), a simple method derived from natural water sampling (Thomas and Théraulaz, 1994) can be applied for the definition of the final grab sampling procedure. In order to estimate the spatio-temporal variability, field measurement of simple parameters is performed during a pre-sampling programme. Grab sampling using either a field portable sampling line (strain, pipe and pump) or a flask fixed at the end of a pole is done at different locations and times, and on-site conductivity measurement and UV absorption spectrum acquisition are carried out. Conductivity characterizes the mineral matrix of wastewater and the UV spectrum gives quantitative and qualitative information on both dissolved organic absorbing substances and on the particulate fractions (suspended solids and colloids). The results can be used for the estimation of variability and for the final choice of sampling procedure (precise location and date), depending on the sampling objective.

1.2.3.3 Assistance for Automatic Sampling

Automatic sampling is very often used for the evaluation of treatment efficiency of a treatment plant. In this case, the sampling programme is defined according to the objectives of the monitoring. The location is easy (inlet and outlet of treatment plant, discharge or mixture) and the sampling starts at one given moment to end generally one day afterwards. In some other applications, automatic sampling is planned for the survey of nonpermanent events, such as the study of overflows or discharge impact on a receiving medium. In order to be sure that sampling is carried out only if, for example, threshold limits are passed for some parameters, the automatic sampler can be equipped with a multiprobe for the continuous measurement of given parameters (temperature, pH, dissolved oxygen, conductivity, turbidity). If a value exceeds the limit (alarm status), the sampling period starts and a message is sent to the operator for planning further complementary analysis in the laboratory. This interesting function is however limited by the measured parameters (no information on organic pollution). In some cases, the sampling container can be automatically drained out and washed for another sampling phase, if the alarm is not validated (some automatic samplers work each day and drain after 24 h, before restarting).

An adaptation of this method is available for combined sewer overflows monitoring. The automatic sampler starts only when an overflow occurs. This is detected by the measurement of the water table height on the overflow system, giving at the same time an estimation of the discharge volume. The same way can be envisaged for the monitoring of bypass flow to storage tanks in the case of heavy rain, for industrial wastewater.

1.2.3.4 Remote Sensing and Sampling

Starting from the previous configuration with physico-chemical measurement devices, other sensors can be added such as an optical analyser for the acquisition of UV absorption spectra for the estimation of qualitative and quantitative parameters (see Chapter 1.5). Moreover, a field data logger coupled with a transmission procedure (through internet or cellular phone), can be used for the automatic management of the system. If a threshold limit is exceeded, or if a given UV spectrum shape is obtained [corresponding to a (high) polluted state, for example], the operator is warned and can decide to manually start sampling from the internet or cellular phone. This is very useful because the person–machine interaction includes the validation of the protocol. Moreover, a warning message can be sent before the limit is passed, from the increasing trend of some parameters. Therefore, the operator is able to start remote sampling when he or she decides. This procedure is a simplification of the previous SCADA (supervisory control and data acquisition) system, largely used for more complex industrial environments.

REFERENCES

APHA (2005) Standard methods for the examination of water and wastewater. American Public Health Association (APHA), American Water Works Association (AWWA) and Water Environment Federation (WEF) (Eds), New YorK.

Baurès, E., Berho, C., Pouet, M.-F. and Thomas, O. (2004) In situ UV monitoring of wastewater: a response to sample aging. *Water Sci. Technol.*, 49(1), 47–52.

Degrémont (2005) *Mémento technique de l'eau*, 10th Edn. Paris.

Dieu, B. (2001) Application of the SCADA system in wastewater treatment plants. *ISA Trans.*, 40(3), 267–281.

Eckenfelder, W.W. (2001) *Industrial Water Pollution Control*. McGraw Hill Series in Water Resources and Environmental Engineering, 3rd Edn. McGraw Hill, Boston.

European Commission (1991) Council Directive of 21 May 1991 concerning urban waste water treatment (91/271/EEC). European Commission, Brussels.

European Commission (1993) Council Directive of 14 June 1993 concerning hygiene of foodstuffs treatment (93/43/EEC). European Commission, Brussels.

ISO 5667 (1980) Water quality – Sampling – Part 1: Guidance on the design of sampling programmes. International Standardization Organization, Geneva.

ISO 5667-2 (1991) Water quality – Sampling – Part 2: Guidance on sampling techniques. International Standardization Organization. Geneva.

ISO 5667-3 (2003) Water quality – Sampling – Part 3: Guidance on the preservation and handling of water samples. International Standardization Organization, Geneva.

ISO 5667-10 (1992) Water quality – Sampling – Part 10: Guidance on sampling of wastewaters. International Standardization Organization, Geneva.

Metcalf and Eddy (2003) *Wastewater Engineering, Treatment and Reuse*, 4th Edn. McGraw Hill, Boston.

Muttamara, S. (1996) Wastewater characteristics. *Resour. Conserv. Recycl.*, 16, 145–159.

Seldon, J. (2004) Sampling and limits are your environmental fingerprints. *Metal Finish.*, 11, 24–33.

Sophonsiri, C. and Morgenroth, E. (2004) Chemical composition associated with different particle size fractions in municipal, industrial and agricultural wastewaters, *Chemosphere*, 55, 691–703.

Thomas, O. and Théraulaz, F. (1994) Analytical assistance for water sampling. *Trends Anal. Chem.*, 13(9), 344–348.

Thomas, O. and Pouet, M-F. (2005) Wastewater quality monitoring: on line/on site measurement. In: *The Handbook of Environmental Chemistry*, Vol. 5, Part O, D. Barcelo, Ed. Springer, Berlin, pp. 245–272.

WEF (1996) Characterization and sampling of wastewater in Operation of Municipal Wastewater Treatment Plants – MOP 11. Water Environment Federation, Alexandria, VA, pp. 475–507.

1.3
Standard Methodologies

Estelle Dupuit

1.3.1 INTRODUCTION

The monitoring of process effluents and wastewater discharges is required under implementation of the Industrial Pollution Prevention and Control (IPPC) Regulations (96/61/EEC Directive) and the Urban wastewater Treatment Regulations

Wastewater Quality Monitoring and Treatment Edited by P. Quevauviller, O. Thomas and A. van der Beken

(91/271/EEC Directive) (see Chapter 1.1). These have put pressure on the water and wastewater treatment industries with respect to discharge requirements. Traditionally, the quality of treated wastewater is defined by the measurement of global parameters such as biological oxygen demand (BOD), chemical oxygen demand (COD), total organic carbon (TOC), total suspended solids (TSS), etc. (Bourgeois *et al.*, 2001). For example, the COD level is required to be 125 mg l^{-1} (as O_2) to meet the discharge standards applied in European Union countries (Table 1.3.1). In the last few years, more specific parameters, such as total nitrogen, total phosphorus, polycyclic aromatic hydrocarbons, absorbable organic halogens, etc., and a list of dangerous substances have appeared, e.g. in the context of the Water Framework Directive (2000/60/EC).

With respect to the analyses, all countries use nationally or internationally recognised methods. There is a trend in the direction of accepting quick test methods or on-line instrumentation.

This chapter provides background information on what a standard method is, what the different names used are and what national or international organisation is involved. It also reviews the standard methods for monitoring global or specific parameters and describes the different tools developed to trend the quality of wastewater measurements and consequently harmonise the results obtained within the European Union particularly in support of EC regulations (compliance with EC Directives), standardisation (pre-normative research) and calibration means (transfer standards in metrology, CRMs in chemistry, see Chapter 1.6).

1.3.2 DEFINITIONS AND SOURCES

1.3.2.1 Definition

ISO/IEC Guide 2:1996 defines a standard as a document, established by consensus and approved by a recognised body, that provides, for common and repeated use, rules, guidelines or characteristics for activities or their results, aimed at the achievement of the optimum degree of order in a given context.

Four major types of standards may be cited:

- Fundamental standards which concern terminology, metrology, conventions, signs and symbols, etc.
- Standards which define the characteristics of a product (product standard) or of a specification standard which service (service activities standard) and the performance thresholds to be reached (fitness for use, interface and interchangeability, health, safety, environmental protection, standard contracts, documentation accompanying products or services, etc.).
- Organisation-related standards which deal with the description of the functions of the company and with their relationships, as well as with the modelling of the

Table 1.3.1 Minimum requirements for discharges from urban wastewater treatment plants (Tables 1 and 2 of Annex 1 of Directive 91/271/EEC)

Parameters	Concentration (mg l^{-1})	Minimum percentage of reduction[a]	Reference method of measurement
Biological oxygen demand (BOD5 at 20 °C) without nitrification[b]	25	70–90	Homogenised, unfiltered, undecanted sample. Determination of dissolved oxygen before and after 5-day incubation at 20 ± 1 °C, in complete darkness. Addition of a nitrification inhibitor
Chemical oxygen demand (COD)	125	75	Homogenised, unfiltered, undecanted sample potassium dichromate
Total suspended solids (TSS)	35 (>10 000 p.e.)	90[c] (more than 10 000 p.e.)	Filtering of a representative sample through a 0.45 μm filter membrane. Drying at 105 °C and weighing
	60 (2000–10 000 p.e.)	70 (2000–10 000 p.e.)	Centrifuging of a representative sample (for at least 5 min with mean acceleration of 2800–3200 *g*), drying at 105 °C and weighing
Total phosphorus	2 (10 000–100 000 p.e.) 1 (>100 000 p.e.)	80	Molecular absorption spectrophotometry
Total nitrogen[d]	15 (10 000–100 000 p.e.) 10 (>100 000 p.e.)[e]	70–80	Molecular absorption spectrophotometry

[a] Reduction in relation to the load of the influent.

[b] The parameter can be replaced by another parameter: total organic carbon (TOC) or total oxygen demand (TOD) if a relationship can be established between BOD5 and the substitute parameter.

[c] This requirement is optional. Analyses concerning discharges from lagooning shall be carried out on filtered samples; however, the concentration of TSS in unfiltered water samples shall not exceed 150 mg l^{-1}.

[d] Total nitrogen means: the sum of total Kjeldahl-nitrogen (organic N + NH_3), nitrate (NO_3^-)-nitrogen and nitrite (NO_2^-)-nitrogen.

[e] Alternatively, the daily average must not exceed 20 mg l^{-1} N. This requirement refers to a water temperature of 12 °C or more during the operation of the biological reactor of the wastewater treatment plant. As a substitute for the condition concerning the temperature, it is possible to apply a limited time of operation, which takes into account the regional climatic conditions. This alternative applies if it can be shown that paragraph 1 of Annex 1.D is fulfilled.

activities (quality management and assurance, maintenance, value analysis, logistics, quality management, project or systems management, production management, etc.).

- Test methods and analysis standards which measure characteristics (standard methods) (www.wssn.net).

In the USA, standard method is a joint publication of the American Public Health Association (APHA), the American Water Works Association (AWWA) and the Water Environment Federation (WEF). The regulatory method authorised by the Environmental Protection Agency and referenced in the Code of Federal Regulation (CFR title 40) is the EPA method. In France, it is known as the normalised method. In this chapter, 'standard method' refers to a document which outlines the procedures used to analyse impurities and characteristics in air, ground and water.

In particular, a standard method is defined as a published procedure that contains details for measuring a specific analyte (or analytes) in a specified medium (e.g. water, soil, air, etc.) and, where applicable, matrix (subcategories of media such as drinking water, groundwater, industrial or municipal wastewaters, etc.). Methods may apply to sample preparation, instrumental analysis (including both field and fixed-site laboratory analyses) of environmental samples, QA/QC procedures, etc., and to to a wide variety of analytes including organic and inorganic chemicals, radioactive isotopes, microbiological and macrobiological organisms. A standard method consists of providing the pertinent information necessary to compare the attributes among methods and determine which, if any, best meet user-specific needs. This includes the determinative technique employed, major instrumentation required, metadata (e.g. accuracy, precision, detection level, rates of false positive and false negative conclusions, etc.), interferences, relative cost and some summarised procedural information.

1.3.2.2 Sources of International, Regional and National Standardisation

Standards are drawn up at international, regional and national level. The coordination of the work at these three levels is ensured by common structures and cooperation agreements.

International Standardisation Organisation (ISO)

Founded in 1947, the International Standardisation Organisation (ISO) is a worldwide federation of national standards bodies, currently comprising over 125 members, one per country. The mission of ISO is to encourage the development of standardisation and related activities in the world in order to facilitate international exchanges of goods and services and to achieve a common understanding in the intellectual, scientific, technical and economic fields. Its work concerns all the fields of standardisation, except electrical and electronic engineering standards, which fall within the scope of the IEC.

ISO counts over 2800 technical work bodies (technical committees, subcommittees, working groups and ad hoc groups). To date, ISO has published over 11 000 International Standards.

ISO has its central offices in Geneva, Switzerland. The transposition of ISO standards into the national collections is voluntary: it may be complete or partial.

A large number of international organisations are in liaison with ISO and participate to varying degrees in their work. Several of these organisations have themselves standardisation activities in their own area of interest, which are recognised at international level. In a number of cases, the results of the standardisation work of these organisations are fed directly into the ISO system and appear in International Standards published by ISO. However, some of these organisations themselves publish normative documents, and these must be taken into account in any review of international standardisation.

Pan American Standards Commission (COPANT)

COPANT is a civil, nonprofit association. It has a complete operational autonomy and unlimited duration. The basic objectives of COPANT are to promote the development of technical standardisation and related activities in its member countries with the aim of promoting the industrial, scientific and technological development in benefit of an exchange of goods and the provision of services, while facilitating cooperation in the intellectual, scientific and social fields. The Commission coordinates the activities of all institutes of standardisation in the Latin American countries. The Commission develops all types of product standards, standardised test methods, terminology and related matters. The COPANT headquarters are in Buenos Aires, Argentina.

European Committee for Standardisation (CEN)

Founded in 1961, CEN draws up European standards and regroups 18 European standards institutes. CEN has witnessed strong development with the construction of the European Union. Its headquarters is located in Brussels, Belgium. A Technical Board is in charge of the coordination, planning and programming of the work which is conducted within the work bodies (technical committees, subcommittees, working groups), the secretariats of which are decentralised in the different EU member states. CEN, which counts over 250 technical committees, has published some 2400 documents, including 2100 European standards. Over 9000 documents are under study.

1.3.2.3 National Standardisation

Each country possesses its own national standardisation system. The central or most representative national standards body (Table 1.3.2) participates within the regional or international bodies.

Table 1.3.2 List of national organisations for standardisation (www.wssn.net)

Country	Acronym	National members of ISO
Algeria	IANOR	Institut algérien de normalisation
Argentina	IRAM	Instituto Argentino de Normalización
Armenia	SARM	Department for Standardisation, Metrology and Certification
Australia	SAI	Standards Australia International Ltd–Australian National Committee of the IEC
Austria	ON	Austrian Standards Institute
Belgium	IBN	The Belgian Institution for Standardisation
Bolivia	IBNORCA	Instituto Boliviano de Normalización y Calidad
Brazil	ABNT	Associaçao Brasileira de Normas Técnicas
Brunei Darussalam	CPRU	Construction Planning and Research Unit, Ministry of Development
Canada	SCC	Standards Council of Canada
Chile	INN	Instituto Nacional de Normalizacion
China	SACS	State Administration of China for Standardisation
Colombia	ICONTEC	Instituto Colombiano de Normas Técnicas y Certificación
Costa Rica	INTECO	Instituto de Normas Técnicas de Costa Rica
Croatia	DZNM	State Office for Standardisation and Metrology
Czech Republic	CSNI	Czech Standards Institute
Denmark	DS	Dansk Standard
Ecuador	INEN	Instituto Ecuatoriano de Normalización
El Salvador	CONACYT	Consejo Nacional de Ciencia y Tecnología
Ethiopia	QSAE	Quality and Standards Authority of Ethiopia
Finland	SFS	Finnish Standards Association
France	AFNOR	Association française de normalisation
Germany	DIN	Deutsches Institut für Normung
Greece	ELOT	Hellenic Organisation for Standardisation
Guatemala	COGUANOR	Comisión Guatemalteca de Normas
Hong Kong, China	ITCHKSAR	Innovation and Technology Commission
Hungary	MSZT	Magyar Szabványügyi Testület
Iceland	STRI	Icelandic Council for Standardisation
India	BIS	Bureau of Indian Standards
Indonesia	BSN	Badan Standardisasi Nasional
Iran, Islamic Republic	ISIRI	Institute of Standards and Industrial Research of Iran
Ireland	NSAI	National Standards Authority of Ireland
Israel	SII	The Standards Institution of Israel
Italy	UNI	Ente Nazionale Italiano di Unificazione
Jamaica	JBS	Bureau of Standards, Jamaica
Japan	JISC	Japan Industrial Standards Committee
Kenya	KEBS	Kenya Bureau of Standards
Korea, Republic of	KATS	Korean Agency for Technology and Standards
Kyrgyzstan	KYRGYZST	State Inspection for Standardisation and Metrology
Latvia	LVS	Latvian Standard
Lithuania	LST	Lithuanian Standards Board

Table 1.3.2 (*Continued*)

Country	Acronym	National members of ISO
Luxembourg	SEE	Service de l'Energie de l'Etat, Organisme Luxembourgeois de Normalisation
Malaysia	DSM	Department of Standards Malaysia
Malta	MSA	Malta Standards Authority
Mexico	DGN	Dirección General de Normas
Moldova, Republic of	MOLDST	Department of Standardisation and Metrology
Morocco	SNIMA	Service de normalisation industrielle marocaine
Netherlands	NEN	Nederlands Normalisatie-Instituut
New Zealand	SNZ	Standards New Zealand
Nicaragua	DTNM	Dirección de Tecnología, Normalización y Metrología
Norway	NSF	Norges Standardiseringsforbund
Oman	DGSM	Directorate General for Specifications and Measurements
Peru	INDECOPI	Instituto Nacional de Defensa de la Competencia y de la Protección de la Propiedad Intelectual
Philippines	BPS	Bureau of Product Standards
Poland	PKN	Polish Committee for Standardisation
Portugal	IPQ	Instituto Português da Qualidade
Russian Federation	GOST-R	State Committee of the Russian Federation for Standardisation, Metrology and Certification
Saudi Arabia	SASO	Saudi Arabian Standards Organisation
Singapore	PSB	Singapore Productivity and Standards Board
Slovakia	SUTN	Slovak Standards Institution
Slovenia	SIST	Slovenian Institute for Standardisation
South Africa	SABS	South African Bureau of Standards
Spain	AENOR	Asociación Española de Normalización y Certificación
Sri Lanka	SLSI	Sri Lanka Standards Institution
Sweden	SIS	Standardiseringen i Sverige
Switzerland	SNV	Swiss Association for Standardisation
Syrian Arab Republic		The Syrian Arab Organisation for Standardisation and Metrology
Thailand	TISI	Thai Industrial Standards Institute
Trinidad and Tobago	TTBS	Trinidad and Tobago Bureau of Standards
Turkey	TSE	Türk Standardlari Enstitüsü
Uganda	UNBS	Uganda National Bureau of Standards
Ukraine	DSTU	State Committee of Standardisation, Metrology and Certification of Ukraine
United Arab Emirates	SSUAE	Directorate of Standardisation and Metrology
United Kingdom	BSI	British Standards Institution
United States	ANSI	American National Standards Institute
Uruguay	UNIT	Instituto Uruguayo de Normas Técnicas
Venezuela	FONDONORMA	Fondo para la Normalización y Certificación de la Calidad
Vietnam	TCVN	Directorate for Standards and Quality

1.3.3 STANDARD METHODS OF MAIN PARAMETERS

EU water directives include guidance on the selection of appropriate monitoring methodologies, frequency of monitoring, compliance assessment criteria and environmental monitoring. The quality of the treated wastewaters must be better than reference values for parameters such as BOD, COD, TSS and even the global nitrogen and total phosphorus. These provisions are of great importance but the chosen parameters are not easy to measure without sampling, storage and laboratory analysis.

1.3.3.1 Biological Oxygen Demand

The determination of BOD is an empirical test in which standardised laboratory procedures are used to determine the relative oxygen requirements of wastewater, effluents and polluted waters. It is defined as the potential for removal of oxygen from water by aerobic heterotrophic bacteria which utilise organic matter for their metabolism and reproduction. In fact, the BOD values indicate the amount of biodegradable organic material (carbonaceous demand) and the oxygen used to oxidise inorganic material such as sulfides and ferrous iron. It also may measure the oxygen used to oxidise reduced forms of nitrogen (nitrogenous demand) unless their oxidation is prevented by an inhibitor.

The BOD test has its widest application in measuring waste loading to treatment plants and in evaluating the BOD removal efficiency of such treatment systems. BOD has been determined conventionally by taking a sample of water, aerating it, placing it in a sealed bottle, incubating for a standard period of time at $20 \pm 1\ ^\circ C$ in the dark, and determining the oxygen consumption in the water at the end of incubation (NF EN 1899-1 and 2 standards). According to the American standard (EPA method 405.1), the incubation time is 5 days and the BOD values based on this standard are called BOD5 for short, whereas the incubation time is 7 days in the Swedish standard and the abbreviation is BOD7. The conventional BOD test has certain benefits such as being a universal method of measuring most wastewater samples, and furthermore, no expensive equipment is needed (Liu and Mattiasson, 2006).

Indeed, BOD5 is an indicator of biological activity and provides an indication of the eventual degradation of the organic waste. This parameter is therefore a suitable measurement in biological treatment processes (Guwy *et al.*, 1999). It has, however, the limitation of being time consuming, and consequently it is not applicable to on-line process monitoring. Thus, it is necessary to develop an alternative method that circumvents the weakness of the conventional BOD test described above (Liu and Mattiasson, 2006). Since the BOD5 test takes 5 days it is of no use in automated control systems and often other automatic/on-line measurements are used in its place, such as the so called short-term BODs based on respirometric techniques, COD, TOC, fluorescence and UV absorbance (Guwy *et al.*, 1999). Fast determination of

BOD could be achieved by the biosensor-based methods (Liu and Mattiasson, 2006). These alternative techniques are presented in the following paragraphs.

1.3.3.2 Chemical Oxygen Demand

The COD test is now widely used as a means of measuring the organic strength of domestic and industrial waste, often replacing BOD as the primary parameter in wastewater. It is based upon the fact that most organic compounds can be oxidised by the action of strong oxidising agents under acid conditions (Bourgeois *et al.*, 2001).

The measurement of COD is carried out on the basis of the 'closed reflux, colorimetric method' described in water quality norms (NF T90-101/ISO 6060:1989/EPA method 410.3). Sample, blanks and standards in sealed tubes are heated in an oven or block digestor in the presence of dichromate at 150 °C. After 2 h, the tubes are removed from the oven or digestor, cooled and measured on a UV/VIS spectrophotometer at a wavelength of 600 nm. Chlorides are quantitatively oxidised by dichromate and represent a positive interference. Mercuric sulfate is added to the digestion tubes to complex the chlorides.

The described method essentially consists of measuring the amount of oxygen required. It takes into account any substance or element presenting a reducing character. Some reducer salts [nitrites, sulfides and iron(II)] are also oxidised but the equivalence dissolved organic carbon (DOC) values are known (Table 1.3.3). Moreover, the aromatic hydrocarbons and the pyridine are not completely oxidised. Some very volatile organic compounds are not oxidised because of evaporation. In addition, the not ramified aliphatic compounds are oxidised only with the presence of sulfuric acid – sulfuric silver.

Organic matter is converted to carbon dioxide and water regardless of the biological assimilability of the substances. For example, glucose and lignin are both oxidised completely.

This method is applicable to water whose DCO is higher than 30 mg l^{-1} and whose chloride concentration (expressed as ion chloride) is lower than 2000 mg l^{-1}.

Table 1.3.3 COD equivalence of some reducer salts (Berne and Cordonnier, 1991. Reproduced by permission of Editions TECHNIP Paris)

Compound	Ion	COD (mg O_2 mg^{-1})
Cyanide	CN^-	1–2.9
Thiocyanate	SCN^-	0.6–1.5
Sulfide	S^{2-}	2
Sulfur	S°	1.5
Thiosulfate	$S_2O_3^{2-}$	0.57
Tetrathionate	$S_4O_6^{2-}$	0.5
Sulfite	SO_3^{2-}	0.2

The maximum value of DCO which can be given, under the defined conditions, on a sample not diluted, is 700 mg l^{-1}.

The major advantage of the COD test is that the results can be obtained within a relatively short time (approximately 2 h instead of 5 days for the BOD5). In the case where there is no change in wastewater quality and no evolution time, a correlation between COD and BOD values can be established. DCO values are correct only when the effluent is completely biodegradable and not have reducer salt. Nevertheless, when used in conjunction with BOD, the COD test can provide an indication of the biodegradability of the wastewater by calculating the BOD/COD ratio. It can also be helpful in relation to toxic conditions.

One of the main limitations of the COD test is its inability to differentiate between biodegradable and biologically inert organic matter on its own. Therefore, the use of chemicals such as acid, chromium, silver and mercury produce liquid hazardous waste which requires disposal (Bourgeois *et al.*, 2001). It is interesting to develop alternative methods without toxic reagents (biosensors, optical sensors, etc.).

1.3.3.3 Total Organic Carbon

Different forms of carbon can be found in wastewater (Figure 1.3.1), such as mineral or organic, volatile or not. The relevant parameter for the global determination of the organic pollution is the TOC.

Two main techniques are usually used for the conversion of organic carbon to carbon dioxide for TOC determination. In the first one, called wet chemical oxidation (WCO), oxidation is performed at low temperature by UV light and the addition of persulfate reagent, after removal of inorganic carbon by acidification and aeration. The second uses a catalyst at high temperature (650–900 °C) and is known as high temperature catalytic oxidation (HTCO).

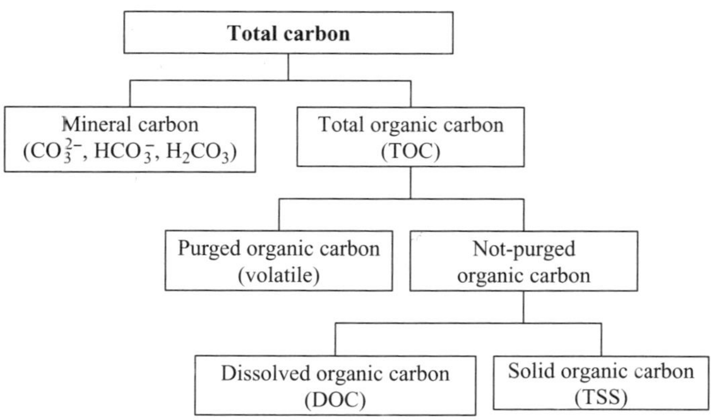

Figure 1.3.1 Different forms of carbon (Ministère de l'aménagement du territoire et de l'environnement, 2000)

Significant differences and conflicting results between the two techniques have been shown (Thomas *et al.*, 1999). As a result, both methods are still being investigated and their accuracy is still subject to controversy (Bourgeois *et al.*, 2001). The use of TOC is difficult in a wastewater treatment plant because of the lack of correlation between TOC and BOD. In fact TOC only measures the content of organic compounds, not other substances that may contribute to BOD (APHA, 1992).

1.3.3.4 Total Suspended Solids

The TSS (in mg l^{-1}) is measured by weighing after filtration or centrifugation and drying at 105 °C (NF T90-105 standard). The centrifugation method is used when filtration is not applicable because of a high risk of clogging of filters.

The decanted solids correspond to the TSS which decant during a time fixed conventionally at 2 h. The decanted solids (in cm^3 l^{-1}) are measured by direct reading of the volume occupied at the bottom of a decantation cone.

The colloidal solids represent the difference between the TSS and decanted solids. The particle size roughly lies between 10^{-8} mm and 10^{-2} mm.

In addition, the TSS are constituted of mineral solids and organic solid, or suspended volatile solids. Organic solid can be determined by the calcination test to 180 °C (NF T90-029 and NF EN 872 standards), but could not be very precise due to partial or total decomposition of certain salts (bicarbonates, chlorides, nitrates, etc.).

1.3.3.5 Specific Organic Compounds: Phenols

Phenols belong to the base, neutral and acid organics family. Two methods are usually used: extraction coupled with gas chromatography analysis (ISO 8165-1:1992, 40 CFR Part 136, Appendix A, method 625) and extraction with colorimetry (ISO 6439: 1990, EPA method 420.1).

The first method is applicable to the determination of extractable organics in municipal and industrial discharges.

A 1 l aliquot of sample is adjusted to pH >11 and extracted in a separatory funnel with three 60 ml portions of methylene chloride or with 200–500 ml methylene chloride in a continuous extraction apparatus. The pH of the sample is then adjusted to <2 and the extraction procedure is repeated. The extracts are concentrated with a Kuderna–Danish concentrator fitted with a three-ball Snyder column. The final volume is adjusted to 1 ml. The organic priority pollutants are determined in the extracts by capillary column or packed column gas chromatography – mass spectrometry. The interferences are contaminants from glassware or compounds that are co-extracted with sample.

The second method determines phenolic compounds in drinking, surface and saline waters or domestic and industrial wastes. Phenolic materials react with 4-aminoantipyrine in the presence of potassium ferricyanide at high pH to form a stable reddish-brown coloured antipyrine dye. The amount of colour produced is proportional to the concentration of phenolic materials. However, the colour response of all phenolic compounds is not equivalent and the results (which are compared against pure phenol standards) represent the minimum concentration of phenolic compounds in the sample.

Interferences from sulfur compounds are eliminated by acidifying the sample to pH <4 with phosphoric acid and aerating briefly by stirring and adding copper sulfate. Oxidising agents can oxidise phenolics, causing results to be low. The presence of oxidising agents is rested for with potassium iodide strips. If present, they are removed when sampling by adding ferrous ammonium sulfate in excess.

1.3.3.6 Mineral Compounds: Total Nitrogen and Total Phosphorus

Total nitrogen in water corresponds to nitrate and nitrite compounds. They are analysed by colorimetric method with an automated hydrazine reduction (NF EN ISO 11732, EPA method 352.1).

This method is applicable to drinking and surface water, and domestic and industrial wastes. The applicable range of this method is 0.01–10 mg l^{-1} nitrate–nitrite nitrogen. Nitrate is reduced to nitrite with hydrazine sulfate and the nitrite (that originally present plus reduced nitrate) is determined by diazotising with sulfanilamide and coupling with *N*-(naphthyl)-ethylenediamine dihydrochloride to form a highly coloured azo dye which is measured colorimetrically. Sample colour that absorbs in the photometric range used for analysis will interfere. The apparent NO_2^- and NO_3^- concentrations varied ±10 % with concentrations of sulfide ion up to 10 mg l^{-1}.

Phosphorus is also analysed by colorimetry (NF EN 1189, ISO 6878:1998, EPA method 365.1). This method is based on reactions that are specific for the orthophosphate ion. Thus, depending on the prescribed pretreatment of the sample, the various forms of phosphorus that may be determined are given in Figure 1.3.2.

Ammonium molybdate and antimony potassium tartrate react in an acid medium with dilute solutions of phosphorus to form an antimony-phosphomolybdate complex. This complex is reduced to an intensely blue-coloured complex by ascorbic acid. The colour is proportional to the phosphorus concentration. The applicable range is 0.01–1 mg P l^{-1}.

Only orthophosphate forms a blue colour in this test. Polyphosphates (and some organic phosphorus compounds) may be converted to the orthophosphate form by manual sulfuric acid hydrolysis. Organic phosphorus compounds may be converted to the orthophosphate form by manual persulfate digestion. The developed colour is measured automatically.

No interference is caused by copper, iron, or silicate at low concentrations. However, high iron concentrations can cause precipitation, and subsequent loss, of phosphorus. The salt error for samples ranging from 5 to 20 % salt content was found to be less than 1 %. Arsenate is determined similarly to phosphorus and should be considered when present in concentrations higher than phosphorus.

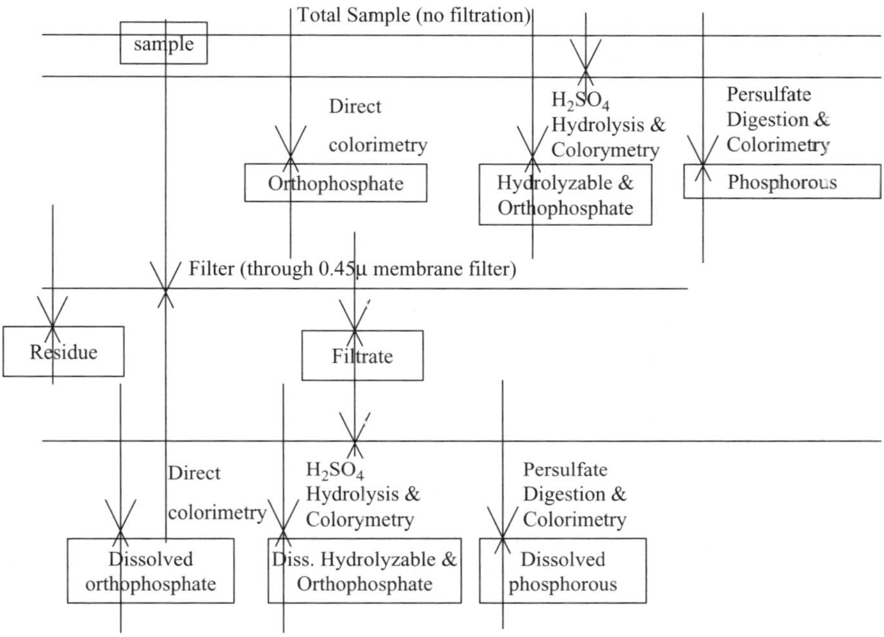

Figure 1.3.2 Analytical scheme for differentiation of phosphorus forms (EPA method 365.1)

1.3.4 IMPROVEMENT IN QUALITY OF WASTEWATER ANALYSIS

Due to the demand for reliable and comparable methods, performance requirements have been established at national and international level by implementation of accreditation systems, QA guidelines and standards (e.g. ISO 9000 and EN 45 000 series), organisation of interlaboratory studies, proficiency testing and production of laboratory and certified reference materials (Anklam *et al.*, 2002; see also Chapter 1.6). Indeed, any method proposed to become official must be validated in a collaborative trial study, resulting in defined method performance characteristics, while the framework for the design and conduction of such collaborative trial studies as well as the statistical evaluation are also defined in appropriate protocols (Horwitz, 1995). Any method that has been successfully validated according to these protocols can be recognised as an official method for use in legal cases or for international trade purpose. In addition to these performance criteria, economical and prevention strategy aspects have also lately become important in method development. Demands for fast and efficient procedures (consumption of chemicals and materials) and the ability for automation are highly desired.

The objective of the method validation is to demonstrate that the defined system (which may include various steps in the analytical procedure, and may be valid for a restricted matrix) produce acceptably accurate, repeatable and reproducible results for a given property. Depending upon the intended purpose of the analysis, different validation parameters have to be evaluated.

1.3.4.1 Tools for Establishing and Controlling Robust Analytical Processes

To define the performance characteristics of a method, two validation schemes can be used.

The first concerns in-house studies based on a detailed investigation and evaluation of one single analytical procedure by:

- Studying its applicability for a range of matrices by checking its compliance to various acceptance criteria (e.g. within-laboratory, within-day repeatability and within laboratory, between-day reproducibility).
- Studying its accuracy for a range of matrices by comparing it with an already validated and robust analytical procedure (in France, XP T 90-210) or a certified reference material (CRM).

The second way of assessing the performance of analytical methods is to compare them within the frame of interlaboratory studies (NF ISO 5725). The comparison of different techniques as applied in different laboratories allows the detection of errors due to a particular method, or part of a method (e.g. insufficient extraction, uncontrolled interferences), or due to a lack of quality control within one laboratory. The participation in such interlaboratory studies may then help in establishing the state of the art in a particular field of analysis and to improve the quality of the measurements (Quevauviller, 2002). An example for such interlaboratory study is given in Chapter 1.6. These interlaboratory studies can have different purposes:

- To validate one single analytical procedure or sampling plan applied by different laboratories and to derive typical performance characteristics (e.g. repeatability, reproducibility, and accuracy).
- To compare different analytical procedures or sampling plans applied by different laboratories to identify systematic errors.
- Both of the above described types can be organised as the so-called 'step by step' approach. This approach consists of a series of interlaboratory studies following the different steps of the analytical process.

These data provide information on the expected precision (within laboratory standard deviation), possible systematic error (bias), recovery values (on the basis of spiking

Table 1.3.4 Parameters determined through performance and validation studies

Term	Description
Specificity	The probability of obtaining a negative result, given that there is no analyte present
Linearity	Proportionality of the signal to the amount of reference material, demonstrated by the calculation of a regression line with the adequate statistical method
Range	Range of analyte concentrations over which the method is considered to perform in a linear manner
Accuracy	The closeness of agreement between a test result and the accepted reference value (ISO 3534-1)
Trueness	The closeness of agreement between the average value obtained from a large series of test results and an accepted reference value (ISO 3534-1)
Detection limit	Minimum level the presence of an analyte can be measured with a given certainty (e.g. 95 %) (DIN 32645)
Quantification	Minimum level the analyte can be quantified with a given certainty (e.g. 95 %) (DIN 32645)
Robustness	Stability of the method with respect to deliberate variations in the method parameters

measurements), applicability, and interference with other compounds and/or matrix components during analysis and best calibration approaches (Table 1.3.4).

1.3.4.2 Tools for Establishing On-line Sensors/Analysing Equipment in Water

A new project funded by the European Commission, 'European Testing and Comparability of On-line Sensors (ETACS)' has recently been initiated. The purpose of the project is to develop generic laboratory and field test protocols to facilitate acceptance of validated on-line sensors/analyser and increase market capabilities. This project was funded under the EC Standards, Measurement and Testing Programme. This work has been progressed within the ISO TC 147/WG2 and underpins the draft international standard (ISO/DIS 15839).

This objective is to initialise a process, which will establish a validation scheme, which will have the form of a test protocol. The standard is applicable to most sensors/analysing equipment by defining (scope of draft of ISO/CD 15839):

- on-line sensors/analysing equipment;
- the terminology describing performance characteristics of on-line sensors/analysing equipment;
- the test procedures (for laboratory and field) used to evaluate the performance characteristics of on-line sensors/analysing equipment;

The instrument testing is organised in two parts:

- Laboratory based tests to ensure that instruments perform to the required specifications.
- Field trials over a few-months period, to ensure that the instruments then work on a real application (Jacobsen and Lynggaard-Jensen, 1998).

The instrument performance standards are modular specifications built from the relevant sections of a number of ISO and CEN standards. Therefore, the content of a test protocol should be based on the typical performance characteristics of *in situ* on-line sensors, which include linearity, response time, lower detection limit and repeatability (Table 1.3.5) (Lynggaard-Jensen, 1999).

The purpose of the project is to develop generic test protocols both in the laboratory and field to facilitate acceptance of validated on-line sensors/analysers and increase market capabilities. In these cases, the different tests, the definitions and the information/materials can be described by the scheme shown in Figure 1.3.3.

The properties of sensors, the results of laboratory tests and the results of field tests have to be written in a report. Technical aspects such as the principle of measurement, reliability, accuracy and detection limit and the intrinsic properties of the sensors (single or multiparameter, need for external sampling and filtration, etc.) dictate whether or not the technology can be accepted as a standard method by the end user and the relevant authorities.

Table 1.3.5 Performance characteristics of on-line *in-situ* sensors/analysers. (Reprinted from Talanta, so, Lynggaard-Jensen, Trends in monitoring of wastewater systems, pp. 707–716, Copyright 1999, with permission from Elsevier)

Performance characteristics	Laboratory test	Field test
Linearity (range)	X	
Lowest detectable change	X	
Selectivity	X	
Limit of detection	X	
Limit of quantification	X	
Response times	X	X
Dead (lag) time	X	
Rise and fall times	x	
Ruggedness	X	
Trueness/bias	X	X
Repeatability	X	
Reproducibility	X	
Up time		X
Drift		X
Memory effects	X	

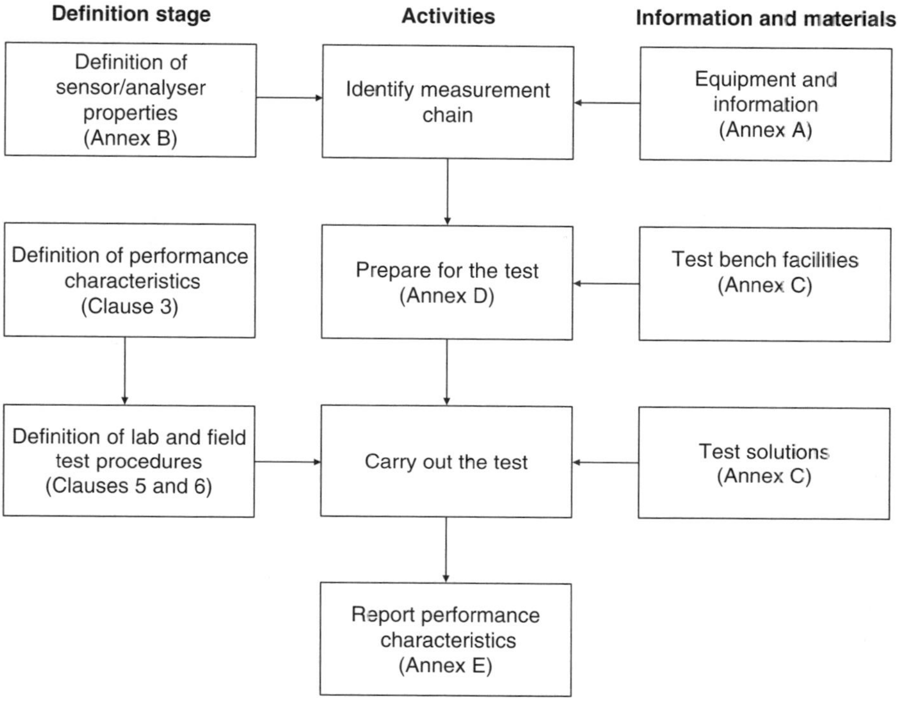

Figure 1.3.3 Diagram of the overview of the test activities (draft ISO/CD 15839). The terms and definitions taken from ISO 15839:2003, Figure 1 Overview of Test, are reproduced with permission of the International Organization for Standardization ISO. This standard can be obtained from any ISO member and from the website of ISO Central Secretariat at the following address: www.iso.org. Copyright remains with ISO

1.3.5 CONCLUSIONS

A standard method is defined as a published procedure that gives details to measure specific analyte(s) in specific medium. Each country publishes these procedures by specifics organisations. In the USA, they are published by the Environmental Protection Agency as regulations (Title 40 of the Code of Federal Regulation) and in France, the standard method can be found in AFNOR books.

There are many standard methods to measure water parameters cited by the Directives. This chapter has presented the most current ones. Standard techniques for the measurement of global parameters, such as BOD, COD and TOC, pose some problems to the end user and the legislator because of their performance characteristics. Theses techniques have been designed as off-line methods, requiring sample collection and retrospective laboratory analysis. The quality water directives include guidance on the selection of the appropriate monitoring methodologies, frequency of monitoring, compliance assessment criteria and environmental monitoring

(Bourgeois *et al.*, 2001). In order to comply with the regulation, there is a general trend for using continuous monitoring and automated measuring techniques (Environmental Agency, 2001).

On-line sensors and other analytical tests in continuous or sequential mode would facilitate process control and plant operation strategy. Nevertheless, the water industry remains slow in taking up new technologies because of the lack of recognised and standardised methods or instruments that would satisfy all their practical requirements (Jacobsen, 1999). A project, ETACS, has been initiated to facilitate acceptance of validated on-line sensors/analysers. This project will define the procedures to control the performance characteristics of these sensors.

Associated with the performance characteristics, other factors, such as cost of ownership, ease of use and sensor placement, will influence the consumer's choice.

REFERENCES

Anklam, E., Stroka, J. and Boenke, A. (2002) *Food Control*, **13**, 173–183.

APHA (1992) *Standard Methods for the Examination of Water and Wastewater,* 18th Edn. Washington, DC.

Berne, F. and Cordonnier, J. (1991) *Traitement des eaux*. Editions TECHNIP, Paris.

Bourgeois, W., Burgess, J.E. and Stuetz, R.M. (2001) *J. Chem. Technol. Biotechnol.*, **76**, 337–348.

Environmental Agency (2001) Proposal to extend the environment Agency's monitoring Certification Scheme (MCERTS) to continuous Water Monitoring Systems.

Guwy, A.J., Farley, L.A., Cunnah, P., Hawkes, F.R., Hawkes, D., Chase, M. and Buckland, H. (1999) *Water Res.*, **33**(14), 3142–3148.

Horwitz, W. (1995) *Pure Appl. Chem.*, **67**, 331–343.

Jacobsen, B.N. (1999) *Talanta*, **50**(4), 77–723.

Jacobsen, H.S. and Lynggaard-Jensen, A. (1998) On-line measurement in wastewater treatment plants: sensor development and assessment of comparibility of one-line sensors, In: *Monitoring of Water Quality*. Elsevier, Amsterdam, pp. 89–102.

Liu, J. and Mattiasson B. (2002) *Water Res.*, **36**(15), 3786–3802.

Lynggaard-Jensen, A. (1999) Trends in monitoring of wastewater systems. *Talanta*, **50**(4), 707–716.

Ministère de l'aménagement du territoire et de l'environnement (2000) Principaux rejets industriels en France, bilan de l'année 2000.

Quevauviller, P. (2002) *Quality Assurance for Water Analysis*, Water Quality Measurement Series, John Wiley & Sons, Ltd.

Thomas, O., El Khorassani, E., Thouraud, E. and Bitar, H. (1999) *Talanta*, **50**(4), 743–749.

1.4
Alternative Methods

Olivier Thomas

1.4.1 CONTEXT AND DEFINITION

The needs of water and wastewater quality monitoring increase but the technical means and the financial resources are limited. The classical way based on sampling and analysis is a rather complex, time consuming and expensive solution, but

Wastewater Quality Monitoring and Treatment Edited by P. Quevauviller, O. Thomas and A. van der Beken

essential for some applications, as, for example, in a regulatory context. For other purposes like early warning, end users ask frequently for more simple procedures such as the use of sensors or on line systems for real time information.

1.4.1.1 Limits of the Sampling/Analysis Procedure

The main procedure for wastewater quality monitoring is based on the following steps:

- sampling (grab or integrated, with time or flow);
- conservation, storage (usually at low temperature);
- transportation;
- and laboratory analysis (immediate or postponed).

This general procedure, often completed with sample pretreatment and on site-flow measurement, is well established and several standards define the different steps (for example ISO standards, see Chapter 1.3), but there exist limits or drawbacks, with regard to some monitoring objectives.

The monitoring objectives for water and wastewater can be numerous but a way to present them is to follow the definition of the three modes of monitoring specified by the European Water Framework Directive (European Commission, 2000): surveillance monitoring to assess long-term changes; operational monitoring to provide extra data on water bodies at risk or failing to meet the environmental objectives of the Water Framework Directive (WFD); and investigative monitoring to determine the causes of such failure where they are unknown. For wastewater, the appliance of the general procedure for the two last modes, leads to the following limitations:

- The first limit of the general procedure is related to the delay, from sampling to results. Generally, and depending on the type of analysis, a delay of at least 1 or 2 weeks is required for the results. This delay can be shortened if needed, but with a high cost increase. Even so, a delay of several days can be problematic in some cases (operational and investigative monitoring).
- The second limit is the relevance of results with regard to the monitoring objectives. The quality parameters, either aggregate [biological oxygen demand (BOD), chemical oxygen demand (COD), toxicity, . . .] or specific [total organic carbon (TOC), nitrogen forms, organics, . . .] cannot be analyzed for each sample, and a choice has to be made for each monitoring program. Usually only a few of parameters are selected, limiting thus the possibility of investigation if needed.
- Another limit is the economical frame of the procedure. The operative costs increase with the number of control points and measurement/analysis, and with the

use of automatic samplers. The choice of some analysis (screening of metals or organics, trace analysis, ...) can represent an important fraction of the operative costs.

- Obviously there may be some other drawbacks with the general procedure, such as the lack of reactivity in case of accidental (industrial) pollution (see Chapter 4.2) or the study of discharge impact in a receiving medium (see Chapter 5.1).

In this context, the use of alternative methods, mainly for measurement and analysis, give an opportunity to improve the general procedure, for example as it has been shown in Chapter 1.2, for the sampling assistance.

1.4.1.2 Evolution of Wastewater Quality Monitoring

Before considering the definition and characteristics of alternative methods, let us consider the evolution of wastewater quality monitoring.

In the 1970s, on-line analyzers were proposed for remote measurement, namely for industrial applications. The first TOC meters, COD meters and nutrients meters were adapted from instrumental procedures designed for the laboratory, and thus not fitted for an automatic use on raw wastewater, without continuous checking. This is the reason why the success of these devices was limited, due particularly to sample line clogging and electronic trouble shooting. However, sensors designed for process control as flow meters, oxy-meters or sludge blanket detectors were well accepted and are always used. The 1980s corresponded to the development of other sensors, such as multiprobe systems, for temperature, pH, conductivity and oxygen measurement, and of turbidimeters for turbidity measurement in surface and tap water and for suspended solids estimation for wastewater, with limited success for this latter use. Since the 1990s, a lot of new methods and devices have been designed for on-site/on-line wastewater quality monitoring. Designed with efficient sampling line and adapted fluidic part, they offer a real possibility for on-site automatic measurement, namely for treatment processes control (Thomas, 1995; Bourgeois *et al.*, 2001; Vanrollegem and Lee, 2003; Thomas and Pouet, 2005).

Figure 1.4.1 gives an example of the context and evolution of ammonium analysis in water and wastewater. Not less than four reference methods exist for laboratory analysis and at least seven alternative ones, mainly for on-site/on-line measurement. Reference methods are based on either simple procedures using classical laboratory material, or instrumental techniques, with photometric detection. Alternative methods are mainly adapted from reference or standard methods but are also based on other principles.

Besides on-site/on-line monitoring, a lot of screening tools or methods have been commercially available for the last decade, thanks to progress in biodetection and biosensors development. This development has been largely studied and promoted through European Commission funded research projects (Dworak *et al.*, 2005).

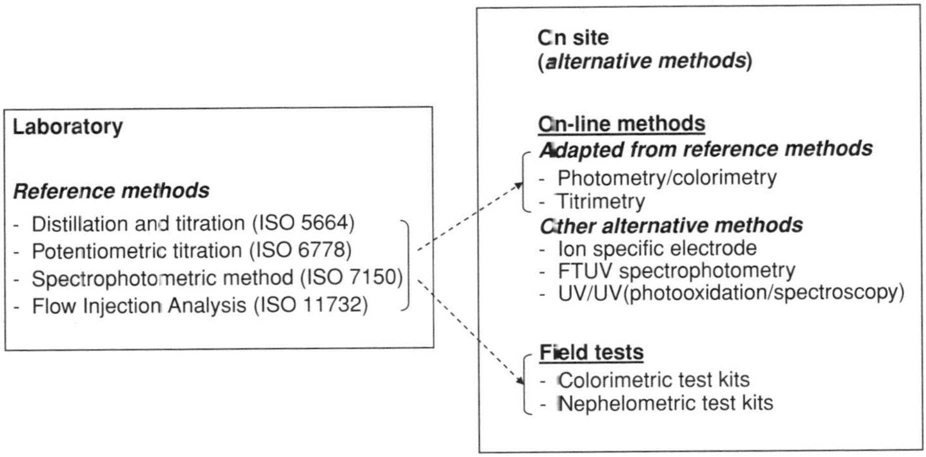

Figure 1.4.1 Laboratory and on-site (alternative) methods for ammonium measurement (ISO 5664, 1984; ISO 6778, 1984; ISO 7150-1, 1984; ISO 7150-2, 1986; ISO 11732, 1997)

1.4.1.3 Definition of Alternative Methods

An alternative (or alternate for US) method is defined by US Environmental Protection Agency (EPA) as 'any method of sampling and analyzing for an air or water pollutant that is not a reference or equivalent method but that has been demonstrated in specific cases–to EPA's satisfaction–to produce results adequate for compliance monitoring'. This definition can be refined by considering that an alternative method must give comparable results with regard to the use of reference method, as for equivalent method. The latter is defined by the US EPA as 'any method which has been demonstrated to be an acceptable alternative to normally used reference methods'.

The urban wastewater treatment European directive (European Commission, 1991) in its Annex I-D-1 states that 'Alternative methods . . . may be used provided that it can be demonstrated that equivalent results are obtained'; the equivalence of results being related to the use of reference methods (see Chapter 1.1).

An extension of the above definition can be proposed with the integration of complementary (emerging) tools used for biological monitoring or other characterization of wastewater. Thus, an alternative method is either a method of sampling and analyzing, giving comparable results to the ones of a reference method for compliance monitoring, or a method complementary to reference or other equivalent or alternative method, giving information not available otherwise. For example, some ready-to-use test kits can be considered as alternative methods, as well as some bioassays or nonparametric methods based on UV spectrophotometry (see later).

In any case, an alternative method must have complementary specific characteristics justifying its use mainly on site, in order to avoid the delay between the sampling transportation and the final result. The general characteristic is that the method must be fit for purpose, i.e. adequate for compliance monitoring or for water quality diagnosis with qualitative measurement. Thus, it must give rapid results and also be as simple as possible, robust, and reliable. From an economic point of view, an alternative method has to be cost effective, considering both investment and operational costs. Other considerations, such as portability or automation can be envisaged. Sensitivity, not included in the above characteristics, is rather dedicated to a reference (standard) method than to alternative ones.

Before considering the types of alternative methods, a comparison between emerging tools, a new concept accompanying the implementation of the European Water Framework Directive, and alternative methods has to be made. The concept of emerging tools concerns new methods and procedures for the chemical and biological monitoring of water quality (Allan *et al.*, 2006). For chemical monitoring, emerging tools are: (i) passive samplers; (ii) on-line, *in-situ* and laboratory-based sensors and biosensors; and (iii) immunoassays. For biological monitoring, emerging tools are: (i) biomarkers; (ii) whole-organisms bioassays; and (iii) biological early warning systems. Thus, emerging tools must be considered as alternative methods, being not reference ones and giving either quantitative parameters or complementary information.

Finally, an alternative method must be retained as a reference one. Numerous examples can be found for wastewater quality monitoring, such as the toxicity measurement based on the use of *Vibrio fisheri* (ISO 113483, 1998).

1.4.2 TYPES OF ALTERNATIVE METHODS FOR WASTEWATER QUALITY MONITORING

Alternative methods can be grouped in to several classes, depending on their principle and their objectives (type of parameter). The three first groups include methods for the rapid measurement of concentrations or parameter values and the two other groups include methods giving qualitative results.

1.4.2.1 Transposition of Reference Methods

Methods of this group give quantitative results and are characterised by the simplification of reference methods, with respect to their potential on-site use, either in developing an automated procedure or in size reduction of instruments. For example, a flow injection system with an automatic sampling, feeding a fast reaction/detection line or a colorimetric test kit with a simple colored scale can be designed on the same analytical scheme as a reference method. Actually, only tests kits are considered as alternative methods, because the procedure of automated systems developed for the laboratory, generally does not differ from the reference method. An ISO standard (ISO 17381, 2003) states on the selection and application of ready-to-use test kit methods in water analysis, very often used for wastewater quality monitoring.

Other systems adapted from standard methods are based on size reduction and electronic integration. For example, microchromatographic systems or simple spectrophotometric devices with optical fiber can be used on site because they are portable and more easy to use than laboratory instruments.

1.4.2.2 Alternative Methods Based on Other Principles

This group of methods giving quantitative results is more important than the first one, because the same parameters can be measured/estimated by several methods which differ in their principle from the one(s) of reference methods. Because reference methods are exhaustive (in terms of interferences treatment) and thus sometimes complex, many alternative methods of this group are based on simple systems. The two main families of methods of this group are the optical sensors and biosensing systems.

The optical methods are not colorimetric ones because no reagent is needed for the measurement and the measurement can be carried out at several wavelengths. They are easy to implement and to adapt for on-line or off-line systems. The most used optical method is the estimation of total suspended solids (TSS) from turbidity, measured either by nephelometry or by absorptiometry for higher concentration (>100 mg/l). Even if the correlation is sometimes poor, due to the interferences of the colloidal fraction, turbidity can give acceptable results, after calibration, principally for treated wastewater. Another optical method for wastewater quality monitoring is UV spectrophotometry. A lot of substances (principally organic) absorb in the UV region and several applications are available from UV absorption measurement. The simplest is the UV254 absorbance value (for a 1 cm pathlength) or the SAC (spectral absorption coefficient), but the exploitation of the whole spectrum gives more relevant information as for example, the estimation of TOC or the measurement of surfactants, phenols or nitrate (Thomas and Constant, 2004).

Biosensing-based systems are increasingly numerous, rather simple to use, but unfortunately not sufficiently validated for wastewater quality monitoring, except for discharge survey (see Chapter 5.1). For example, the number of parameters which can be measured by immunoenzymatic test kits, particularly micropollutants, is very high.

Electrochemical systems can be added to this group for the measurement of mineral ions including metals, providing interference compensation (with ionic strength buffer for example).

1.4.2.3 Modeling, Software Sensors

This group of methods aims at giving quantitative results from mathematical models. The principle of measurement is thus very different from the one of the reference method. Considering that wastewater composition is complex and some parameters difficult to obtain and sometimes not directly measurable, the estimation of a given parameter can be calculated from simpler parameters and the use of more or less complex mathematical models. Even if these methods seem interesting, very few applications are available. For example, the estimation of BOD for a pulp and paper mill, is possible from physico-chemical parameters (conductivity, pH, COD, etc.) and from production parameters (pulp production, paper production, etc.) with the use of a multilinear regression (Oliveira-Esquerre *et al.*, 2004a) or a neural network (Oliveira-Esquerre *et al.*, 2004b). The estimation of wastewater nitrifiable nitrogen, nitrification and denitrification rates, using oxido-reduction potential and dissolved oxygen dynamics has also been proposed (Sperandio and Queinnec, 2004). This approach, drawn from process control and automation, is rather complex and not actually applied to wastewater monitoring.

1.4.2.4 Qualitative Alternative Methods

This group of methods does not give quantitative results for physico-chemical or biological parameters. The results are more qualitative (presence–absence, classification, tests, etc.) and complete the usual characterisation of wastewater. In this group are placed the nonparametric measurements, for which the knowledge of wastewater composition is not indispensable and can be replaced by characterisation of properties of wastewater (variability, treatability, etc.). Even if these properties can be estimated from physico-chemical parameters, alternative procedures can be proposed from the direct use of analytical factors. This last point is the basic principle of the nonparametric measurement, which, as for a nonparametric statistical test, does not require to be related to a given parameter (respectively, a given statistical law) (Baurès, 2002). This means that there exists a qualitative relationship between the analytical factor and the information to be given. For the purpose, UV spectrophotometry based methods have been developed for the variability estimation (Thomas and Pouet, 2005), a rapid treatability test of chemical and petrochemical wastewater (Castillo *et al.*, 1999) and the global characterisation of industrial wastewater matrix (Muret *et al.*, 2000). Chapter 4.2 presents some applications of nonparametric methods.

1.4.2.5 Toxicity Evaluation and Related Methods

Considering the importance of the knowledge of wastewater toxicity and more generally of biological monitoring, either in the sewer to protect the treatment plant biological processes or in the discharge to prevent toxic effects in the receiving medium, several complementary methods are available, such as whole-organism bioassay and biological early warning systems (Allan *et al.*, 2006). However, considering that the composition of a sewer is generally toxic for the majority of biological methods, the actual application is for wastewater discharge monitoring (see Chapter 5.1).

Chapter 1.5 presents extended information on biosensors and biological monitoring for assessing water quality.

1.4.3 USE OF ALTERNATIVE METHODS

Alternative methods can be used anywhere, but preferably on site. They are effectively useful only if they are affordable, reliable and produce data that are of comparable quality between times and locations (Greenwood *et al.*, 2004), but also if they give rapidly relevant information necessary for decision making such as screening, incidents and accident detection, monitoring compliance process monitoring or specific knowledge. A review on these alternative methods for wastewater quality monitoring has been recently published (Thomas and Pouet, 2005).

The majority of alternative methods are for chemical monitoring, but emerging tools open the way for improving biological monitoring, particularly for wastewater discharges (Allan *et al.*, 2006).

1.4.3.1 Ready-to-Use Methods

A ready-to-use method, also named field method, is an analytical method that is ready-made for use, and may be employed in the field with no need for a laboratory (ISO 17381, 2003). It is very often a colorimetric test kit method based on a simplification and size reduction of a reference method, applied to the measurement of the main parameters (N and P forms, some metals, etc.). It can also be a rapid method based on another principle, such as immunoassay test kits, for the measurement of emergent pollutants (pesticides, pharmaceuticals, etc.). Unfortunately, the matrix complexity of wastewater often limits the reliability of these results.

Different types of ready-to-use methods are available. The simplest ones give semiquantitative results using a colored scale with test sticks or reagent. More quantitative results are available with field dosage from drops counting, but above all with photometric cuvette tests with a colorimeter and the use of the Beer–Lambert law.

The ISO standard on the selection and application of ready-to-use test kit methods in water analysis (ISO 17381, 2003) aims to set up criteria for the choice and evaluation of ready-to-use methods for water and wastewater chemical monitoring. Annex B2 gives an application for the determination of nitrogen nutrients (ammonium, nitrite and nitrate) in wastewater, as an important part of the control of sewage treatment plants.

1.4.3.2 Handheld Devices

These methods complementary to the previous use a handheld instrument but no reagent is generally needed. Largely used for the main physico-chemical parameters of water quality (temperature, conductivity, pH, dissolved oxygen), often integrated in a same instrument equipped with a multi-probe, they also concern other parameters, measured by optical or electrochemical techniques. Handheld turbidimeters are sometimes used for TSS estimation. Field portable UV spectrophotometers give estimation of aggregate (TOC, COD, TSS, etc) or specific parameters (nitrate, sulphide, phenol, chromium, etc) (Thomas, 2004). They also provide useful information in the frame of non parametric methods for wastewater variability measurement and incident or accident detection for example (see section 4.2). Ion specific electrodes are proposed for nutrients measurement (nitrate, ammonium for example). The list will increase with the development of future biosensing based handheld devices.

All these devices are easy to use, with simple calibration, and more and more data storage and integrated traceability procedure.

1.4.3.3 On-line Sensors/Analyzers

Recent reviews have been published on the topic (Bourgeois *et al.*, 2001; Vanrolleghem an Lee, 2003; Bonastre *et al.*, 2005; Thomas and Pouet, 2005). On-line sensors/analyzers are often installed in an industrial context, for treatment plant protection and process monitoring. Actually, such systems group on-line equipment (rarely in line) but more often off-line equipment, the advantage being the maintenance facility due to the existence of a rapid sample loop. In fact, contrary to the previous alternative methods, on-line sensors/analyzers are part (the most important) of a measurement chain including data validation and transmission. As for ready-to-use methods, there exists an international standard (ISO 15839, 2003) for the specification of the test procedures to be used to evaluate the performance characteristics of on-line sensors/analyzing equipment. Considering the complexity of the evaluation, it is recommended to check the performance characteristics first at laboratory level and then on field (see Section 1.5.4). One key point of the on line sensors/analyzers use is the concept of availability of the measurement, which represents the percentage of time of the full measurement period during which the measurement chain is available for making measurements (ISO 15839, 2003). This period includes all specified automatic or manual maintenance but also all measurement chain stops due to trouble shooting. For example, a study of TOC measurement availability from four TOC analyzers, carried out over 1 year in a petrochemical site, has shown that the mean availability is about 80 % of the time, with good maintenance, representing each year about 50 % of the investment cost of the analyzers (Thomas *et al.*, 1999). However, such effort is necessary regarding the protection of the wastewater treatment plant.

1.4.3.4 Other Systems

The last group of alternative methods concern principally those for biological monitoring, including emerging tools not considered in previous groups. These are biomarkers, whole-organism tests and biological early warning systems for biological monitoring (Allan *et al.*, 2006). These tools, many being under validation, even if they are commercially available, are actually designed for water bodies monitoring and very few for wastewater. However, considering their complementary nature with reference and other alternative methods, there are several new methods for biological monitoring. Further developments will be devoted to direct application to wastewater quality. Meanwhile, a lot of emerging tools can already be used for discharge toxicity monitoring, such as bioassays and biological early warning systems (see Chapter 5.1). Other emerging tools designed for chemical monitoring are passive samplers, immersed in a stream, for the selective adsorption and concentration of micropollutants. A recent review (Vrana *et al.*, 2005) has pointed out the huge development of this approach for water quality monitoring. Even if only a few applications exist for wastewater quality monitoring with analysis of polar organic compounds (Alvarez *et al.*, 2005) or trace metals and organic micropollutants (Petty *et al.*, 2004), the use of passive samplers appears to be a very promising technique, even if the calibration is difficult as it is strongly dependent on the composition of water. This is the reason why applications deal with wastewater discharge impact.

1.4.4 COMPARABILITY OF RESULTS

The purpose of this section is not to give an exhaustive overview of the tools for quality control and assurance for water quality (the reader will find complete information in Quevauviller, 2002), but rather to stress a simple procedure to check the comparability of results between a reference method and an alternative one (candidate for being recognised as an equivalent method).

There exist very few standards for the purpose. The French experimental standard (AFNOR XP T90-210, 1999) on the evaluation protocol of a physico-chemical quantitative analysis (for water analysis) regarding a reference method, defines some principles and tools for the comparability of methods. Considering the complexity of the problem, this standard is still experimental, and discussions still exist. However, the principles of this standard have been chosen for the evaluation procedure for comparing two methods intended for the detection or quantification of the same target group or species of microorganisms (ISO 17994, 2004). ISO 17994 provides the mathematical basis for the evaluation of the average relative performance of two (quantitative) methods against chosen criteria of equivalence. Another international standard (ISO 11726, 2004) describes procedures for validating alternative (quantitative) methods of analysis for coal and coke either directly by comparison with the relevant international standard method or indirectly by comparison with reference materials that have been exhaustively analysed using the relevant international standard method.

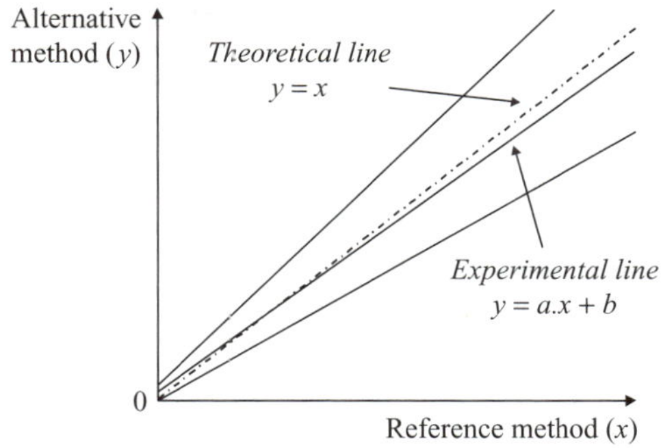

Figure 1.4.2 Comparison between reference and alternative methods

The principles of comparison are simple and schematically based on two steps:

- The first step aims to calculate the analytical characteristics of the two methods (reference and alternative), including the reproducibility for a given value (from a standard solution). A first comparison is carried out on the average values, from a Fisher–Snedecor test. If the test is conclusive (if the two values are not statistically different), the second step can be performed.
- Then, the equivalence between methods must be statistically verified by plotting the results (Figure 1.4.2) and checking the coordinates of the experimental regression line [comparison of the slope and intercept values which must be not statistically different from, respectively, 1 and 0, values of the theoretical line ($y = x$)]. For the purpose a Student test is carried out.

An example is given in Table 1.4.1, showing the results of the Student test of a comparison from real urban and industrial wastewater (grab samples) for the measurement of total Kjeldahl nitrogen (TKN). Reference and alternative methods are, respectively, standard NF EN 25663 and UV/UV procedure (Roig *et al.*, 1999) for TKN. The regression line between the estimated (by the alternative method) and measured values (by the reference method) is: $TKN_{est} = 0.96\ TKN_{ref} + 0.86$ ($R^2 = 0.98$). The results obtained from the comparison of the slope and intercept values to, respectively, 1 and 0, show that the alternative method can be considered as equivalent.

In fact, the scientific decision must be determined by other considerations, such as the improvement of the alternative method if it brings some consistency advantages regarding the reference methods (very cheap, rapid, etc.), and the acceptability of the procedure (Figure 1.4.3).

Once the equivalence between methods is confirmed, the validation procedure results given for on-/off-line instruments (permanent measurement) must be completed, taking into account the sampling procedure is different for a laboratory method and a permanent measurement. For example, considering that regulation

Table 1.4.1 Results of Student test (95 % confidence interval) for TKN measurement by UV (method described in Roig *et al.*, 1999)

Student test	Values
Slope δ	0.9643
Y intercept γ	0.8609
S_δ	0.02
S_γ	0.63
Degree of freedom	55
$t_{0.975}$	2.01
$\delta - t_{0.975}{}^{*}S_\delta$	0.92
$\delta + t_{0.975}{}^{*}S_\delta$	1.0045
$\boldsymbol{\delta - t_{0.975}{}^{*}S_\delta < 1 < \delta + t_{0.975}{}^{*}S_\delta}$	**0.92 < 1 < 1.0045**
$\gamma - t_{0.975}{}^{*}S_\gamma$	−0.405
$\gamma + t_{0.975}{}^{*}S_\gamma$	2.127
$\boldsymbol{\gamma - t_{0.975}{}^{*}S_\gamma < 0 < \gamma + t_{0.975}{}^{*}S_\gamma}$	**−0.405 < 0 < 2.127**

constraints require 24 h composite sampling before laboratory analysis, the challenge is to obtain equivalent results with this procedure and with permanent measurement. In this case, the results to be compared are the mean values for each measurement during the permanent acquisition, with the reference value of the corresponding composite sample (Thomas and Pouet, 2005).

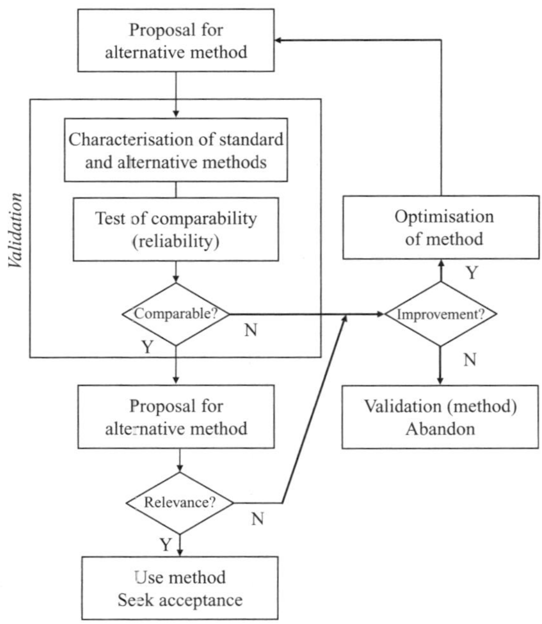

Figure 1.4.3 Validation procedure of a candidate alternative (equivalent) method (adapted from Bruner *et al.*, 1997)

Finally, the international standards already cited (ISO 17381, 2003; ISO 15839, 2003) should be considered for the general evaluation of ready-to-use test kits methods and on-line systems. Other procedures can also be cited (Battelle, 2002, 2004), including works in progress in the frame of the European project Swift-WFD (www.swift-wfd.com).

REFERENCES

AFNOR XP T90-210 (1999) Qualité de l'eau – Protocole d'évaluation d'une méthode alternative d'analyse physico-chimique quantitative par rapport à une méthode de référence.

Allan, I.J., Vrana, B., Greenwood, R., Mills, G.A., Roig, B. and Gonzalez, C. (2006) *Talanta*, **69**, 302–322.

Alvarez, D.A., Stackelberg, P.E., Petty, J.D., Huckins, J.N., Furlong, E.T., Zaugg, S.D. and Meyer, M.T. (2005) *Chemosphere*, **61**, 610–622.

Battelle (2002) Generic verification protocol for long-term deployment of multiparameter water quality probes/sondes. http://www.epa.gov/etv/pdfs/vp/01_vp_probes.pdf.

Battelle (2004) Generic verification protocol for portable technology for detecting cyanide in water. http://www.epa.gov/etv/pdfs/vp/01_vp_cyanide.pdf.

Baurès, E. (2002) La mesure non paramétrique, un nouvel outil pour l'étude des effluents industriels: application aux eaux résiduaires d'une raffinerie. PhD thesis, Université Aix Marseille III, France.

Bonastre, A., Ors, R., Capella, J.V., Fabra, M.J. and Peris, M. (2005) *Trends Anal. Chem.*, **24**(2), 128–137.

Bourgeois, W., Burgess, J.E. and Stuetz, R.M. (2001) *J. Chem. Technol Biotechnol.*, **76**, 337–348.

Bruner, L.H., Carr, G.J., Curren, R.G. and Chamberlain, M. (1997) *Comm. Toxicol.*, **6**, 37–51.

Castillo, L., El Khorassani, H., Trebuchon, P. and Thomas, O. (1999) *Water Sci. Technol.*, **39**(10–11), 17–23.

Dworak, T., Gonzalez, C., Laaser, C. and Interwies, E. (2005) *Environ. Sci. Pol.*, **8**, 301–306.

European Commission (1991) Council Directive of 21 May 1991 concerning urban wastewater treatment (91/271/EEC).

European Commission (2000) Council Directive of 23 October 2000 establishing a framework for Community action in the field of water policy (2000/60/EC).

Greenwood, R., Roig, B. and Allan, I.J. (2004) Draft report: operational manual, overview of existing screening methods (available at: http://www.swift-wfd.com).

ISO 5664 (1984) Water quality – Determination of ammonium – Distillation and titration method.

ISO 6778 (1984) Water quality – Determination of ammonium – Potentiometric method.

ISO 7150-1 (1984) Water quality – Determination of ammonium – Part 1: Manual spectrometric method.

ISO 7150-2 (1986) Water quality – Determination of ammonium – Part 2: Automated spectrometric method.

ISO 11732 (1997) Water quality – Determination of ammonium nitrogen by flow analysis (CFA and FIA) and spectrometric detection.

ISO 11348-3 (1998) Water quality – Determination of the inhibitory effect of water samples on the light emission of *Vibrio fischeri* (Luminescent bacteria test) – Part 3: Method using freeze-dried bacteria.

ISO 15839 (2003) Water quality – On line sensors/analysing equipment for water: specifications and performance tests.

ISO 17381 (2003) Water quality – Selection and application of ready-to-use test kit methods in water analysis.

ISO 11726 (2004) Solid mineral fuels – Guidelines for the validation of alternative methods of analysis.

ISO 17994 (2004) Water quality – Criteria for establishing equivalence between microbiological methods.

Muret, C., Pouet, M.F., Touraud, E. and Thomas, O. (2000) *Water Sci. Technol.*, **42**(5–6), 47–52.

Oliveira-Esquerre, K.P., Seborg, D.E., Bruns, R.E. and Mori, M. (2004a) *Chem. Engin. J.*, **104,** 73–81.

Oliveira-Esquerre, K.P., Seborg, D.E., Mori, M. and Bruns, R.E. (2004b) *Chem. Engin. J.*, **105**, 61–69.

Petty, J.D., Huckins, J.N., Alvarez, D.A., Brumbaugh, W.G., Cranor, W.L., Gale, R.W., Rastall, A.C., Jones-Lepp, T.L., Leiker, T.J, Rostad, C.E. and Furlong, E.T. (2004) *Chemosphere*, **54**, 695–705.

Quevauviller, Ph. (2002) *Quality Assurance for Water Analysis*. Water Quality Measurements Series. John Wiley & Sons Ltd, Chichester.

Roig, B., Gonzalez, C. and Thomas, O. (1999) *Anal. Chim. Acta*, **389**, 267–274.

Sperandio, M. and Queinnec, I. (2004) *Water Sci. Technol.*, **49**(1), 31–38.

Thomas, O. (1995) *Métrologie des eaux résiduaires.* Tec et Doc: Paris; Cebedoc: Liège.

Thomas, O. and Constant, D. (2004) *Water Sci. Technol.*, **49**(1), 1–8.

Thomas, O. and Pouet, M.-F. (2005) Wastewater quality monitoring: on-line/on-site measurement. In: *The Handbook of Environmental Chemistry*, 5, part O, Barcelo, D., (Ed.). Springer-Verlag: Berlin, pp. 245–272.

Thomas, O., El Khorassani, H., Touraud, E. and Bitar, H. (1999) *Talanta*, **50**, 743–749.

Vanrollegem, P.A. and Lee, D.S. (2003) *Water Sci. Technol.*, **47**(2), 1–34.

Vrana, B., Mills, G.A., Allan, I.J., Dominiak, E., Svensson, K., Knutsson, J., Morrison, G. and Greenwood, R. (2005) *Trends Anal. Chem.*, **24**(10), 845–868.

1.5
Biosensors and Biological Monitoring for Assessing Water Quality

Carmen Rebollo, Juan Azcárate and **Yolanda Madrid**

1.5.1 INTRODUCTION

The implementation of wastewater treatment procedure (WWTP), including sewerage systems, WWTP and effluent quality control and potential reuse, and the control of environmental impacts on the receiving waters imply the availability of a considerable amount of analytical data in order to facilitate the management of water resources and the decision-making processes.

Wastewater Quality Monitoring and Treatment Edited by P. Quevauviller, O. Thomas and A. van der Beken

These needs are derived basically from the three following points:

- Normative requirements. Within the EU environmental policy, the Water Framework Directive is likely to cause an important change regarding water quality monitoring. Additionally, other European Directives have been developed, concerning protection of water against the harmful effects of particular substances, the quality of water dedicated to different uses and the obligation of wastewater treatment to achieve a degree of performance and effluent quality. This quality control has to be carried out as analytical measurements.
- Operation and maintenance needs (O&M). In WWTP and sewerage, analytical data are essential for the monitoring process, detecting changes in the process, following the process evolution, better understanding the process and for performance evaluation. The monitoring of raw water is also needed as an alarm system to protect biological processes, during water-clean up, which could be easily damaged by uncontrolled industrial discharges.
- Research and development (R&D). The increasing use of mathematical models for designing and operation of sewer networks and WWTP demands also lots of raw analytical data in order to validate the model itself for a specific site. For research purposes in the environmental field, to assess the aquatic ecosystem status, etc., analytical data are also important.

In order to satisfy these needs on a permanent basis, treatment plant managers, environmental authorities as well as consumers and polluters require the implementation of rapid and accurate analytical measuring techniques. On-line systems, such as sensors, biosensors and other analytical tools in continuous or sequential mode, offer as main advantages faster response, lower cost and easier automatization compared with classical laboratory methodologies. Besides, on-line monitoring provides more detailed information than that obtained from composite samples, because it takes into consideration time-dependent variations.

However, on-line methods have limitations. Although they are normally rapid and inexpensive, currently only a narrow range of parameters can be measured automatically, satisfying the required quality and sensitivity criteria within a reasonable cost. Thus, it is not always possible to carry out continuous monitoring of the required analytes (direct parameters), and often it is necessary to use indirect parameters, correlated to the former or even global pollution indicators. In addition, accuracy and reliability are often lower than laboratory methods. In most cases, a combination of field analysis, laboratory analysis and on-line monitoring is the best choice.

Within the broad range of on-line monitoring devices, special reference should be made to biosensors considering recent advances in technology and applications to the environmental field. A wide range of applications have been described in the literature, both as screening techniques and for the determination of specific compounds. Despite this variations in biosensor-related methods, a common definition could be

Table 1.5.1 Main biosensor applications in the monitoring of wastewater systems

Area	Objective	Measured parameter
Sewer system	Pollution load	BOD, biodegradability
	Industrial discharges	Pesticides, phenols, heavy metals, solvents, toxicity
Wastewater treatment plants	Alarm systems	Toxicity
	Process control	BOD, O_2 consumption
Environmental monitoring	Effluent quality/effluent reuse	Microbiological pollution, BOD
	Aquatic ecosystem evolution	Chlorophyll, global chemical parameters

'an analytical device composed of a biological recognition element directly interfaced to a signal transducer, which together relate the concentration of an analyte or group of related analytes to a measurable response' (Allan *et al.*, 2006). The different types of biosensor and the classification criteria will be discussed below. The term biosensor, in a wide sense, could include not only the determination of chemical species but also the determination of biological populations through the changes of chemical or physical properties. This type of on-line technique is referred to within the text as biological monitoring.

The main potential applications in which biosensors could offer special advantages are listed in Table 1.5.1.

1.5.2 BIOSENSORS

1.5.2.1 Definition and Classification

Despite the wide variation in biosensors and biosensor-related techniques that have been introduced, the widely accepted definition for these devices remained fairly constant. A biosensor can be described as an analytical device composed of a biologically active material directly interfaced to a signal transducer.

Biosensors for environmental applications have employed a wide variety of biological recognition systems (isolated enzymes, intact bacterial cells, mammalian and plant tissue, antibody and bioreceptor proteins) coupled to a similarly wide range of signal transducers (Allan *et al.*, 2006). In a broad sense, biosensors can be divided into three categories according to the biological recognition mechanism: biocatalytic-, bioaffinity- and microbe-based systems. These biological recognition systems have been linked to electrochemical, optical and acoustic transducers.

The biocatalytic-based biosensors for environmental applications are based on the use of enzymes that can act following two operational mechanisms. The first one involves the catalytic transformation of a pollutant (typically from a nondetectable form to a detectable form). The second mechanism involves the detection of pollutants that inhibit or mediate the enzyme activity.

Bioaffinity-based biosensors for environmental applications depend on the use of antibodies and antigenes to measure a wide variety of substances ranging from complex viruses and micro-organisms to simple pesticide molecules and industrial pollutants. The key reagents in these types of biosensors are antibodies, which are soluble proteins, produced by the immune system in response to infection by foreign substances (called antigens).

The fundamental concept behind immunoassays is that antibodies prepared in animals can recognize and bind with relatively affinity and specificity to the antigen that stimulated their production. The binding forces involved in the specific interaction between antibodies (Ab) and antigens (Ag) are of a noncovalent, purely physicochemical nature: hydrogen bonds, ionic bonds, hydrophobic bonds and van der Waals interactions. Since these interactions are weaker that the covalent bonds, an effective Ab–Ag interaction requires the presence of a large number of these interactions and a very close fit between the Ab and Ag.

Antibodies are glycoproteins produced by lymphocite B cells, usually in conjunction with T-helper cells, as part of the immune system response to foreign substances. Antibodies (also known as immunoglobulins) are found in the globulin fraction of serum and in tissue fluids and they are able to bind in a highly specific manner to foreign molecules. There are five classes of immunoglobulins: IgG, IgM, IgA, IgD and IgE. The predominant immunoglobulin in serum is IgG which has an approximate molecular weight of 160 000 Da. All five classes of immumoglobulins share a common basic structure comprised of two light chains and two heavy chains linked by disulfide bonds and noncovalent forces. The antibody molecule usually is represented as a Y-shaped structure.

Immunoassays can be classified as competitive and noncompetitive. Because most low-molecular-weight organic pollutants in the environment have distinguishing optical or electrochemical characteristics, the detection of stoichiometric binding of these compounds to antibodies is typically accomplished with the use of competitive binding assay formats. Competitive immunosensors rely on the use of an antigen tracer that competes with the analyte for a fixed and limited number of antibody binding sites. As antigen tracer radioisotopes, enzymes, liposomes, fluorophores or chemiluminescent compounds are commonly used.

For affinity-based biosensors, this is typically accomplished in several ways. In one method, the antigen tracer competes with analyte for immobilized antibody binding sites. In another format, the antigen is immobilized to the signal transducer while free binding sites on the antibody, which has been previously exposed to the analyte, bind to the surface-immobilized antigen. The third commonly used format requires an indirect competitive assay and relies on the use of an enzyme-labelled antigen tracer. In this format, the assay is completed in two ways. First, the enzyme tracer competes with the analyte for immobilized antibody binding sites. Then, after removal of the unbound tracer, a nondetectable substrate is catalytically converted to a detectable product.

Immunosensors are becoming the most popular type of biosensors for environmental applications.

Micro-organism-based biosensors for environmental monitoring and toxicity assessment use devices with sensitivity over a broad spectrum rather than highly specific ones. As the array of contaminants is wide and the threat unknown, the choice of cellular rather than molecular systems is more suitable. Whole cell biosensors probably offer the greatest technological changes among the existing alarm systems.

In contrast to the previous biosensors, which exploit only one combination, namely, enzyme/substrate or Ag/Ab, microbial biocatalysts are living cells, i.e. complete organisms with multiple biochemical pathways governed by multiplicity of enzymes, which thus offer the greatest potential of investigation. Therefore, microbial sensors share the property of presenting a wide spectrum of response to toxicants with vertebrates and invertebrates. These types of biosensors use three mechanisms.

For the first mechanism, the pollutant is a respiratory substrate being mainly applied to the measurement of biological oxygen demand (BOD). Another mechanism used for micro-organism-based-biosensors involves the inhibition of respiration by the analyte of interest. In this case, these devices might be most applicable for general toxicity screening or in situations where the toxic compounds are well defined, or where there is a desire to measure total toxicity. Biosensors have also been developed with the use of genetically engineered micro-organisms (GEMs) that recognize and report the presence of specific environmental pollutants.

Biological recognition systems have been linked to several types of transducers: electronic, optical and acoustic. Electronic transduction is the most applied in biosensors being classified in potentiometric, amperometric and conductimetric biosensors. The potentiometric transducers are based on the use of ion-selective membranes that make these devices sensitive to various ions, gases and enzyme systems. The enzymatic modification of ion-selective electrodes by covalent binding of the enzymes to the membrane surface is a common procedure for the development of biosensors with high sensibility, stability and fast response. Most potentiometric biosensors for detection of environmental pollutants have used enzymes that catalyse the consumption or production of protons.

Amperometric biosensors typically rely on an enzyme system that catalytically converts electrochemically nonactive analytes into products that can be oxidized or reduced at a working electrode. The electrode is maintained at a specific potential and the current produced is linearly proportional to the nonelectroactive enzyme substrate. The enzymes typically used are oxidases, peroxidases and dehydrogenases. Despite efficient electron transfer from redox enzymes with corresponding electron carrier molecules, few redox enzymes can transfer electrons directly to a metal or semiconductor electrode. Several molecular interfaces that enhance electron transfer from redox enzymes on the electrode surface have been developed. Electron mediators such as ferrocene and its derivates and Meldona Blue have been successfully applied into enzyme sensors.

An optical biosensor incorporates a biologically active material which alters its optical properties, reversibly and selectively in response to the analyte, usually a chemical species. Due to the diversity of optical methods, a vast number of optical transduction techniques can be used for biosensor development. These include adsorption, fluorescence, phosphorescence, chemiluminescence, polarization, rotation and interference. The choice of a particular optical method depends on the nature of the application and the desired sensitivities. The biologically active material can be a catalyst immobilized at the surface of a single fibre, waveguide or fibre bundle that converts the analyte to a detectable species or an Ab with excellent selectivity via Ab–Ag recognition, enabling measurement. However, most cases require the sample to be taken into the instrument. Optical fibre sensors form a large subset of the family of optic sensing and measurement techniques of particular relevance because they offer the ability to perform *in situ* and remote measurements.

Fibre optics serve analytical sciences in several ways. They enable optical spectroscopy to be performed on sites inaccessible to conventional spectroscopy, over large distances or even in several spots along the fibre. Fibres are available now with transmissions over a wide spectral range. However, the transmission capabilities of most fibres are optimized for the telecomunication purposes in the range of 800–1600 nm. In an optical fibre sensor, the fibre forms the coupling optics, and transmits the light from the light source to the modulation zone, where the properties or the light are modulated in response to a change in an external parameter, which can be physical, chemical or biological. The light is then transferred to the detector, where the perturbation in the light characteristics is converted into an electrical signal. The advantage of optical fibre sensors over conventional sensor systems have been well documented and are:

- immunity to electromagnetic interferences;
- electronic isolation is possible to apply these sensors in wet environments;
- transmission of light over long distances, enabling remote or distribute sensing due to the low losses achievable in optical fibres;
- chemical immunity to corrosion enabling use in hostile environments.

In Figure 1.5.1 a summary of different biosensors is given.

1.5.2.2 Environmental Applications of Biosensors

Although a wide range of biosensors have been developed for water monitoring, most of the work has been performed at research level and relatively few of these devices have been introduced into commercial markets. Incorporation of biosensors and, in general, field methods into environmental measurements reduces problems related to sample transportation and time consumption of the analytical measurement.

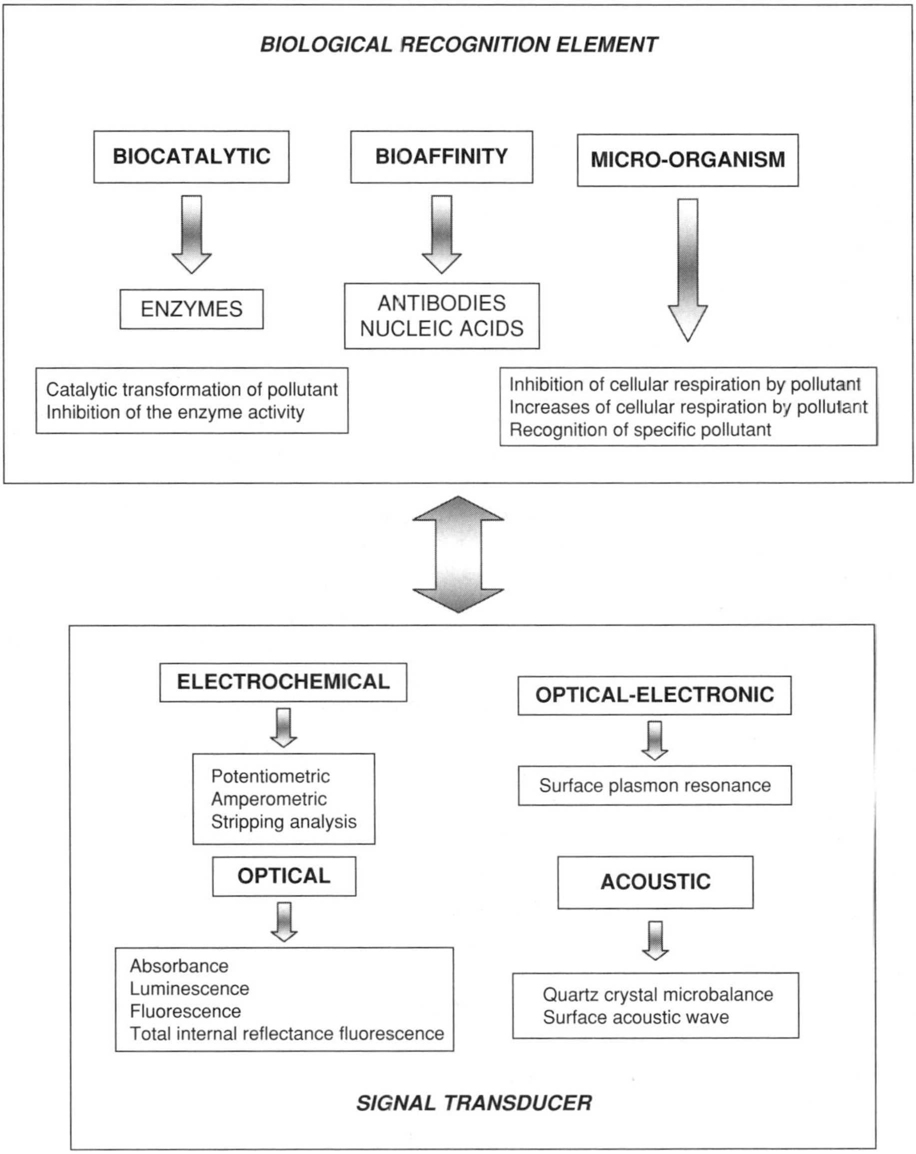

Figure 1.5.1 Summary of different biosensors

In recent years a variety of biosensors, based on some of the mechanisms previously mentioned, have been reported. They are intended for on-line or *in-situ* monitoring of some parameters including global pollution indicators and single compounds or classes of compounds. Most of them are dedicated to the determination of BOD, the direct or indirect measurement of toxicity and, less frequently, to specific compounds such as pesticides, phenols, heavy metals, etc.

On-line Biochemical Oxygen Demand analysis

The estimation of the organic load is a key parameter in conventional wastewater treatment for assessing the environmental effect (oxygen depletion) caused by a wastewater discharge into a receiving aquatic system. One of the parameters used to determine the organic load in a water sample is the BOD. The BOD is an indicator of the amount of biodegradable organic compounds found in a water sample. The conventional analysis carried out in the laboratory involves the determination of oxygen consumption after a 5-day incubation (BOD5) of the water sample with an inoculum of micro-organisms (in case of urban wastewater samples the inoculum is not needed because a micro-organism population already exists).

Biosensors that detect biodegradable organic compounds as BOD are the most widely used micro-organism-based sensors. Different equipment is already commercially available. Also, the use of this device has been incorporated into standard methods in Japan.

The availability of a device that provides a reliable and continuous estimation of BOD on-line is of great interest for WWTP operation, since the BOD loading changes on a timescale of hours and the conventional analytical method takes 5 days from sample collection to final result. Thus, different rapid techniques, with BOD data generation within a short time (typically 15 min–1 h) have been implemented. This technology is advantageous for process control purposes.

The principle of BOD (biodegradation) explains the convenience of using micro-organism-based biosensors for its continuous monitoring. Instead of measuring the dissolved oxygen concentration at the initial and end-point of the test, the use of micro-organisms interfaced to signal transducers allows the measurement of the biodegradation by means of the rate of organic compound metabolism and results obtained in short time frame can be correlated to BOD5.

The instruments commercially available basically consist of an on-line bioreactor in which a population of micro-organisms (biomass) is aerated until it reaches the endogenous respiration stage. When a wastewater sample is added, the micro-organisms begin to degrade it rapidly, causing an increase in oxygen uptake rate and a decrease in dissolved oxygen (DO) compared with the level during endogenous respiration. When the organic matter in the sample is consumed, the micro-organisms return to the endogenous stage. Precise DO consumption measurements during the degradation phase correlate closely with the BOD5 of the sample. The unit usually includes a monitor, DO sensor, temperature sensor, heater and aeration, sample tank, mixing vessel, nutrient tank, agitator and air pump.

The main differences between microbial BOD sensors rest on the characteristics of the bioreactor. In some types of biosensors, the microbial population is immobilized in synthetic membranes (Rasgoti *et al.*, 2003) or in supporting materials, like small plastic rings, to provide the growth surface for micro-organisms inside the reaction chamber (ISCO, 2004). Other kinds of analysers use a suspension of activated sludge as reactive biomass. Recently, techniques based on continuous availability of active micro-organisms from an integrated chemostat independent of the reaction cell have been successfully evaluated (Diez-Caballero, 2000).

Micro-organism-based biosensors can be used either in routine control of BOD in wastewater treatment processes (elimination rates, effluent quality, etc.) or as part of alarm systems where the inhibition of respiration caused by toxic compounds is measured. The use of biosensors in toxicity assessment is described next.

Toxicity analysis

It is widely accepted that routinely used chemical monitoring and analysis methods only detect a limited fraction of the toxic compounds that may be present. There is a need of rapid, easy and inexpensive methods that can be used as alarm systems for aquatic environmental monitoring. Toxicity data can be used as an exclusion parameter, as a binary yes/no response in order to discard the chemical analysis of nontoxic wastewater samples One solution is based on the use of biological systems that indicate that a harmful condition exists, even though it cannot be assigned to a particular substance. Each type of organism will show a specific sensitivity for various pollutants or pollutant mixture. Different biological recognition systems have been used for toxicity assessment including enzymes, antibodies, bacteria, plants, invertebrates and fish.

Many of these tests are time consuming and the use of higher organisms such as fish prevent the method from being automated. Biosensors which exploit only one combination, enzyme/substrate or Ag/Ab, cannot offer the broad response spectrum to toxicants achieved by living cells with multiple biochemical pathways governed by numerous enzymes. Consequently, in the last few years interest in bacterial screening tests has increased. Bacterial biosensor measurements rely mainly on the determination of oxygen consumption using a respirometer or on measuring optical properties, such as luminescence.

Inhibition of microbial respiration by the analyte of interest is one of the mechanisms used in microbial biosensors. The oxygen consumption can be measured both electrochemically by means of an oxygen electrode or optically with an optrode. In the last case the sensor uses optical fibres as signal transducer. Several references on the use of a luminescent ruthenium complex, whose luminescent intensity depends on the oxygen concentration of the sample in contact with the sensing film can be found in the literature. This type of sensor has been evaluated by measuring the inhibition effect of heavy metals on the respiration of micro-organisms in activated sludge.

For this type of measurement some of the equipment uses pure micro-organism cultures, activated sludge from a wastewater treatment plant or even GEMs, which recognize the presence of specific environmental pollutants. Some commercially available automated equipment for toxicity evaluation by means of respirometric methods are: Toxiguard, Biox 1000T, Toxalarm, Rodtox.

Other techniques for the determination of toxicity are based on the direct measurement of optical properties. Toxicity tests based on the bioluminescence inhibition of *Vibrio fisheri* have been frequently used, because it is a well-known organism, well introduced and standardized. These tests offer rapid, easy handling and cost effective responses, and a large database for many chemicals is available.As standard ISO 11340 protocols exist for this assay, many commercial devices are available. Commercial instruments such as Microtox® (Azur Environmental) for toxicity measurements use the freeze-dried marine bacteria stored in a cooled area within the instrument. A standard amount is rehydrated and mixed with the water sample. It is widely used in the laboratory (Araujo *et al.*, 2005). Two bioluminescent inhibition assays from Merck, Toxt Alert 10 and Toxt Alert 100, are also based on the inhibition of *V. fisheri*. Toxt Alert 100 is a portable device with no temperature control and uses freeze-dried bacterial reagents and Toxt Alert 100 uses liquid–dried bacterial reagent and the incubation takes place at controlled temperature (Farré *et al.*, 2002). Other commercial equipment for toxicity measurements are Eclox (Aztec Environmental & Control Ltd) and Aquanox (Randox Laboratories). These use an enhanced chemiluminescent reaction; a free radical reaction for the oxidation of luminol in presence of horse radish peroxidase enzyme using *p*-iodophenol as an enhancer and to stabilize the reaction.

Research in the field of whole-cell biosensors had led to many systems which may be used to quantify general toxicity, cytotoxicity and genotoxicity. Bacterial or yeast cells may be immobilized onto screen-printed electrodes (e.g. the CellSense biosensor), in solution or added to the sample with measurement undertaken by fluorescence or luminescence. Biosensors with a range of standard micro-organisms are available, e.g. *V. fischeri*, activated sluge, *Pseudomonas putida*, *Bacillus subtilis*, *Escherichia coli* (Freitas dos Santos *et al.*, 2002) and genetically modified cells including a fluorescent or luminescent reporter (Philip *et al.*, 2003).

An optical fibre biosensor, for on-line monitoring of toxic effluents, measures the rate of hydrolysis of fluorescein diacetate (FDA) by micro-organisms which is proportional to their metabolic rate, thus indicating the sample toxicity. An amperometric biosensor with *E. coli* for the determination of toxicity in textile and tanneries industry wastewater has been reported (Farré, 2001).

Chemical substances detection

Biosensors for water monitoring cover a broad range of substances. Pesticides and chlorinated compounds are a hardly biodegradable group of pollutants for which numerous sensing schemes have been presented. The area of optical fibre immunosensors is fast growing.

An example of a biosensor environmental application whose mechanism involves the catalytic transformation of a pollutant from a nondetectable form is the use of cholinesterase biosensor for the determination of pesticides, such as carbaryl, aldicarb, carbofuran and dichlovos. These compounds can be detected because they inhibit the enzyme activity (Marty *et al.*, 1995). This type of biosensor in many cases requires the use of substrates, cofactors and mediators and acts in an irreversible way. Also, interference from other compounds that can inhibit the enzyme activity (i.e. heavy metals) can be expected. However, for some classes of compounds they show good sensitivities in the $\mu g\ l^{-1}$ to $ng\ l^{-1}$ range.

Enzyme-based biosensors represent potential alternatives to the analysis or screening of phenolic compounds in wastewater samples. Phenols can be detected by means of the enzyme tyrosinase through the electrochemical reduction of quinone intermediates or through oxygen comsumption with an electrode. For example, portable amperometric biosensors using two enzymes, cellobiose dehydrogenase and quinoprotein-dependent glucose dehydrogenase, have been used to analyse catechol (Nistor *et al.*, 2002).

Using the well known Ag/Ab system, some biosensors configured as waveguides have been developed for the detection of many pesticides such as isoproturon, antibiotic and endocrine disrupting chemicals (Tschemaleak *et al.*, 2005).

Few applications have been dedicated to the determination of inorganic pollutants. Biosensors intended for heavy metal detection primarily use enzymes or GEMs as biological recognition elements (Holmes, 1994).

1.5.3 BIOLOGICAL MONITORING

1.5.3.1 Microbiological Contamination

Biological monitoring in the field of wastewater management is mainly related to the control of pathogen micro-organisms and indicators of faecal pollution. The application of these on-line techniques is likely to play an increasing role in the near future due to hygienic requirements set in new environmental regulations and the growing trend to reuse treated wastewater. The secondary uses of recycled water (irrigation, street cleaning, industrial uses, etc.) on many occasions implies a potential contact with human beings or with the human food chain.

In the last few years, a great development of rapid techniques for microbiological control has occurred. These new procedures could represent an advantageous alternative to conventional methods of detection by culture and colony counting, which usually are laborious and whose results cannot be expected within less than 3–5 days. The major part of the rapid detection methods has been developed in the food, drug and cosmetic industry for raw materials characterization, hazard control of critical points of the processes and sterility of final products. Nevertheless, the stricter hygienic requirements in environmental regulations, particularly in the water field, the greater the analytical needs. Consequently, the potential application of rapid methods in natural water and wastewater monitoring has raised an increasing interest.

Depending on the principle on which the techniques are based it is possible to differentiate four main types:

- ATP luminescence;
- electric properties (impedance, amperometry);
- enzyme immunoassay;
- DNA hybridization (PCR).

ATP luminescence

This is based on the quantification of a cellular component, ATP, by means of an enzymatic reaction that uses ATP as co-substrate. The enzyme is luciferase and the other co-substrate is D-luciferine (Stanley *et al.*, 1989).

$$\text{ATP} + \text{D-luciferine} + O_2 \text{—(luciferase)} \rightarrow \text{AMP} + \text{PPi} + \text{D-oxiluciferine} + CO_2 + \text{light}$$

The time required for the test is in the range of 20 s, to reach maximum light emission, but sometimes a pre-incubation step is needed. The main disadvantage of this determination is the low sensitivity of portable luminometers (on-line monitoring), that are approximately of 10^5 CFU ml^{-1} (total number of Colony Forming Units) and the lack of specificity. The procedure just allows the evaluation of total micro-organisms (bacterial load).

The future of ATP luminescence test utilization in environmental monitoring of natural and wastewaters is restricted to the assessment of global microbiological pollution but the potential of the technique (luminometer) to work on-line supposes a great advantage for the continuous surveillance of disinfection processes.

Electric properties

Biological monitoring of bacterial populations can be based on the measurement of changes in the electric properties of a medium caused by the metabolism activity of the micro-organisms. The final measurement can be conductance, impedance or capacitance of the culture media or of a second solution (indirect measurement) and detection times are significantly shorter than in conventional methods. Other advantages over classical procedures are the avoiding of dilutions, elimination of agar plates and nearly continuous measurement of micro-organism growth.

This rapid technique has great opportunities for automatization and sensitivity is high. However, the selective enrichment on culture media is an essential precondition of this procedure and thus, the specificity depends on the selectivity of the culture media employed.

The required utilization of culture media, membrane filtration step and incubation conditions similar to the conventional analytical procedures compromise the application of this technique for on-line monitoring of water samples although some automatic devices (Pless *et al.*, 1996) have been marketed for laboratory use. Many applications have been developed for food samples and drinks, including detection of coliform bacteria, clostridia, salmonella, *E.coli* and total microbial activity.

More specificity is achieved in amperometric measurement of electroactive compounds. The technique is rapid and sensitive and can be applied to the detection and quantification of pathogens in environmental samples. For example, detection of 4-AP produced by enzymatic hydrolysis of 4-APGal by the bacterial enzyme β-D-galactosidase can lead to a rapid determination (less than 10 h) of low concentrations of *E. coli* (Pérez *et al.*, 2001). However, optimization of the assays is needed to increase the practical applicability in on-line monitoring, particularly regarding the automatization of filtration and incubation steps.

Immunoassays

The use of immunoassays for biological monitoring is a particular application of the bioaffinity-based biosensors that are widely used in the control of environmental pollutants. The principle of operation consists in the specific recognition and binding of a bacterial antigen by antibodies. The final measurement can be carried out by means of several techniques (colorimetry, fluorimetry, luminescence, electrochemistry, etc.). The format of the immunoassay (reusable vs disposable, direct vs indirect competitive techniques, final measurement, etc.) will determine the test sensitivity.

Specificity is high, allowing the detection of different pathogen organisms, and time required to complete the test is considerably lower than for conventional procedures. Once again, the necessary previous stages of isolation and enrichment can jeopardize the applicability for on-line monitoring of an aquatic environment.

A chemiluminescence enzyme immunoassay kit has been marketed (GEM Biomedical, Inc.) for the qualitative detection of *E. coli* O157 (competitive ELISA technique) in less than 7 h. This assay is based on the ability of purified and highly specific antibodies against *E. coli* O157 adsorbed onto a solid phase to detect the organism in a previously enriched sample.

DNA sequence sensors

The principle of these biosensors is the bioaffinity of nucleic acids. The hybridization of DNA sequences from the target organism with complementary sequences allows an extraordinary specificity in the identification. The signal transduction technologies include luminescence measurements or electrochemical biosensors for detecting DNA sequences (Cheng *et al.*, 1998).

Their applicability to on-line monitoring is still limited due to problems of isolation and processing the micro-organism of interest in order to amplify the selected DNA sequences before the hybridization step. The development of multistep genetic analysis devices, including amplification through PCR technique, in the biochip technology (Wooley *et al.*, 1996) represents significant promise for use in environmental monitoring. Besides the application in biomonitoring of organisms of hygienic/environmental interest and difficult to determine through conventional laboratory procedures, such as viruses, these biosensors, also known as genosensors, can be useful for the detection of chemically induced DNA damage.

1.5.3.2 Algae Monitoring

Together with pathogen and indicator micro-organisms, we should also mention the use of biosensors on biological monitoring targeted to the control of phytoplankton. Examples of this application are the recycling of WWTP final effluent for ponds recharge or in cases of eutrophication risks in the receiving water bodies. Chlorophyll 'A' on-line analysers based on luminescent measurements or immunoassay (antibodies directed toward *Alexandrium affine* – red tide – in sea water) (Nakanishi *et al.*, 1996) have been described and the former are at present commercially available.

Natural chlorophyll fluorescence can be used to measure presence of algae based on spectrofluorimetry to detect the effects of pollutant on algae or algal blooms (Europto, 1995).

1.5.4 FUTURE TRENDS

Legislation determines what parameters have to be monitored for assessing water quality. In most cases, these parameters are determined by the existing laboratory analytical methods which are quite often expensive, slow and tedious. This justified the interest in developing methods which allow continuous measurement to monitor the species at the point of discharge, in external environment or for real-time on-line process control. On-line systems including sensors and biosensors and other continuous systems can be considered as an alternative to the classic analytical methods for determining biological and chemical parameters.

One of the most important shortcomings in biosensor development is the validation of these devices under field conditions. Although most of them have shown good performance at the laboratory or pilot scale (wastewater treatment plant), discrepancies in results when the biosensors are applied at the field scale, especially in continuous flow, have been reported by several researchers and explain the small number of automated biological systems in use in biomonitoring.

In fact, a biosensor is not just a simple association between a biocatalyst and the transducer but a device which is affected by different interferences, requiring perhaps thermostatic control, addition of nutritive solutions, adjustment of pH, salinity, exposure to light and elimination of suspended solids. All these parameters need to be carefully controlled in field applications (sometimes this is a difficult task) in order to assure the quality of the data produced by these systems.

Another problem is related to the measurement systems (specially the optical instrumentation). In order to perform *in-situ* analysis it is advisable to design small instruments to make them cheaper and more compact. Battery-operated instruments based on solid-state technology (e.g. excitation with LED or laser diodes, silicon photodiode detection, etc.) would be a potential solution for obtaining portable instruments.

Therefore validation of such devices in field conditions and development of a robust and portable instrumentation is a priority to include biosensors and other continuous analytical systems in biomonitoring of water and to help to improve protection of the aquatic environment. Otherwise these systems will remain mostly within the academic and research frame. Only the systems which are fast, simple, cheap and validated will have commercial success. This aim obviously cannot be achieved without the cooperation of the biologists, engineers, statisticians and electrical engineers. This interdisicplinary cooperation is absolutely necessary to ensure success.

REFERENCES

Allan, I., Vrana, B., Greenwood, R., Mills, G., Roig, B. and Gonzalez, C. (2006) *Talanta*, **69**(2), 302–322.

Araujo, C.V., Nascimiento, R.B., Oliveira, C.A., Strotmann, U.J. and da Silva, E.M. (2005) *Chemosphere*, **28**, 1277–1281.

Cheng, J., Sheldon, E.L., Wu, L., Uribe, A., Gerrue, L.O., Carrino, J., Heller, M. and O'Conell, J.P. (1998) *Nature Biotechnol*, **16**, 541–546.

Diez-Caballero, T. (2000) *Ingen. Quím.*, **6**, 119–125.

Europto (1995) Air toxics and water monitoring. *SPIE*, **2503**.

Farré, M., Pasini, O., Alonso, M.C., Castillo M. and Barceló, D. (2001) *Anal. Chim. Acta*, **426**, 155–165.

Farré, M., Kloter, G., Petrovic, M., Alonso, M., Jose Lopez de Alda, M. and Barceló, D. (2002) *Anal. Chim. Acta*, **456**, 19–30.

Freitas dos Santos, L., Defrenne, L. and Krebs-Brown, A. (2002) *Anal. Chim. Acta*, **456**, 41–54.

Holmes, D.S. (1994) *Environ. Geochem. Health*, **16**, 229–233.

ISCO (2004) Publicity data. STIP ISCO GmbH.

Marty, J.L., García, D. and Rouillon, R. (1995) *Trends Anal Chem.*, **14**, 329–333.

Nakanishi, K., Masao, A., Sako, Y., Ishida, Y., Muguruma, H. and Karube, I. (1996) *Anal Lett.*, **9**, 1247–1258.

Nistor, C., Rose, A., Farré, M., Stoica, L., Wollenberger, U., Ruzgas, T., Pfeiffer, D., Barceló, D., Gorton, L. and Emmneus, J. (2002) *Anal. Chim. Acta*, **456**, 3–17.

Pérez, F., Tryland, I., Mascini, M. and Fiksdal, L. (2001) *Anal. Chim. Acta*, **427**, 149.

Philip, J., Balmand, S., Hajto, E. and Bailey, M.J. (2003) *Anal. Chim. Acta*, **487**, 61–74.

Pless, P., Futschik, K. and Schopf, E. (1996) *J. Food Protect.*, **57**(5), 369–376.

Rasgoti, S., Kumar, A., Mehra, N.K., Makhijani, S.D., Manoharan, A., Gangal, V. and Kumnar, R. (2003) *Biosensors Bioelectr.*, **18**, 23–29.

Stanley, P.E., McCarthy, B.J. and Smither, R. (Eds) (1989) *ATP-Luminiscence: Rapid Methods in Microbiology.* Blackwell, Oxford, vol. 26.

Tschernaleak, J., Proll, G. and Gauglitz, G. (2005) *Talanta* , **65**, 313.

Wooley, A.T., Hadley, D., Landre, P., Demello, A.J., Mathies, R.A. and Noarthrup, M.A. (1996) *Anal Chem*, **68**, 4081–4086.

1.6
Reference Materials

Philippe Quevauviller, Christian Dietz and **Carmen Cámara**

1.6.1 INTRODUCTION

Pollutants continuously discharged into the environment within the borders of the enlarged European Community present a significant risk to or via the aquatic environment, including the risks of affecting waters used for the abstraction of drinking

Wastewater Quality Monitoring and Treatment Edited by P. Quevauviller, O. Thomas and A. van der Beken

water. The closing of water cycles is here an essential part of sustainable water resource management, requiring protection of surface waters from especially problematic compounds, which are difficult to remove, toxic, endocrine disrupting or affecting the organoleptic quality of the resulting drinking water. Impacts are both direct and indirect, through degradation products, causing acute and/or chronic toxicity and/or long-term effects via bioaccumulation in aquatic food chains. The characterization of the physico-chemical state of the aquatic environment should include its dynamic aspects, the interrelation among the different environmental substrates and the integration of the information concerning all these factors.

The current Water Framework Directive (WFD) is the major Community instrument for the control of point and diffuse discharges of dangerous substances. Decision no. 2455/2001/EC of 20 November 2001, amending water policy directive 2000/60/EC, defines priority hazardous substances, subject to cessation of emissions, discharges and losses into water. Their respective concentrations in the aquatic environment are aimed to be set back to values close to zero within a timeframe of not more than 20 years.

Wastewater Treatment Plants play a key role in sustainable water resource management, requiring protection of surface waters from all compounds which are difficult to remove and/or toxic. Sound decisions on wastewater treatment procedures should be based on accurate chemical measurements, which may be verified by various means, e.g. proficiency testing (AOAC, 1992)or use of Certified Reference Materials (Quevauviller and Maier, 1999; Stoeppler *et al.*, 2001). Various Certified Reference Materials (CRMs) are available for the quality assurance of water analyses, as discussed in detail in a separate volume of the present Series (Quevauviller, 2002). However, discussions in the frame of a workshop dedicated to reference materials for water analysis have highlighted the lack of materials representative of wastewater composition (Quevauviller, 1998). Indeed, the quality control of trace element determinations in wastewater can hardly be fully demonstrated by the use of CRMs of different water matrices. Recent developments made within a project carried out through the Standards, Measurements and Testing Programme (follow-up of the BCR Programme, European Commission) have allowed the verification of the feasibility of preparation of real wastewater reference materials through an interlaboratory trial and to certify wastewater reference materials for their trace element content. This chapter gives an overview on CRM requirements, with specific details related to the wastewater CRM project.

1.6.2 TYPES OF REFERENCE MATERIALS

A Reference Material (RM) may be defined as a material or substance with one or more property values that are sufficiently homogeneous and well established to be used for calibration of an apparatus, assessment of a measurement method, or assigning values to materials. A CRM is situated above those in the traceability hierachy and are RMs accompanied by a certificate, with property values that are certified by a procedure that establishes its traceability to an accurate realization of the unit in which the property values are expressed, and for which each certified value is accompanied by an uncertainty at a stated level of confidence (ISO, 1993). CRMs are designed to verify and improve the quality of environmental chemical analyses in various matrices; they are essential tools in the chain of traceability ensuring comparable analytical data between laboratories, across borders, and through time.

Various types of RMs are used in analytical chemistry for different objectives (e.g. internal quality control, interlaboratory studies). RMs used for internal quality control purposes are often referred to as Laboratory Reference Materials (LRMs) or Quality Control Materials (QCMs). As described later, LRMs are used as a means to compare results from one laboratory with another (in the frame of interlaboratory studies) and/or monitor method reproducibility (through control charts), whereas CRMs enable the results to be linked to those of known standards at the international level, and to verify the accuracy of a method at any desired moment.

RMs can be:

- Pure substances or solutions used for the calibration and/or the identification of given parameters, or aimed at testing part or totality of an analytical procedure (e.g. raw or purified extracts, spiked samples, etc.).
- Materials with a known composition, aimed at the calibration of certain types of measurement instruments. In the case of CRMs, calibrating solutions have to be prepared gravimetrically by specialized laboratories.
- Matrix reference materials, representing as much as possible the matrix analysed by the laboratory. In the case of LRMs, the materials may be prepared by the laboratory for internal quality control purposes (e.g. establishment of control charts) or for use in interlaboratory studies. CRMs are certified for specific parameters and are reserved for the verification of a measurement procedure. The certification is based on specific procedures that are described in the following sections.
- RMs that are operationally defined. The assigned or certified values are directly linked to a specific method, following a strict analytical protocol.

CRMs are expensive items. Their production and certification are very costly (typically several hundred thousands euros). Hence, they should in principle be reserved for the verification of the accuracy of analytical procedures and not for daily use (e.g. routine internal control of a laboratory). Two further disadvantages of using CRMs for certain purposes result from the compromises that have to be accepted by the end user. One is the additional material manipulation to achieve the necessary homogeneity and stability for a CRM. The other is the fact that the matrix of any CRM never matches that of real samples to be analysed 100 %. The user must decide whether the resulting deviation can be accepted within the Quality Assurance process.

1.6.3 REFERENCE MATERIAL REQUIREMENTS

Major requirements for the preparation of RMs are related to their representativeness, homogeneity and stabilty over long-term storage. The following sections describe general rules to be followed for the preparation of water matrix-CRMs, with details that are specific to wastewater matrices. Examples of other type of water RMs are described in the literature (Quevauviller, 2002), illustrating that tailor-made preparation procedures have to be adapted for each type of material and that they have to fit the purpose of the analytical work.

Correct conclusions on the performance of an analytical method or a laboratory require the use of one or several RMs with a composition as close as possible as the samples routinely analysed by the laboratory. This means that a RM should, in principle, pose similar analysis difficulties, i.e. induce the same sources of error, to those encountered when analysing real samples. Requirements for the representativeness of a RM imply in most cases a similarity of matrix composition, concentration range of substances of interest, binding states of the analytes, occurrence of interfering compounds, and physical status of the material.

In many cases, a 'perfect' similarity of CRMs with natural samples cannot be entirely achieved. The material should be homogeneous and stable to guarantee that the samples provided to the laboratories are similar, and compromises have often to be made at the stage of preparation to comply with this requirement. Some important parameters, and characteristics of real samples [e.g. coagulation of colloids, oxidation of iron (II), etc.], may change. Unstable compounds or matrices cannot be easily stabilized or their stabilization may severely affect their representativeness. The degree of acceptance of these compromises will depend upon the producer and the user's needs. For example, the preparation of 'natural' groundwater RMs has been demonstrated to be feasible for the certification of trace element contents, whereas sets of artificial RMs had to be prepared for the certification of major elements owing to the instability of some constituents (e.g. nitrates, ammonia) in natural samples (Quevauviller *et al.*, 1999). Both natural and artificial samples (matching the matrix of 'natural' samples) actually corresponded to compromises in comparison with the samples collected for monitoring purposes, but they fulfilled the customer's needs with respect to quality control. Users should, in any case, be informed about the real status of the sample, its treatment and possibly the treatment that has to be applied to bring the sample to a state that is more representative of a natural sample.

1.6.4 PREPARATION

The preparation of a CRM comprises a series of steps to be carried out, from preproduction steps, such as the establishment of the need for a new CRM, and the planning of a certification campaign to post-production processes, such as storage and selling of a new material (Quevauviller, 2002). Details of these steps with respect to wastewater CRMs will be discussed in the following sections.

1.6.4.1 Collection

The amount of collected sample has to be adapted to the aim of the analysis, and to various parameters such as the size of the current sample intakes, the stability, the frequency of use and the potential market (for CRMs). It is sometimes better to prepare a limited batch of samples to respond to the needs for a given period (e.g. 5 years) and to prepare a new batch of material when new requests are made to respond to needs of modern analytical techniques or to changes in regulations. The collected amount may vary from some litres for the preparation of LRM (used for internal QC) to some cubic metres for materials to be used in interlaboratory studies or for the production of CRMs. The producer should be equipped to treat the appropriate amount of material without substantially changing its representativeness.

With respect to wastewater, the chemical composition, even from the same sampling point, can vary considerably, depending on the time and date when the samples are taken. Considering the variability of wastewater samples according to their origin, a wide range of metallic concentrations has to be covered. In the above- mentioned BCR project, a feasibility study was undertaken, focusing on three types of samples: urban wastewater containing relative low and high levels of metals and an industrial wastewater (Segura *et al.*, 2000). The urban wastewater sample was collected in the Wastewater Treatment Plant of the city of Madrid, which deals with the wastewater coming from the centre of the city and whose influent is almost entirely of urban origin. The sample was collected with a magnetic drive pump without metal parts in contact with the solution, in an existing canal after the screening treatment and before the sand removal processes (raw wastewater), when the wastewater organic load was medium–high. Two industrial wastewater samples were collected in a sewer from an industrial area, with a medium flow of 0.9 m^3 s^{-1}, collecting the effluent of different types of industries. The industrial wastewater sample was taken in an easy access site with turbulent flow in order to facilitate the sample homogenization and to get representative samples. Details on the composition of the collected materials are given elsewhere (Segura *et al.*, 2000). The samples were collected in pre-cleaned high-density polyethylene containers; 25 litres of each sample was collected in high density polyethylene containers (previously cleaned by leaching with reagent grade nitric acid 5 % and rinsing with ultrapure water), acidified (pH below 2) with (70 %) $HNO_3$and homogenized by stirring for a period of 16 h.

1.6.4.2 Sample Treatment

Typical operations for the preparation of water reference materials include the stabilization, possible filtration and homogenization. The stabilization step is one of

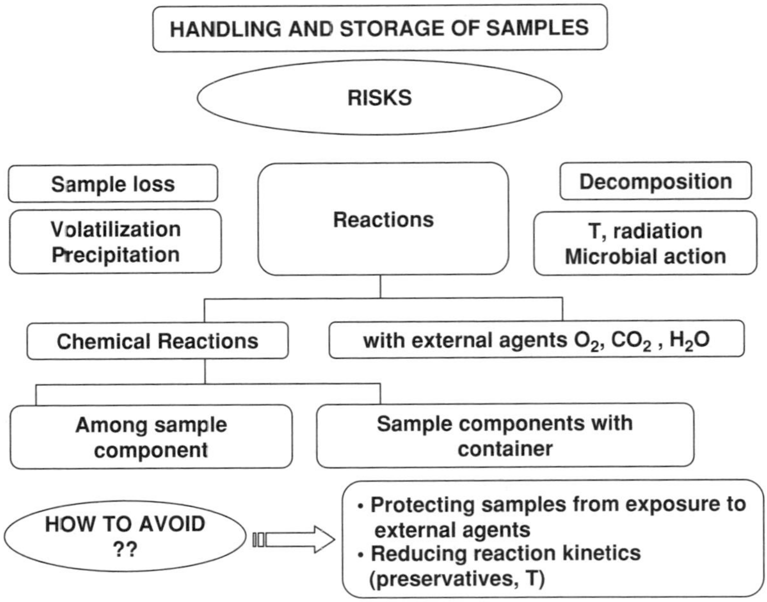

Figure 1.6.1 Risks and solutions during sample treatment

the most critical steps that may affect the material representativeness. This step is, however, mandatory to ensure the long-term stability of the material. Stabilization has to be adapted to each particular case (matrix, type of substance) and should in principle be studied systematically before proceeding to the treatment of the bulk sample. Synthetic solutions containing mixtures of conservative pure substances are generally stable and do not require stabilization. Conversely, natural samples are often very unstable, in particular for compounds that are sensitive to long-term temperature variations or prone to chemical changes (e.g. carbon dioxide, pH of low conductivity samples, metal speciation, etc.).

Figure 1.6.1 gives an overview of possible risks to be taken into account during sample pretreatment and storage when preparing aqueous RMs.

A material may be used as reference only if on each occasion of analysis an identical portion of sample is available. Therefore, when a material is stabilized, it has to be homogenized to guarantee a homogeneity that is sufficient within and between each bottle/vial for the certified properties (Quevauviller and Maier, 1999). Homogenization is not the most difficult problem for water samples (in comparison to solid materials). Regarding wastewater materials, acidification ($\approx$pH < 2 with HNO_3) is, in general, necessary to ensure a proper stability of the samples. Though this treatment may affect the representativeness of the RMs, it is considered to reflect the best compromise in comparison to 'real samples', which can hardly be stabilized over a long-term period.

A general scheme for sample pretreatment when dealing with liquid samples is given in Figure 1.6.2.

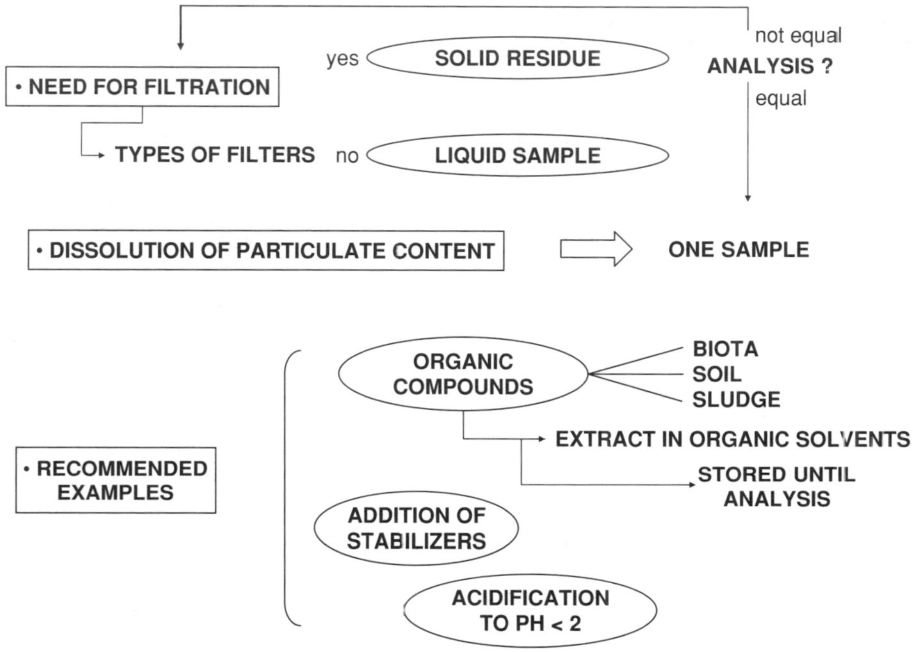

Figure 1.6.2 Sample treatment strategy for liquid sample preparation

The samples processed using the above described certification campaign were filtered in a continuous operation. Due to the original low element contents, they were spiked with selected elements (As, Cd, Cr, Cu, Fe, Mn, Ni, Pb, Se and Zn) at different concentration levels; this spiking was necessary in order to ensure vaild evaluation of data comparison among the laboratories participating in the exercise. The exact spiking levels are given in the literature (Segura *et al.*, 2000). The samples were then prefiltered through on-line prefilter cartridges (pore size 1.2 μm) and thereafter filtered by means of cartridges (pore size 0.5 μm) placed after a peristaltic pump. The filtration was performed in continuous operation to avoid a prolonged stay of the water sample in the tubing. The sample flow rate was about 90 ml min^{-1}. The bottling operation is described below.

1.6.5 STORAGE AND TRANSPORT

The parameters related to the homogeneity and stability of the RM are implicitly linked to the vial used for the long-term storage. Containers used for the storage of water RMs can be sealed ampoules or glass bottles (generally in polyethylene or polycarbonate, more rarely in glass). It is generally recommended to protect the materials from light and amber glass or high-density polymers has generally been used (Table 1.6.1). In cases where risks of contamination from the walls of the flasks

Table 1.6.1 Examples of recommended storage conditions for selected samples

Conditions	Adequate samples	Not recommended samples
Freezing (−20 °C)	Samples with high enzymatic activity (e.g. lever) Unstable analytes	Fruits and vegetables Aqueous samples
Cooling (4 °C)	Soil, minerals Liquid samples Fruits and vegetables	Samples with possible biological activity
Ambient temperature (20 °C)	Dry powders or granulates Minerals Stable analytes	Fresh food Biological fluids
Dryer	Hygroscopic samples	Samples with higher hygroscopy than the drying material

are suspected (e.g. from glass), silica may be recommended. In such a case, the ampoule has to be stored in a closed light-tight tube to avoid any exposure to light and shocks.

The storage temperature should be appropriate for ensuring sufficient stability of the RM. Low temperatures are often recommended but are not always necessary. As previously highlighted, cooling of materials may sometimes affect some parameters, e.g. precipitation of dissolved compounds. Aqueous samples are normally not frozen for storage due to the high risk of analyte interconversion, e.g. from one metal-organic species to another.

Storage conditions, as well as the selected transport means, should be derived from a well-designed stability study that has been adapted to each type of matrix and parameter. A preliminary study on various storage conditions (different temperatures and flask types) is often recommended, in particular for the preparation of CRMs. Adding preservatives during the preparation of a RM may be done in order to reduce decomposition by altering pH, redox conditions, solubility or by converting species to other more stable ones. Careful selection of suitable reactives is mandatory, as the preservatives shall not interfere with subsequent analytical measurements. Another approach often used to avoid ongoing biological activity is sterilization by means of radiation. General requirements for electron beam, X-ray, ^{60}Co and ^{137}Cs irradiators, though designed for medical products, and guidance in qualifying product for radiation sterilization and validating the sterilization process can be found in ISO 11137 Standard concerning the sterilization of healthcare products. The transport has to be performed in the shortest possible time window. Express distribution systems are expensive and must be used in particular cases (e.g. microbiological samples that are only stable for some hours or 1 or 2 days). The material should in principle be accompanied by a form to be sent back to the organizer of the interlaboratory tests or the producer (for a CRM), indicating the status of receipt of the material. Temperature indicators may be added to the sample in order to detect high temperatures that possibly occurred during transport.

In the case of the wastewater RM example (Segura *et al.*, 2000), bottling was carried out in 120 ml Pyrex ampoules (washed twice with demineralized water and dried at 60 °C). They were manually filled with 100 ml of wastewater using a 50 ml plunger pump 'Dispensette'. The cylinder of the pump was constructed of borosilicate glass protected with Teflon and the plunger protected with PSA to avoid contamination. During the bottling procedure the wastewater was continuously homogenized under inert Ar gas in order to ensure a good homogenization before bottling and prevent physical and chemical changes and microbiological contamination from contact with the atmosphere. Filled ampoules were loaded manually onto the carriage of an automatic sealing machine and were automatically moved to a flame warming and sealing station for closing. The storage of water in Pyrex ampoules has less risk of leaking during transport of the material than polyethylene bottles but particular care is required for opening them; in addition, large volumes cannot be stored in such ampoules. The choice of ampoules was preferred over polyethylene bottles since difficulties have been experienced with other water CRMs in the past, mainly due to leaking problems during transport; ampoules are considered to be safer in this respect for the purpose of CRM storage. Three ampoules of the same kind of sample were packed and identified on an outer bag. Samples were dispatched at ambient temperature for the homogeneity, stability study and intercomparison exercise, to the coordinator and the rest of the participants of the intercomparison campaign.

1.6.6 HOMOGENEITY CONTROL

During a chemical analysis, the sample intake of a given material can only be used once since it is generally destroyed during the analysis. The amount of material in a bottle or an ampoule has, therefore, to be sufficient to carry out several determinations. Moreover, the producer has to guarantee that the material is similar from the first vial prepared to the last one. Therefore, the homogeneity of the material should be verified between vials (in the case of water samples – for solid samples, a within-vial check is also necessary) of a same batch to guarantee that no significant difference may occur between sample intakes taken from different vials. The (in)homogeneity may be estimated by comparing the coefficients of variation of repeated measurements on samples from different vials with those of repeated measurements of samples taken from a single vial (which, in the case of water analysis, are considered as the uncertainty of the analytical method). The analytical method used for a homogeneity study should be sufficiently precise (suitable repeatability and reproducibility). A high level of trueness is usually not required since the interesting parameter is, in this case, the existing difference between the samples.

Continuing with the example of wastewater RMs, elements selected for homogeneity and stability checking with the analytical techniques used were: Cr, Mn, Ni, Cu, Zn, Cd, Pb by inductively coupled plasma mass spectromelry; Fe by FAAS; As, Se by hydride generation atomic fluorescence spectroscopy (Segura *et al.*, 2000). After samples were received for the feasibility study, particulate matter appeared in two samples even when they were stored at $-20\,^{\circ}C$. To evaluate if this particulate matter had some influence on the trace metal content, these samples were analysed without and with sample treatment. For the latter procedure a 8 ml aliquot of wastewater samples was treated with 1 ml of (sub-boiled) HNO_3 and 1 ml (30 %w/v) H_2O_2 in a microwave oven and then analysed by the techniques mentioned above. The results obtained showed that the presence of particulate matter did not significantly affect the metal content in solution. Therefore, the following stability and homogeneity studies were performed by analysing the samples without any further treatment.

Statistical tests were applied for the homogeneity testing. The within bottle variability calculated as the coefficient of variation (CV_{wb})was tested by 10 replicate determinations in one ampoule of the three tested solutions. The samples (10 randomly selected ampoules of each solution) were analysed in triplicate by random order in the most repeatable way (sample day, same equipment, same analyst). The results were presented as the between bottle coefficient of variation (CV_{bb}). The estimation of the uncertainty U_{CV} of the coefficient of variation (CV) was calculated as follows:

$$U_{CV} = CV/(2n)^{1/2}$$

where n is the number of replicates.

Statistical data showed that no significant differences at the 95 % confidence level could be detected for all the elements tested. On the basis of the results obtained, it was concluded that the sets of the samples used were homogeneous (Segura *et al.*, 2000). As an example, Figure 1.6.3 shows the homogeneity pattern for Cu in the three wastewater samples.

1.6.7 STABILITY CONTROL

The composition of a RM and the studied parameters should remain stable over the entire utilization period of the material. The extent of the study of the temporal stability will depend upon the use of the material. If a material is to be used in a short-term interlaboratory trial (e.g. 6 months), its stability should only be verified for the duration of the exercise. Additional studies may be needed, e.g. to simulate conditions that may be encountered during the transport of the material (e.g. severe climatic conditions with temperature changes). In the case of a CRM, the stability study has to be planned over some years. The stability (or instability) has to be studied or known before producing the RM on a large scale, and it has to be verified on the entire batch of material (taking a given number of samples randomly over the whole batch). Analyses for studying the stability of a CRM may start at the beginning of the storage period and after various intervals, e.g. 1, 3, 6, 12 months or more, if necessary.

One of the ways to study the stability of (water) CRMs is to use samples stored e.g. at $+4\,^{\circ}C$ as reference for studying samples stored at e.g. $+20\,^{\circ}C$ and $+40\,^{\circ}C$.

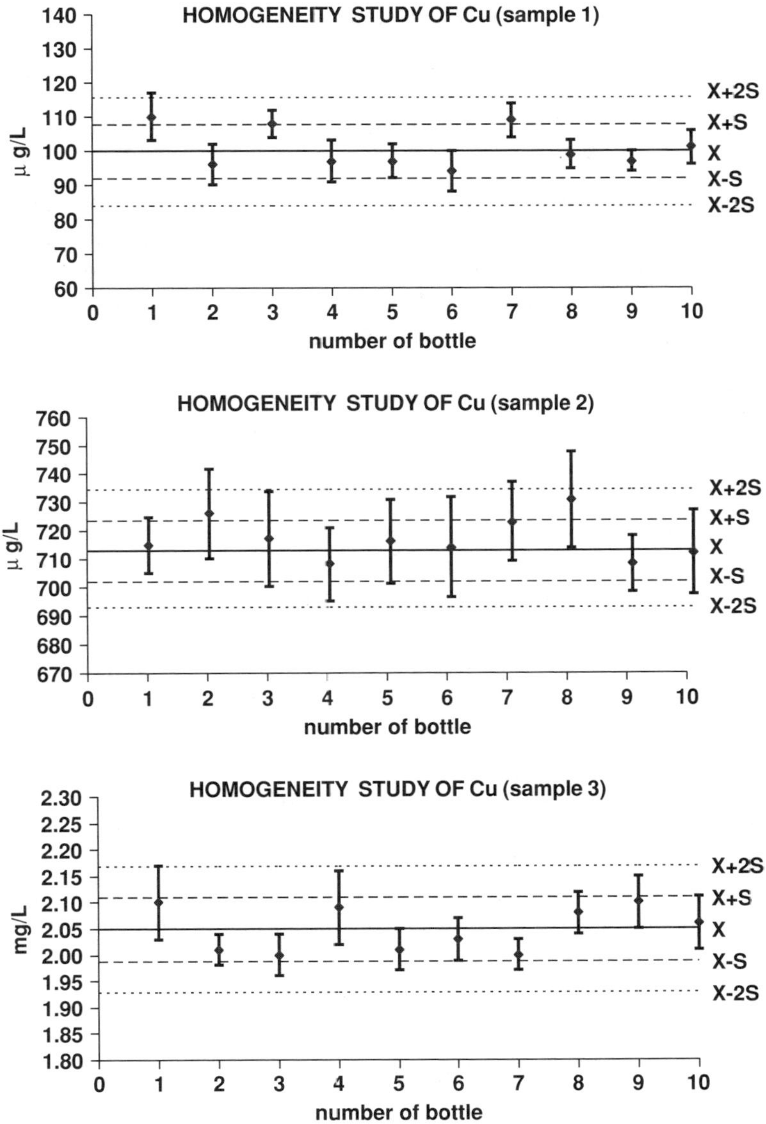

Figure 1.6.3 Homogeneity study for Cu in the three wastewater samples

The ratios (R_T) of the mean values (X_T) of, e.g. five measurements carried out at +20 °C and +40 °C, respectively, and the mean value (X_{Ref}) of five determinations carried out at the same period of analysis on the samples stored at +4 °C, are calculated:

$$R_T = X_T/X_{Ref}$$

The total uncertainty U_T is obtained from the coefficient of variation (CV) of *n* measurements carried out at each temperature:

$$U_T = \left(CV_T^2/n + CV_{Ref}^2/n\right)^{1/2} \cdot R_T$$

This approach overcomes possible variations that are only due to the analytical method (reproducibility). Indeed, these variations are in principle similar, at a given period, for the analysis of CRMs stored at the reference temperature and those stored at +20 or +40 °C. In the ideal case, the ratios R_T should be equal to 1. In practice, random errors on measurements allow one to estimate that the CRM is stable if the expected value 1 is between the values of $(R_T - U_T)$ and $(R_T + U_T)$.

Examples are shown in Figures 1.6.4 and 1.6.5 for the stability study of wastewater RMs stored in the above-described conditions (stability of As and Ni at +20 °C). Results showed no significant variation within the tested time for the 10 elements investigated even in sample 2 where the major formation of particulate matter was observed (Segura *et al.*, 2000). So it was concluded that the samples were stable over the tested period. The formation of particulate matter had no influence on the metal content and sample stability and homogeneity. These particulate matters may have been due to small colloids and dissolved humic matter that passed through the filters. Although the organic particulate matter did not interfere with trace metal analysis at low pH, potential inhomogeneities are introduced that may interfere with the analysis especially when electrochemical techniques like ASV (anodic stripping voltametry) are used.

The reference to samples stored at low temperature may, however, have limitations. This approach is not applicable to the study of water samples in which some compounds may precipitate at low temperature without the possibility of redissolving them in a reproducible manner upon warming of the sample. This feature was apparrently not detected for wastewater samples during this campaign.

1.6.8 PROCEDURES TO OBTAIN CERTIFIED/ REFERENCE VALUES

1.6.8.1 Certification of Reference Materials

There is no true value of any characteristic, state or condition that is defined in terms of measurement or observation. Change of the procedure for measurement or observation will always produce a new number. Therefore the operationally defined reference values are used, a best estimate of the true value provided on a certificate of analysis, or report of investigation where all known or suspected sources of bias have been fully investigated. The certification of RMs has to follow strict rules that are described in the ISO Guide 35 (ISO, 1989). Various approaches may be followed

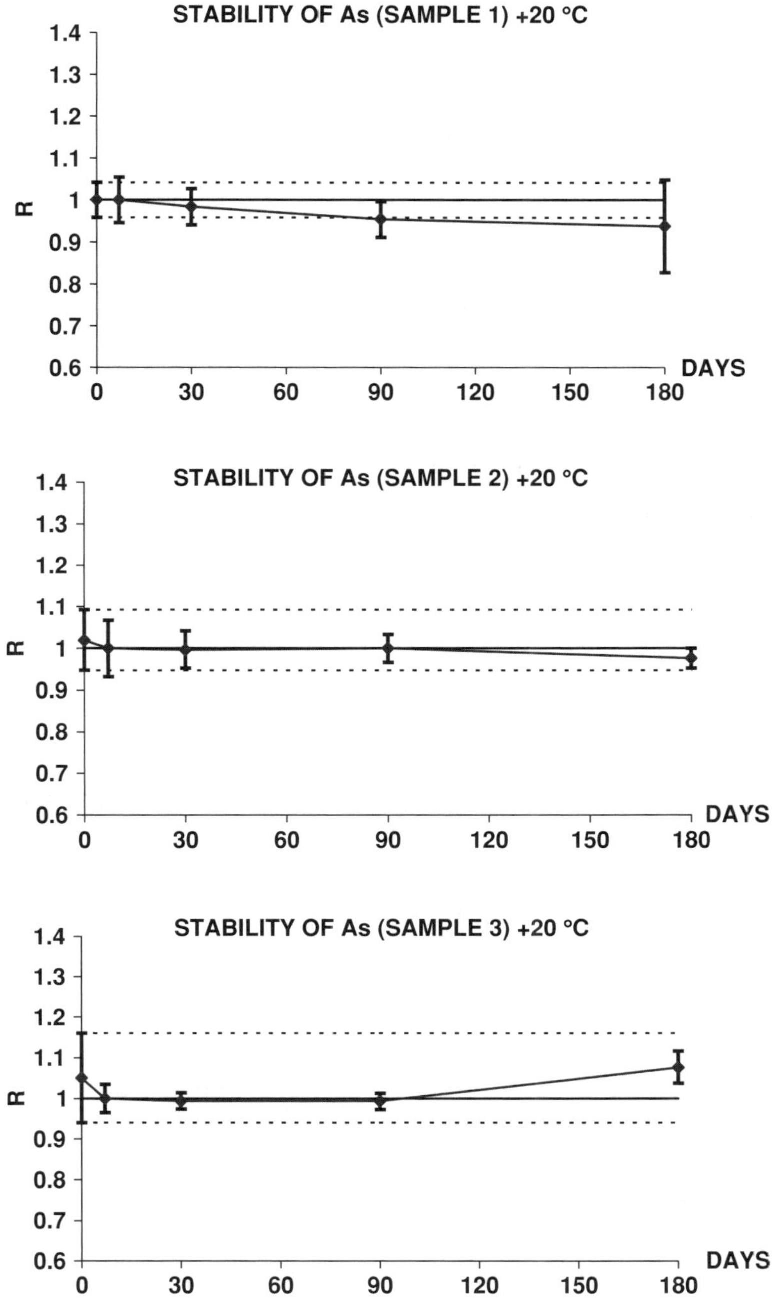

Figure 1.6.4 Stability control for As during production of BCR-713 wastewater RM

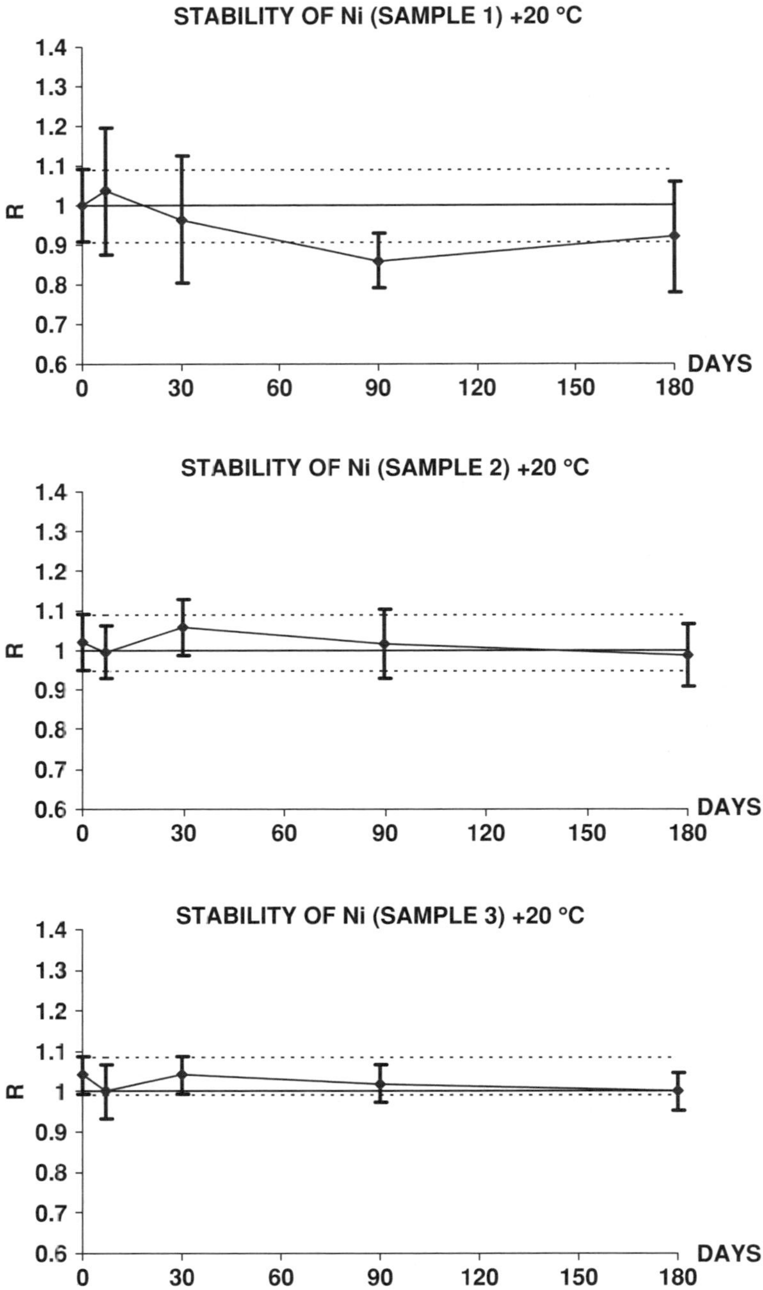

Figure 1.6.5 Stability control for Ni during production of BCR-713 wastewater RM

in relation to the types of properties and matrices to be certified. With respect to calibrating solutions of pure substances, the certification relies on the identification of compounds, the evaluation of their purity and stoichiometry, and gravimetric measurements. Matrix CRMs cannot be certified on the basis of gravimetric methods since the samples are generally analysed after partial or total transformation of the matrix. In this case, three different approaches exist:

- Certification in a single laboratory, using a so-called 'definitive method' applied by one or more independent analysts.
- Certification in a single laboratory, using one or more reference methods applied by one or more independent analysts.
- Certification through interlaboratory studies, using one or more independent methods, if possible including 'definitive methods'.

In all cases, only experienced laboratories should take part in the analytical work. The first two approaches, based on the use of 'definitive methods' by a single laboratory do not eliminate risks of systematic errors related to the human factor (manipulation error). A supplementary confirmation by interlaboratory testing – even limited – is therefore recommended. For some chemical parameters (mainly inorganic), so-called direct methods (not requiring external calibration), e.g. gravimetry, titrimetry, volumetry, etc., or 'definitive' methods are available, e.g. isotope dilution mass spectrometry. The certification of matrix RMs using a single 'definitive' method (e.g. for trace elements) does not give the user, who does not apply this technique in his routine work, a good estimate of the uncertainty obtained with more classical techniques. Moreover, the application field of these methods is limited with respect to the types of matrices and parameters that may be certified. These techniques do not yet exist for the certification of organic or organometallic compounds for which the certification through interlaboratory studies remains the most adopted method.

Certifications based on interlaboratory studies are organized following the same basic principles that classical interlaboratory studies [see details on their organization in Quevauviller (2002)] but they only involve specialized laboratories. All the participating laboratories should, in principle, have demonstrated their capabilities in preliminary exercises. The organizer should also work according to well-defined rules and his ability to organize such exercises should be recognized. The best way to check the reliability of participating laboratories is to request them to demonstrate their performance in interlaboratory improvement schemes. This approach has been followed by the European Commission's BCR programme for all new RMs that had to be certified for the first time, in particular the matrix CRMs (Quevauviller and Maier, 1999).

In each interlaboratory study, detailed instructions and forms to submit results are prepared, requesting each participant to demonstrate the quality of the performed

Table 1.6.2 Summary of the techniques used in the interlaboratory trial

Element	Techniques
Cr	ZETAAS, ICPAES, ICPMS, HRICPMS, INAA
Fe	FAAS, ICPAES, ICPMS, HRICPMS
Mn	ZETAAS, ICPAES, ICPMS, HRICPMS, INAA
Ni	ZETAAS, ICPAES, ICPMS, HRICPMS
Cu	FAAS, ICPAES, ICPMS, HRICPMS
Zn	FAAS, ICPAES, ICPMS, HRICPMS, INAA
As	HGAAS, HGAFS, ICPMS, HRICPMS, INAA
Se	HGAAS, HGAFS, ICPMS, HRICPMS
Cd	ZETAAS, ICPAES, ICPMS, HRICPMS
Pb	ZETAAS, ICPAES, ICPMS, HRICPMS

[1] FAAS, flame atomic absorption spectrometry;
[2] HGAAS, hydride generation atomic absorption technique;
[3] HRICPMS, high resolution inductively coupled plasma mass spectrometry;
[4] ICPAES, inductively coupled plasma emission spectrometry;
[5] ICPMS, inductively coupled plasma mass spectrometry.
[6] INAA, instrumental neutron activation analysis;
[7] ZETAAS electrothermal atomic absorption spectrometry with Zeeman background correction.

analyses, in particular the validity of calibration (including the calibration of weighing scales, volumetric flasks, etc., the use of calibrants of suitable purity and known stoichiometry, sufficiently pure solvents and reagents, etc.). Absence of contamination should also be demonstrated by blank measurements, and yields of chemical reactions (e.g. derivatization) should in principle be accurately known and demonstrated. All precautions should be taken to avoid losses (e.g. formation of insoluble or volatile compounds). If results of totally independent methods such as isotope dilution mass spectrometry, atomic absorption spectrometry and voltammetry (between-method variations) for trace element determinations by laboratories working independently (between-laboratory variations) are in good agreement, it can be concluded that the risk of systematic error related to each technique is negligible and that the mean value of the obtained results is the closest approximation of the true value. This principle has been followed for the certification project on trace elements in wastewater (Segura *et al.*, 2004), in which 16 European laboratories participated, using the different techniques summarized in Table 1.6.2.

The certification of a given parameter in a RM leads to a certified value that is typically the mean of several determinations or the result of a metrologically valid procedure, e.g. weighing. The confidence intervals or the uncertainty limits of the mean value have also to be determined. These two basic parameters have to be included in the certificate of analysis. In Table 1.6.3, the values for the 16 elements certified during this campign and their corresponding uncertainties are summarized.

Table 1.6.3 Element concentration and uncertainties for the BCR wastewater CRMs

Elements	BCR-713 Effluent wastewater	BCR-714 Influent wastewater	BCR-715 Industrial effluent wastewater
As	9.7 ± 1.1	18.3 ± 1.6	29 ± 4
Cd	5.1 ± 0.6	19.9 ± 1.6	40 ± 5
Cr	21.9 ± 2.4	123 ± 10	$(1.00 \pm 0.09) \times 10^3$
Cu	69 ± 4	309 ± 23	$(0.90 \pm 0.14) \times 10^3$
Fe	$(0.40 \pm 0.04) \times 10^3$	$(1.03 \pm 0.11) \times 10^3$	$(3.00 \pm 0.27) \times 10^3$
Mn	43.4 ± 3.0	103 ± 10	248 ± 25
Ni	30 ± 5	108 ± 15	$(1.20 \pm 0.09) \times 10^3$
Pb	47 ± 4	145 ± 11	$(0.49 \pm 0.04) \times 10^3$
Se	5.6 ± 1.0	9.8 ± 1.2	29 ± 4
Zn	$(0.22 \pm 0.04) \times 10^3$	$(1.00 \pm 0.1) \times 10^3$	$(4.00 \pm 0.4) \times 10^3$

Supplementary information to be provided to the user is described in the ISO Guide 31 (ISO, 2000a) and covers, in particular:

- Administrative information on the producer and the material.
- A brief description of the material, including the characterization of its main properties and its preparation.
- The expected use of the material.
- Information on correct use and storage of the CRM.
- Certified values and confidence intervals.
- Other not-certified values (optional).
- Analytical methods used for certification.
- Identification of laboratories participating in the certification.
- Legal notice and signature of the certification body.

Other information, potentially useful to the user of the CRM, cannot be given in a simple certificate. Therefore, some producers (e.g. BCR, European Commission) provide the materials with a certification report including details on the information given in the certificate. In particular, this report underlines the difficulties encountered during certification and the typical errors that may occur when analysing the material with current analytical techniques. The overall work described in the certification report should, in principle, be examined by an independent group of experts so that all the possibly unacceptable practice can be detected and removed. The experts should have in-depth knowledge in metrology as well as a good grounding in analytical chemistry; they have to decide whether or not the CRM can be certified. In the framework of BCR (now under the responsibility of the Institute for Reference

Materials and Measurements, IRMM), the certification committee is composed of representatives from EU countries and Associated States, covering a wide field of expertise in chemical, biological and physical measurement sectors.

Another approach that is being used for the certification of RMs is actually based on the voluntary participation of expert laboratories in interlaboratory schemes (e.g. proficiency testing), using various analytical methods applied by different laboratories (Ihnat, 1997). This approach is less prone to control and there are generally no technical discussions of the results but rather robust statistics to detect and remove possible outliers (e.g. based on z-scores). This type of study is certainly useful for evaluating the performance of laboratories/methods but is not generally recommended for certification unless highly skilled laboratories are involved.

1.6.8.2 Assigned Values

With respect to not-certified materials, there is an interest to obtain good reference values (assigned values). The same approach and rules as the ones used for certification to, in principle, needed to obtain good assigned values. A high degree of accuracy for these values is rarely mandatory for a LRM used for routine quality control checks (control charts) but it should be attempted for each RM that is used in method performance studies. Assigned values may be established through measurements carried out in the framework of interlaboratory studies involving experienced laboratories (they hence correspond to 'consensus' values), which is very similar indeed to the approach followed for certification. The main difference between a good assigned value and a certified value is actually linked to the (legally binding) guarantee given by the producer (certificate of analysis) and the procedure used to obtain this guarantee.

1.6.9 TRACEABILITY OF REFERENCE MATERIALS

Traceability is defined as a property of a measurement or the value of a standard whereby it can be related to stated references, usually national or international standards, through an unbroken chain of comparisons all having stated uncertainties (ISO, 1993).

CRMs and traceability are closely connected since certified values and their uncertainty should, in principle, be linked to established references. In theory, the certified value of a CRM should be traceable to the amount of substance of the element or compound of concern.

The establishment of a 'hierarchy' of RMs has been proposed by Pan (1997). The author pinpointed that it is difficult, if not impossible, to trace all matrix CRMs to primary RMs, because of matrix effects, the variety of sample composition and substances, etc. In addition, factors influencing the analytical process (e.g. homogeneity of the CRM) have an effect on the certified values (Figure 1.6.6).

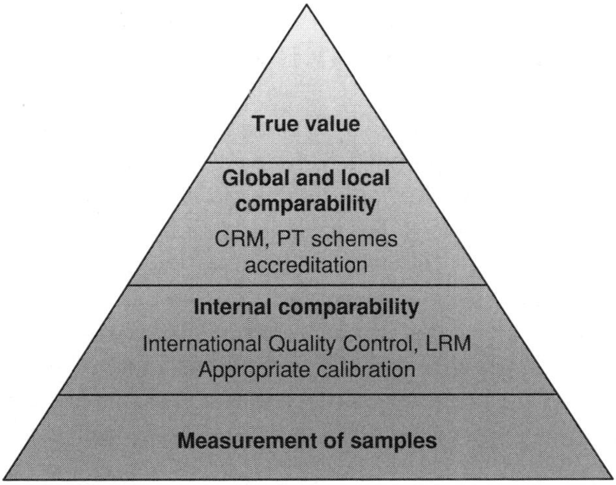

Figure 1.6.6 Traceability hierarchy shows how to achieve results close to the true values

The classification proposed provided the main criteria for establishing a hierarchy in the traceability chain for CRMs:

- metrological quality of methods used for certifying values of the CRM;
- homogeneity and stability;
- calculation of uncertainty;
- metrological competence and recognition of the producer at the national and/or international level;
- demonstration of traceability.

Numerous chemical measurements are carried out, for which RMs cannot readily be prepared owing to their instability (Richter and Dube, 1997). In other cases, RMs may be available but their matrices are significantly different from that of the analysed sample, and the reference used to demonstrate the traceability of the results is then questionable. Some CRMs are directly traceable to SI units and open the possibility of traceability of measurements to these units, e.g. high purity substances, stable isotope calibrants for IDMS, playing the role of primary RMs (Richter and Dube, 1997).

The user of a CRM and of certified values should be informed about all the aspects of traceability that have directed the preparation and certification of the RM, the technical explanations on the rejection of outlying results, the sources of error, the procedures of recovery evaluation (based on a spiking procedure or the analysis of another CRM), the available documentation on the CRMs used to validate the certification methods, etc.

1.6.10 EVALUATION OF ANALYTICAL RESULTS USING A MATRIX CERTIFIED REFERENCE MATERIAL

This section will examine how an analytical result may be evaluated in comparison with the certified value of a matrix CRM. The approach described is adapted from the procedure proposed by Walker and Lumley (1999). The general use of RMs in a validation process of a method is described in detail by them (Walker and Lumley, 1999). The use of a matrix CRM will be based on the evaluation of an analytical result (x) as compared with a certified value (μ) of the CRM. The error on the analytical result (Δ) is calculated using the formula: $\Delta = x - \mu$.

Considering the random errors of the method, the value of Δ will likely not be equal to zero, even if the result is not affected by any systematic error. The greater the random errors (i.e. the poorer the precision), the greater the value of Δ and hence the more difficult to detect the occurrence of a systematic error. The precision is, therefore, a critical parameter that should not be underestimated when evaluating the trueness of a method. Walker and Lumley (1999) distinguish the laboratory internal standard deviation, s_i, characterized by the measurement repeatability of which the estimate should be calculated on the basis of at least seven repetitions of CRM analyses, and the between-laboratory standard deviation, s_e, which is more difficult to estimate. The authors propose several approaches to calculate this latter parameter:

(1) The reproducibility, s_R, may be estimated by replicate analyses (at least 7, preferably up to 20) carried out over a given period of time (if possible over 3 months).

(2) The between-laboratory standard deviation, s_e, may also be estimated in the framework of any method validation interlaboratory study in which the laboratory will know the repeatability values, s_r, and the reproducibility values, s_R, of the method according to the document summarizing the results of the study. The value of s_e will hence be equal to $\sqrt{(s_R^2 - s_r^2)}$.

(3) When the CRM has been characterized in the framework of an interlaboratory study, information on the between-laboratory standard deviation are generally given in the certification report of the material. If the method to be tested is similar to one of those used for the certification of the RM, the value of s_e given in the report may be used.

(4) Predicted values found in the literature may also enable the estimate of s_e. This type of information is available in the agro-food sector but few values comparatively exist in the sector of water analysis.

(5) In the absence of any information, an estimate of s_e may be obtained from the value of s_i according to the formula: $s_e \approx 2s_i$.

The precision σ of an analytical result of a matrix CRM will be calculated by combination of two components:

$$\sigma = \sqrt{\left(\frac{s_e^2 + s_i^2}{n}\right)}$$

where n is the number of replicates of CRM analyses. In general, the value s_i is smaller than the value of s_e (typically by a factor of 2 as indicated above). The fact that n is at least equal to 7 means that s_e will represent the main contribution of σ.

At first sight, it could appear sufficient to base the estimate of the precision σ of a method used by an individual laboratory on the sole value of s_i. However, s_i reflects the random dispersion of results of a series around their mean, which is itself randomly distributed around the CRM certified value with a dispersion that is characterized by the value s_e. Therefore, the combination of s_i and s_e (as indicated above) is used to describe the overall dispersion of the results around the certified value, which is taken as the true value (Walker et Lumley, 1999).

The parameter s_e measures the sources of random errors that cannot be evaluated by replicate analyses in a single laboratory, but however contribute to the result dispersion around the certified value (true or assigned value). An example of random error is the possible variation of the final volume of a sample extract before its introduction in a measurement instrument, without taking care of the variations of ambient temperature. Such volume variations would not be significant for the estimate of the repeatability and would therefore not be considered in the calculation of s_i. However, the same measurements carried out by different laboratories (or by a single laboratory over a given period of time) would be suject to random errors due to variations of the ambient temperature. The effects of such variations would be included in the term s_e.

It is also useful to remember that when a laboratory analyses a matrix CRM, it actually takes an effective part in an 'interlaboratory study' (if the certified values have indeed been measured on the basis of such study). Under these circumstances, it is clearly appropriate that the component s_e of the precision be considered when a laboratory compares its results to CRM values. This is analogous to the comparison of laboratory results in the framework of proficiency testing schemes using z scores [see additional information in Quevauviller (2001)].

If the information on the value s_i is available (e.g. the repeatabilty value s_r of the method as validated through an interlaboratory study), a χ^2 test may then be carried out that will establish whether s_i (measured by the laboratory) is acceptable, i.e. whether the laboratory performs its method with a sufficient precision. However, even if s_i is significantly greater than s_r, if the measured value $s_i^2/\sqrt{n}$ is small in comparison to s_e^2, there will be little or no benefit to repeat a series of measurements of a CRM with the aim to obtain a smaller value of s_i (Walker and Lumley, 1999).

The estimate of the possible occurrence of systematic errors will be based on a statistical test aiming to evaluate whether the value Δ is significantly different from zero. If it is not the case, it is possible to conclude that no systematic error has been demonstrated. A test that is currently used is based on bracketing the value Δ in an interval with limits of $\pm 2\sigma$ in which it is estimated that no systematic error has occurred: $-2\sigma < \Delta < 2\sigma$.

The affirmation that no systematic error has occurred has to be considered with some care. It is indeed possible that errors are left undetected, e.g. in the case of positive and negative errors, which compensate each other. As previously mentioned, the choice of the $\pm 2\sigma$ interval means that the confidence level of this conclusion is about 95 %. The adoption of limits $\pm 3\sigma$ would permit to obtain a confidence level of 99.7 %. This is equivalent to the calculation of z scores used in proficiency testing schemes [as a reminder, $z = (x - X)/\sigma$, the value of σ being based, in this case, on the standard deviation resulting from the test].

It is important that the value of σ be a reliable estimate of the measurement precision. Among the five above-described approaches, procedure (1) implies that at least seven replicate analyses be carried out (which is generally considered sufficient). However, if the method has been previously studied (enabling to be obtained a good estimate of the standard deviation of the measurement for the considered matrix) the number of CRM analyses may be less than seven, although the minimum is to duplicate the analysis. A single analysis may be envisaged where the laboratory is confident in its statistical control. The value of n used for the calculation of σ should obviously reflect the number of replicate analyses effectively carried out on the CRM.

> Walker and Lumley (1999) give an example of application related to water analysis: A water CRM containing certified concentrations of herbicides (LCG 1004) is analysed six times. The certified value of simazine is equal to (26.7 ± 2.0) μg kg^{-1}, and the values obtained by the laboratory are, respectively, 29.4, 24.9, 26.4, 25.7, 22.0 and 23.5, corresponding to a mean concentration of 25.3 μg kg^{-1} and a standard deviation of 2.5 μg kg^{-1}. The adopted value for s_e is 5.2 μg kg^{-1}, based on the measurement of the measurement reproducibility. The value of σ is, therefore, equal to: $\sigma = \sqrt{[(5.2)^2 + (2.5)^2/6]} = 5.3$ μg kg^{-1}.
>
> The calculated value of Δ obtained is: $25.3 - 26.7 = -1.4$ μg kg^{-1}.
>
> It is hence verified that this value responds to the conditions of acceptability of the method, i.e. $-10.6 < 1.4 < 10.6$.

Let us note once more that the validity of the above-described test depends upon the validity of the adopted values for s_i and s_e. If these values are erroneous, the value of σ will be also erroneous, and the test will lead to wrong conclusions.

In some cases, it appears necessary to take into account the uncertainty of the certified value of the CRM (if this uncertainty is significantly different from σ) and to add a term corresponding to an enlarged uncertainty. Further details can be found in the literature (Walker and Lumley, 1999; ISO, 2000a,b).

The error may be expressed in two different ways in the framework of a method validation:

(1) As an absolute value $|x - x_o|$ where a positive error indicates a higher value. Or (more often in the case of method validation):

(2) As a recovery factor, i.e. a fraction or a percentage, x/x_o or $100x/x_o$, where x is the measured value and x_o the certified value. This type of approach is particularly useful when several tests or materials are subject to similar and proportional errors.

1.6.11 REFERENCE MATERIAL PRODUCERS

More than 150 reference material producers exist worldwide, but few of them are dedicated to water analysis. Information on the available materials can be obtained from the searchable VIRM database (http://www.virm.net), a member-led nonprofit organization founded within the 6th EC Framework programme, the COMAR data base, which is jointly operated by the BAM (Berlin, Germany), the LGC (London, UK) and the LNE (Paris, France). It should be noted that the mandatory criteria with respect to production quality (in particular accreditation) are not always fulfilled and that, therefore, it is presently difficult to evaluate the quality of all the materials that are available on the market. Among the major producers, two major organizations cover a large range of CRMs (including water CRMs) and ensure a continuity of the stocks: these are, on the one hand, the BCR in Europe (Institute for Reference Materials and Measurements, European Commission Joint Research Centre, Geel, Belgium) and, on the other hand, the NIST in the USA (National Institute for Standards and Technology, Gaithersburg, MD, USA). These two organizations deliver catalogues that can be obtained free of charge and provide information on the Internet (http://www.irmm.jrc.be/mrm.html for IRMM; http://ts.nist.gov/srm for NIST). Other notable producers for water CRMs are the National Research Council of Canada (Ottawa, Canada), the National Research Centre on CRMs in Pekin (China) and the National Institute for Environmental Sciences in Osaka (Japan). Other organizations produce water (C)RMs for the purpose of proficiency testing schemes in support of laboratory accreditation, e.g. the National Water Research Institute (USA) and the Dutch Ministry of Public and Water Works (The Netherlands).

Various CRMs for the quality control of water analysis, covering different types of matrices (freshwater, estuarine water, seawater, groundwater) are described in Volume 3 of the Water Quality Measurements Series (Quevauviller, 2002). In Table 1.6.4, the currently available CRMs related to wastewater are summarized, excluding the above-discussed BCR materials.

Table 1.6.4 Certified and indicative analyte concentrations of currently available wastewater-related CRMs in Europe

RM code and matrix	Analyte	Value	Provider and contact details
CRM002-100 Activated charcoal water filter	Aluminium	1800 mg kg^{-1} (Noncertified)	RT Corporation http://www.rt-corp.com
	Antimony	2 mg kg^{-1} (Noncertified)	
	Arsenic	30 mg kg^{-1} (Noncertified)	
	Barium	80 mg kg^{-1} (Noncertified)	
	Boron	80 mg kg^{-1} (Noncertified)	
	Cadmium	1 mg kg^{-1} (Noncertified)	
	Calcium	980 mg kg^{-1} (Noncertified)	
	Chromium	36300 mg kg^{-1} (Certified)	
	Cobalt	10 mg kg^{-1} (Noncertified)	
	Copper	96900 mg kg^{-1} (Certified)	
	Iron	1150 mg kg^{-1} (Noncertified)	
	Lead	5 mg kg^{-1} (Noncertified)	
	Magnesium	190 mg kg^{-1} (Noncertified)	
	Manganese	8 mg kg^{-1} (Noncertified)	
	Mercury	5 mg kg^{-1} (Noncertified)	
	Nickel	30 mg kg^{-1} (Noncertified)	
	Potassium	490 mg kg^{-1} (Noncertified)	
	Selenium	4 mg kg^{-1} (Noncertified)	
	Silver	18.3 mg kg^{-1} (Certified)	
	Sodium	480 mg kg^{-1} (Noncertified)	
	Strontium	110 mg kg^{-1} (Noncertified)	
	Thallium	20 mg kg^{-1} (Noncertified)	
	Tin	120 mg kg^{-1} (Noncertified)	
	Titanium	210 mg kg^{-1} (Noncertified)	
	Vanadium	40 mg kg^{-1} (Noncertified)	
RM2 and RM2e Wastewater	Biological oxygen demand	13–216 mg O_2 L^{-1} (Noncertified)	Association Générale des Laboratoires de l'Environnement aglae@nordnet.fr
	Chloride	95–600 mg L^{-1} (Noncertified)	
	Chemical oxygen demand	50–1000 mg O_2 L^{-1} (Noncertified)	
	Conductivity	1150–1530 μS cm^{-1} (Noncertified)	
	Fluorine	0.3–4.5 mg L^{-1} (Noncertified)	
	Potassium	14–35 mg L^{-1} (Noncertified)	
	Suspended solids	11–250 mg L^{-1} (Noncertified)	
	Sodium	71–163 mg L^{-1} (Noncertified)	
	Ammonia	0.6–56 mg N L^{-1} (Noncertified)	
	Nitrite	$<$0.05–3.5 mg N L^{-1} (Noncertified)	
	Nitrate	$<$0.2–150 mg N L^{-1} (Noncertified)	
	Total phosphorous	2–11 mg P L^{-1} (Noncertified)	
	pH	7.1–8 (Noncertified)	
	Phosphate	1.5–5.75 mg P L^{-1} (Noncertified)	
	Sulfate	112–142 mg L^{-1} (Noncertified)	
	Total Kjeldahl nitrogen	6–104 mg N L^{-1} (Noncertified)	
	Total organic carbon	60 mg C L^{-1} (Noncertified)	

Table 1.6.4 (*Continued*)

RM code and matrix	Analyte	Value	Provider and contact details
RM3B Wastewater	Aluminium	85–2000 μg L^{-1} (Noncertified)	Association Générale des Laboratoires de l'Environnement aglae@nordnet.fr
	Arsenic	1.5–80 μg L^{-1} (Noncertified)	
	Boron	300–2850 μg L^{-1} (Noncertified)	
	Barium	65–425 μg L^{-1} (Noncertified)	
	Beryllium	10 μg L^{-1} (Noncertified)	
	Cadmium	1–490 μg L^{-1} (Noncertified)	
	Cobalt	140 μg L^{-1} (Noncertified)	
	Chromium	4.5–3500 μg L^{-1} (Noncertified)	
	Copper	40–12000 μg L^{-1} (Noncertified)	
	Iron	100–2500 μg L^{-1} (Noncertified)	
	Mercury	0.3–50 μg L^{-1} (Noncertified)	
	Manganese	180–1100 μg L^{-1} (Noncertified)	
	Molybdenum	480 μg L^{-1} (Noncertified)	
	Nickel	35–7000 μg L^{-1} (Noncertified)	
	Lead	10–3000 μg L^{-1} (Noncertified)	
	Selenium	<5–85 μg L^{-1} (Noncertified)	
	Tin	500 μg L^{-1} (Noncertified)	
	Titanium	<10–200 μg L^{-1} (Noncertified)	
	Zinc	10–9000 μg L^{-1} (Noncertified)	
RM4B and 60 Wastewater	1,2-Dichloroethane	1–130 μg L^{-1} (Noncertified)	Association Générale des Laboratoires de l'Environnement aglae@nordnet.fr
	Aldrin	0.003–0.10 μg L^{-1} (Noncertified)	
	Anthracene	0.02–0.15 μg L^{-1} (Noncertified)	
	Atrazine	0.1–0.6 μg L^{-1} (Noncertified)	
	Benzene	5–35 μg L^{-1} (Noncertified)	
	Benzo(*a*)anthracene	0.02–0.15 μg L^{-1} (Noncertified)	
	Benzo(*a*)pyrene	0.02–0.15 μg L^{-1} (Noncertified)	
	Benzo(*b*)fluoranthene	0.02–0.25 μg L^{-1} (Noncertified)	
	Benzo(*g,h,i*)perylene	0.02–0.20 μg L^{-1} (Noncertified)	
	Benzo(*k*)fluoranthene	0.02–0.15 μg L^{-1} (Noncertified)	
	Bromodichloromethane	0.95–3 μg L^{-1} (Noncertified)	
	Bromoform	1–5.5 μg L^{-1} (Noncertified)	
	Carbon tetrachloride	0.1–1.5 μg L^{-1} (Noncertified)	
	Chloroform	1–7.5 μg L^{-1} (Noncertified)	
	Chlortoluron	0.1–0.65 μg L^{-1} (Noncertified)	
	Deisopropylatrazine	0.1–0.4 μg L^{-1} (Noncertified)	
	Desethylatrazine	0.05–0.9 μg L^{-1} (Noncertified)	
	Diazinon	0.4 μg L^{-1} (Noncertified)	
	Dibenzo(*a,h*)anthracene	0.02–0.40 μg L^{-1} (Noncertified)	
	Dibromochloromethane	1–5.5 μg L^{-1} (Noncertified)	
	Dieldrin	0.01–0.20 μg L^{-1} (Noncertified)	
	Diuron	0.1–0.9 μg L^{-1} (Noncertified)	
	Ethion	0.2 μg L^{-1} (Noncertified)	
	Fluoranthene	0.02–0.25 μg L^{-1} (Noncertified)	
	Heptachlor	0.009–0.060 μg L^{-1} (Noncertified)	
	Heptachlor epoxide	0.01–0.090 μg L^{-1} (Noncertified)	
	Indeno(1,2,3-*cd*)pyrene	0.02–0.10 μg L^{-1} (Noncertified)	
	Isoproturon	0.08–0.8 μg L^{-1} (Noncertified)	
	Lindane	0.01–0.26 μg L^{-1} (Noncertified)	
	Linuron	0.1–0.65 μg L^{-1} (Noncertified)	
	Methyl(2)fluoranthene	0.02–0.085 μg L^{-1} (Noncertified)	

(*Continued*)

Table 1.6.4 Certified and indicative analyte concentrations of currently available wastewater-related CRMs in Europe (*Continued*)

RM code and matrix	Analyte	Value	Provider and contact details
	Methyl(2)naphthalene	0.02–0.080 μg L^{-1} (Noncertified)	
	PCB 101	0.005–0.75 μg L^{-1} (Noncertified)	
	PCB 118	0.005–0.45 μg L^{-1} (Noncertified)	
	PCB 138	0.005–0.85 μg L^{-1} (Noncertified)	
	PCB 153	0.005–0.90 μg L^{-1} (Noncertified)	
	PCB 180	0.005–0.70 μg L^{-1} (Noncertified)	
	PCB 28	0.005–0.035 μg L^{-1} (Noncertified)	
	PCB 52	0.005–0.35 μg L^{-1} (Noncertified)	
	Propazine	0.1–0.4 μg L^{-1} (Noncertified)	
	Simazine	0.1–0.7 μg L^{-1} (Noncertified)	
	Terbutylatrazine	0.1–0.7 μg L^{-1} (Noncertified)	
	Tetrachloroethylene	0.2–0.80 μg L^{-1} (Noncertified)	
	Toluene	5–40 μg L^{-1} (Noncertified)	
	Total xylene	5–40 μg L^{-1} (Noncertified)	
	Trichloroethylene	1–5 μg L^{-1} (Noncertified)	
RM51 Wastewater	Arsenic	<100–450 μg kg^{-1} dry wt (Noncertified)	Association Générale des Laboratoires de l'Environnement aglae@nordnet.fr
	Cadmium	<100 μg kg^{-1} dry wt (Noncertified)	
	Chromium	<200–5800 μg kg^{-1} dry wt (Noncertified)	
	Copper	<300–1500 μg kg^{-1} dry wt (Noncertified)	
	Mercury	<10–25 μg kg^{-1} dry wt (Noncertified)	
	Nickel	<200 μg kg^{-1} dry wt (Noncertified)	
	Lead	1.1–13 μg kg^{-1} dry wt (Noncertified)	
	Selenium	<200–450 μg kg^{-1} dry wt (Noncertified)	
	Soluble fraction	2.5–40 % dry wt (Noncertified)	
	Zinc	850–7000 μg kg^{-1} dry wt (Noncertified)	
RM5B Wastewater	Anionic surfactants index	500–20000 μg SDS L^{-1} (Noncertified)	Association Générale des Laboratoires de l'Environnement aglae@nordnet.fr
	Phenol index	100–20000 μg C_6H_5OH L^{-1} (Noncertified)	
	Total cyanide index	250 μg CN L^{-1} (Noncertified)	
	Total hydrocarbons index	200–13000 μg L^{-1} (Noncertified)	
VKI-HL1 Wastewater	Aluminium	2.07 μg L^{-1} (Certified)	Eurofins A/S www.eurofins.dk/referencematerials
	Iron	3.03 μg L^{-1} (Certified)	
	Manganese	1.98 μg L^{-1} (Certified)	
	Molybdenum	9.96 μg L^{-1} (Certified)	
	Lead	10.02 μg L^{-1} (Certified)	
	Tin	10.33 μg L^{-1} (Certified)	
	Zinc	0.492 μg L^{-1} (Certified)	
VKI-HL2 Wastewater	Silver	2.06 μg L^{-1} (Certified)	Eurofins A/S www.eurofins.dk/referencematerials
	Barium	2.06 μg L^{-1} (Certified)	
	Cadmium	1.05 μg L^{-1} (Certified)	
	Cobalt	0.52 μg L^{-1} (Certified)	
	Chromium	4.08 μg L^{-1} (Certified)	
	Copper	4.26 μg L^{-1} (Certified)	
	Nickel	2.12 μg L^{-1} (Certified)	
	Strontium	5.11 μg L^{-1} (Certified)	

Table 1.6.4 (*Continued*)

RM code and matrix	Analyte	Value	Provider and contact details
VKI-WW1a	Ammonium Nitrate Phosphate	1.02 mg L^{-1} (Certified) 4.9 mg L^{-1} (Certified) 1.5 mg L^{-1} (Certified)	Eurofins A/S www.eurofins.dk/ referencematerials
VKI-WW2.1	Ammonium Phosphate	10 mg L^{-1} (Certified) 4.97 mg L^{-1} (Certified)	Eurofins A/S
VKI-WW2.2	Nitrate	1 mg L^{-1} (Certified)	Eurofins A/S
VKI-WW3	Total nitrogen Total phosphorus	7.45 mg L^{-1} (Certified) 1.54 mg L^{-1} (Certified)	Eurofins A/S www.eurofins.dk/ referencematerials
VKI-WW4	Chemical oxygen demand Total organic carbon	502 mg L^{-1} (Certified) 204 mg L^{-1} (Certified)	Eurofins A/S www.eurofins.dk/referencematerials
VKI-WW4A	Chemical oxygen demand Total organic carbon	50.4 mg L^{-1} (Certified) 19.8 mg L^{-1} (Certified)	Eurofins A/S www.eurofins dk/ referencematerials
VKI-WW5	BOD5 BOD7	206 mg L^{-1} (Certified) 217 mg L^{-1} (Certified)	Eurofins A/S www.eurofins dk/ referencematerials
VKI-WW6	Suspended solids	239 mg L^{-1} (Certified)	Eurofins A/S

REFERENCES

AOAC (1992) International harmonized protocol for the proficiency testing of (chemical) analytical laboratories. AOAC/ ISO/ REMCO No. 247.

Ihnat, M. (1997) *Fresenius J. Anal. Chem.*, **360**, 308–311.

ISO (1989) ISO Guide 35:1989. Certification of reference materials. General and statistical principles. Geneva, Switzerland.

ISO (1993) *International Vocabulary of Basic and General Terms in Metrology* (VIM), 2nd Edn. BIPM-IEC-IFCC-ISO-IUPAC-IUPAP-OIML. Geneva, Switzerland.

ISO (2000a) ISO Guide 31:2000. Reference materials. Contents of certificates and labels. Geneva, Switzerland.

ISO (2000b) ISO Guide 33:2000. Uses of certified reference materials. Geneva, Switzerland.

Pan, X.R. (1997) *Metrologia*, **34**, 35–39.

Quevauviller, Ph. (1998) *The Analyst*, **123**, 997–998.

Quevauviller, Ph. (2002) *Quality Assurance for Water Analysis*, Water Quality Measurements Series, Vol. 3. John Wiley & Sons, Ltd, Chichester.

Quevauviller, Ph. and Maier, E.A. (1999) *Interlaboratory Studies and Certified Reference Materials for Environmental Analysis – The BCR Approach*. Elsevier, Amsterdam.

Quevauviller, Ph., Benoliel, M.J., Andersen, K. and Merry, J. (1999) *Trends Anal. Chem.*, **18**, 376–383.

Richter, W. and Dube, G. (1997) *Metrologia*, **34**, 13–18.

Segura, M., Cámara, C., Madrid, Y., Rebollo, C., Azcárate, J., Kramer, G.N., Gawlik, B., Lamberty, A. and Quevauviller, Ph. (2004) *Trends Anal. Chem.*, **23**, 194–202.

Segura, M., Madrid, Y., Cámara, C., Rebollo, C., Azcárate, J., Kramer, G. and Quevauviller, Ph. (2000) *J. Environ. Monitor.*, **2**, 576–581.

Stoeppler, M., Wolf, W.R. and Jenks, P. (Eds) (2001) *Reference Materials for Chemical Analysis – Certification, Avalaibility and Proper Usage*. Wiley, Weinheim.

Walker, R. and Lumley, I. (1999) *Trends Anal. Chem.*, **18**, 594–616.

2.1
Sewers (Characterization and Evolution of Sewage)

Olivier Thomas and **Marie-Florence Pouet**

2.1.1 OBJECTIVES OF SEWAGE QUALITY MONITORING

The monitoring of the quality of raw wastewater in sewers is a rather new concern of water authorities. Before the 1990s, the monitoring of wastewater was limited to the inlet of the treatment plant, but in 1991, the urban wastewater treatment European

Wastewater Quality Monitoring and Treatment Edited by P. Quevauviller, O. Thomas and A. van der Beken

directive (Council Directive of 21 May 1991) (European Commission, 1991) stated several new considerations for collecting systems (sewers). They must be designed to collect urban wastewater (domestic and nondomestic, among industrial discharges) with the aim of prevention of leaks, and limitation of pollution of receiving waters due to storm water overflows (Annex I-A of directive). Thus, the main objectives of wastewater monitoring in sewers are the following:

- A better knowledge of wastewater loads and characteristics (mainly origin) for the protection and efficiency of the wastewater treatment plant, complementary to regulatory sampling at inlet/outlet of the plant. Shock loads and toxic effects of pollutants may be avoided.
- The possibility of checking the regulation compliance for nondomestic discharges, mainly industries and other facilities (hospitals, for example), from corresponding sewer branches. This 'through pipe' approach can be a preliminary step for nondomestic reduction load.
- The minimization of impacts of combined sewer overflows (CSOs) on receiving medium in case of unusually heavy rainfall. The knowledge of discharge load leads to a better management of CSOs.
- A complementary knowledge of wastewater characteristics with regard to emergent pollutants.

2.1.2 METHOLOGY

The monitoring of raw wastewater quality, generally involves sampling and laboratory analysis for regulation purpose (at the inlet of a treatment plant). However, some parameters can be measured on site, with handheld or on-line devices.

2.1.2.1 Sampling

Wastewater sampling has been largely discussed in Chapter 1.2. In summary, grab or discrete samples have to be avoided because of the variability with location and time, of sewage composition. Thus, automatic composite sampling usually coupled with flow rate or volume measurement, is better adapted for measuring the daily load in sewer branches or the efficiency of the treatment plant (at the inlet and outlet of the plant in this case). The sampling procedure must be applied with the best practices available, including conservation of samples at low temperature. Depending on objectives, a composite flask or 12 or 24 flasks may be used for an integrated, hourly or bihourly measurement. The choice of sampling points can be decided, either from the HACCP method (see Chapter 1.2) when little information is available, or, directly, for specific objectives like CSOs or nondomestic (industrial) discharges studies. Once the sampling points are located, one or several sampling campaign(s) are planned, depending on the sewer type. For a combined sewer, at least two campaigns have to be organized, one for a dry weather period and another for a wet weather period (if possible with heavy rainfalls, >50 mm per 24 h). The duration of each sampling campaign is generally 24 h, but can be extended to 36 h or 1 week, in case of uncertainty regarding industrial discharges for example. In any case, samples must be carried to the laboratory at least every 24 h.

2.1.2.2 Measurement and Analysis

Several books and reviews cover this topic (Thomas, 1995; Colin and Quevauviller, 1998; Olsson *et al.*, 2002; Fleishman *et al.*, 2003), and some simple recommendations can be proposed. On-site measurement has to be carried out for some parameters, mainly temperature and pH. For other parameters (see Section 2.1.3), rapid measurement and analysis should be done in the laboratory. In the case of field experimentation with several sampling sites possible, for example for the optimization of control points location, on-site measurement can be planned, with field portables devices such as a multiprobe, colorimetric test kits or UV analyser. These handheld systems give in a few minutes field data for parameters such as:

- temperature, pH, conductivity, turbidity (dissolved oxygen) for a usual multiprobe, possibly associated with the automatic sampler;
- N [ammonia, total kjeldhal nitrogen (TKN)] and P (orthophosphate) forms and other specific mineral substances (chloride, sulfide, etc.) for colorimetric test kits;
- Global organic pollution estimation [total organic carbon (TOC), chemical oxygen demand (COD), biological oxygen demand (BOD)], total suspended solids (TSS) and some other specific compounds (phenol, sulfide, nitrate, etc.) for UV sensor.

Except the colorimetric test kits, the other devices can be used either as handheld instruments or as on-line sensors during the sampling period, completing thus the flow or volume measurement system generally placed close to the automatic sampler for integrated sampling proportional to flow rate or volume.

One key point of on-site measurement is the traceability of results, in order to allow the completion and/or comparison of data with results of laboratory analysis from samples.

2.1.2.3 Remote Sensing

Several reviews have been published on the topic (Thomas, 1995; Bourgeois *et al.*, 2001; Vanrolleghem and Lee, 2003). Monitoring of wastewater quality in sewers with on-line devices placed inside the collecting system is difficult, except at the inlet of the treatment plant. On the one hand, there exist few on-line instruments for wastewater quality monitoring, and on the other hand, the environmental conditions for instruments are very severe (humidity and corrosive atmosphere). However, the previous on-line devices (multiprobe, UV analyser) can be completed by oil sensors (based on near infrared reflectance), or more sophisticated instruments like on-line TOC meters. The latter have to be located in a temperature controlled environment (shelter for example), connected to the sewer with a sample fast loop, where wastewater flow speed is very fast, to ensure a good representativity of the sample. Nevertheless, the reliability of the measurement is poor, depending on the maintenance efforts to obtain available measures (validated and when needed). For example, a study of four TOC meters (two on-line and two laboratory) for the wastewater quality monitoring of a petrochemical wastewater treatment plant has shown a difference of about 20 % (Thomas *et al.*, 1999).

2.1.3 PARAMETERS OF INTEREST

A lot of parameters can be considered for raw wastewater quality monitoring in sewers, divided into two main groups: one of usual parameters, often measured for a regulatory purpose; and the other, a group of complementary parameters including the analysis of emergent pollutants and nonparametric (statistical sense) measurements.

2.1.3.1 Usual Parameters

This group has been the same since the beginning of wastewater management almost a century ago or at least for the last 50 years. Except for some organoleptic parameters (colour, odour), they are classified into physico-chemical parameters (temperature, pH, conductivity, dissolved oxygen), chemical parameters, either aggregate [BOD, COD, TSS, total nitrogen (TN), total phosphorus (TP)] or specific (ammonia, nitrate, orthophosphate, etc.), and microbiological ones (mainly faecal coliforms). This classification is however not so simple with regard to some parameters considered either global (aggregate) or specific as total organic carbon (TOC), or TKN (reduced N compounds). Except the physico-chemical group, all other parameters have to be analysed in the laboratory.

2.1.3.2 Complementary Parameters

These are parameters not often measured in wastewater because they are rarely included in a regulated context, but knowledge of them is very important especially for studies related to industrial discharge characterization and control. As for usual parameters, the same classification can be proposed.

Turbidity and redox potential constitute the first group of physico-chemical complementary parameters. They can be measured by sensors, directly (in-line) into the flow or on- off-line.

The second group is that of aggregate parameters, characterizing families of chemical organic substances by way of nonchromatographic techniques, as total petroleum hydrocarbons (TPH), anionic surfactants (methylene blue active substances, MBAS), halogenated organic compounds (adsorbable halogenated organics, AOX) or phenol index. Laboratory analyses are needed for these parameters.

The specific analysis of chemical substances, either minerals (including organo-metallic forms) or organics, constitutes the third group of complementary parameters. There are a lot of substances of interest to be analysed in wastewater, usually in the laboratory by atomic spectroscopy (emission or absorption) for metals, by chromatography (gaseous or liquid) for organics and by chromatography or capillary electrophoresis for mineral and organic ions.

Associated with this group are emergent pollutants, including some potentially toxic substances and their degradation by-products, pharmaceuticals, such as endocrine disruptors (the majority of compounds being pharmaceuticals), pesticides, surfactants, personal care products, etc. (Barcelo, 2005).

A fourth group of complementary parameters, less well known because new and not related to quantitative information (mainly physical result or concentration), includes the so-called nonparametric approach, giving very useful complementary information (Thomas, 1995). The basic principle of the nonparametric measurement (NPM) which, as for a nonparametric statistical test, does not require to be related to a given parameter (respectively, a given statistical law) is the existence of a qualitative relationship between the analytical factor and the information to be given (Baurès, 2002). Thus, the more relevant analytical techniques which can be envisaged are the ones giving multiple responses that are difficult to exploit without extensive knowledge of the phenomenon to be studied. This is the case for all scanning techniques such as spectroscopic techniques (absorptiometry and flurorimetry). UV spectrophotometry is chosen based on its numerous and decades-old existing applications for water and wastewater quality monitoring. From UV spectra to useful information, some basic handling can be envisaged (Vaillant *et al.*, 2002). Derivatives (second often preferable), peak-valley methods, direct comparison and normalization – all these simple transformations can give interesting information. One major application is, however, the exploitation of the presence of isosbestic points (IPs), when several spectra cross together at least at a single point (Pouet *et al.*, 2004). Depending on the condition of the IP appearance, directly from a set of spectra (or after normalization in the case of hidden IPs) the composition of wastewater can vary from one state to another (qualitative conservation) with a possible quantitative conservation when a direct IP occurs. Applications of this nonparametric measurement will be shown in Chapter 4.2 for the calculation of industrial wastewater variability and in Chapter 5.1 for the study of discharges in receiving medium.

2.1.4 EVOLUTION OF SEWAGE

Considering the composition of wastewater, heterogeneous and variable, always changing with inputs of industrial discharges or fresh domestic loads, from upstream to the treatment plant, its evolution is evident but complex, involving, physical, physico-chemical and biological factors. Moreover, the evolution of wastewater depends both on the design principle of the sewer systems (gravity or pressure main) and on the climatic conditions for combine sewers (Nielsen *et al.*, 1992). A lot of studies have been published on the interaction of sewerage and wastewater treatment (Kruize, 1993) and on the role of the sewer as a physical, chemical and biological reactor (Hvitved-Jacobsen *et al.*, 1995). All these studies have been carried out with classical methods for wastewater quality measurement in the laboratory. However, changes in wastewater composition can be appreciated by the measurement of on-site parameters of interest (see above) including the estimation of variability.

2.1.4.1 Physical Factors

The first physical factor is the flow rate ratio in the case of a mixture or discharge, playing a role in the concentration or dilution of pollutants concentration. The main problem is for combined sewers during rain fall, with storm runoff drainage. At the beginning of the event, particulate materials from roads, roofs and parking areas, and also oil, salts, etc., can be carried to the sewer (particularly after a long dry weather period) increasing the pollution load. Then, after flushing, the main phenomenon remains dilution. The effects of storm water in combined sewers vary with the characteristics of the sewers (length, diameters, etc.) and the topography (slope) leading to the equalization of loads in the case of small flow rate and large volumes. In this case, settling of large or dense particles generally occurs, and the settled material can be flushed with the increase of flow if the sewer is combined (collecting both wastewater and storm water runoff). Thus, the wastewater quality of long sewers in a flat area, (partly) combined, presents huge variations and differences between dry and wet periods. Finally, temperature variation (generally an increase) is possible with industrial wastewater of enterprises with cooling open circuits or rejecting hot effluent. A hot temperature leads to the increase of the kinetics of biological and physico-chemical reactions (biodegradation, chemical reactions), mainly by the increase of equilibrium 'constants' (which depend on temperature), but also by the increase solubility of some organics (for example, the solubility of benzene in water increases 20 % up to 1900 mg/l, between 10 °C and 30 °C). A hot temperature leads to the evaporation of solvents for laundry discharge, for example.

2.1.4.2 Physico-chemical Factors

The first physico-chemical factor is the variation of pH responsible for the modification of acidic–basic reactions. Even if wastewater is considered as a buffer medium

Table 2.1.1 Percentage of unionized ammonia with respect to pH and temperature ($pK_a = 9.25$ at 25 °C)

	pH					
Temperature (°C)	6.0	7.0	8.0	9.0	9.25	10.0
10	0.02	0.2	1.8	15.7	24.9	65.0
20	0.04	0.4	3.8	28.4	41.4	79.8
30	0.08	0.8	7.4	44.6	58.8	88.9

considering its composition as a complex mixture, acidic or basic shocks are locally possible with industrial or accidental discharge of concentrated acid or base solutions. One consequence can be, for example, on ammonia equilibrium (Table 2.1.1), with the increase of the toxic form (unionized ammonia) with pH (and temperature). For example, a concentration of ammonia of 10 mg/l at 20 °C and for a pH of 8.0, gives a concentration of the unionized form equal to 0.38 mg/l, which is toxic.

Another physico-chemical factor is the redox potential E_H, fixed by the respective concentration of chemical oxidized and/or reduced substances. As for temperature or pH, variations of redox conditions are related to industrial discharges. A decrease of E_H can give septic conditions (for example, $E_H \leq 40$ mV for pH = 7) leading to odour production and sewer corrosion in the presence of sulfides (Degrémont, 2005).

There are some other physico-chemical factors involved in sewage evolution, like precipitation, due to pH increase (for hydroxides) or exceeding of solubility products in the case of industrial discharge, or complexation by the presence of chelating agents. One last important point is the fate of surfactants, the concentration of which being high in some industrial discharges. Depending on the presence of colloids and on flow conditions, theses substances can be adsorbed on suspended solids, leading to the aggregation of colloids with a decrease of the dissolved amount of dispersants. This phenomenon is responsible for sample ageing (Baurès *et al.*, 2004).

2.1.4.3 Biological Factors

Even if physical and physico-chemical factors of wastewater composition evolution are numerous, the biological ones are more important. Regarding the degradation of organic substances, where used, biological reactions in sewers are principally anaerobic, bur aerobic conditions can be encountered in some gravity sewers. Even if the concentration of dissolved oxygen is very low (<2 mg/l), some aerobic processes may occur as in biological treatment plants. In the pressure part of the sewer or in the case of high organic load, expressed by high oxygen demand values (BOD or COD), no dissolved oxygen is available (nor in the gaseous phase, which is dangerous for operators). Thus, the organic matter can be partially degraded through fermentation reactions, accompanied by the chemical reduction of some minerals, like the sulfate ion to sulfide. This reaction occurs for septic conditions (see above), and leads to odour production and corrosion of the sewer.

Another biological factor is the potential toxicity of a lot of substances, often brought by industrial discharges in sewers, able to cause severe damage in the biological reactors of the wastewater treatment plant (death of active biomass). The toxicity effect depends on the nature and concentration of substances, but also on the existence of an acclimated biomass potentially in contact with wastewater. For example, depending on the organisms, phenol is toxic from concentrations between 10 mg/l and 25 mg/l but concentrations up to 400 mg/l can be treated by biological processes (Bevilacqua *et al.*, 2002). As for the previous factors, the main cause of wastewater quality variation and evolution (except dilution by storm runoff in combined sewers) is the occurrence of shock loads associated with point industrial discharges, the effects of which are important in the case of short sewers or if the discharge is close to the treatment plant.

REFERENCES

Barcelo, D. (2005) *Emerging Organic Pollutants in Wastewater and Sludge. The Handbook of Environmental Chemistry*, vol. 5, parts I and O. Springer-Verlag, Berlin.

Baurès, E. (2002) La mesure non paramétrique, un nouvel outil pour l'étude des effluents industriels: application aux eaux résiduaires d'une raffinerie. PhD Thesis, University of Aix Marseille III.

Baurès, E., Berho, C., Pouet, M.-F. and Thomas, O. (2004) *Water Sci. Technol.*, **49**(1), 47–52.

Bevilacqua, J.V., Cammarota, M.C., Freire, D.M.G. and Sant Anna, G.L. (2002) *Brazilian J. Chem. Engin.*, **19**(2), 151–158.

Bourgeois, W., Burgess, J.E. and Stuetz, R.M. (2001) *J. Chem. Technol. Biotechnol.*, **76**, 337–348.

Colin, F. and Quevauviller, Ph. (Eds) (1998) *Monitoring of Water Quality, the Contribution of Advanced Technologies*. Elsevier, Amsterdam.

Degrémont (2005) *Mémento technique de l'eau*, 10th Edn. Paris.

European Commission (1991) Council Directive of 21 May 1991 concerning urban wastewater treatment (91/271/EEC).

Fleishman, N., Langergraber, G. and Haberl R. (2003) Proceedings of the IWA International Specialised Conference, Vienna, Austria, 21–22 May 2002. *Water Sci. Technol.*, **47**(2).

Hvitved-Jacobsen, T., Nielsen, P.H., Larsen, T. and Aa Jensen, N. (1995) Proceedings of the International Specialised Conference, Aalborb, Denmark, 16–18 May 1994. *Water Sci. Technol.*, **31**(7).

Kruize, R.R. (1993) Proceedings of the International Conference, Amsterdam, The Netherlands, 31 August–4 September 1992. *Water Sci. Technol.*, **27**(5–6).

Nielsen, P.H., Raunkjaer, K., Norsker, N.H. and Hvitved-Jacobsen, T. (1992) *Water Sci. Technol.*, **25**(6), 17–31.

Olsson, G., Jeppsson, U. and Rosen, C. (2002) Proceedings of the IWA International Conference, Malmö, Sweden, 3–7 June 2001. *Water Sci. Technol.*, **45**(4–5).

Pouet, M.-F., Baurès, E., Vaillant, S. and Thomas, O. (2004) *Appl. Spectrosc.*, **58**(4), 486–490.

Thomas, O. (1995) *Métrologie des eaux résiduaires*. Cebedoc, Tec et Doc Lavoisier, Liège, Paris.

Thomas, O., El Khorassani, H., Touraud, E. and Bitar, H. (1999) *Talanta*, **50**, 743–749.

Vaillant, S., Pouet, M.-F. and Thomas, O. (2002) *Urban Water*, **4**, 273–281.

Vanrolleghem, P.A. and Lee, D.S. (2003) *Water Sci. Technol.*, **47**(2), 1–34.

2.2
Sewer Flow Measurement

Charles S. Melching

2.2.1 INTRODUCTION

Sewers are difficult environments in which to obtain accurate discharge estimates for many reasons including rapidly changing flow conditions, surcharge, backwater,

Wastewater Quality Monitoring and Treatment Edited by P. Quevauviller, O. Thomas and A. van der Beken

sediment and debris deposition, turbulence, confined space/hazardous conditions issues, access, variable pipe slope along a reach resulting from differential settlement of individual pipes, and different pipe sizes. Nevertheless, continually increasing environmental concerns and the need to more optimally manage stormwater and wastewater flows have increased the need to accurately monitor flows in storm, sanitary and combined sewers. These concerns are not new, e.g., North Rhine-Westphalia, Germany, issued a decree that the most important detention facilities of the combined sewer network were to be equipped with continuous monitoring devices more than 20 years ago (Weyand, 1996). Fortunately, as the need for measurement devices capable of high accuracy has increased new and improved measurement techniques also have been developed over the last 20 years. This chapter attempts to summarize the accuracy, advantages, and disadvantages of the available techniques.

There are basically two basic types of flow measurement techniques: (1) those that rely on a relation between stage and discharge, e.g., Manning's equation and flumes; and (2) those that estimate average velocity by acoustic or electromagnetic means and multiply this by the cross-sectional area obtained through a depth measurement device and known conduit geometry. Most of these devices have two parts: (1) a primary device that directly interacts with or controls the flowing water; and (2) a secondary device for measuring water depth (Church *et al.*, 1999).

This chapter focuses on the general characteristics of the various measurement techniques of types 1 and 2 and does not provide a direct comparison of the commercially available equipment for flow measurement in sewers that apply the various measurement techniques. Due to the limited space available and the limited number of independent evaluations of measurement equipment, a proper comparison of the equipment cannot be done here. Further, it is not the purpose of this chapter to advocate or criticize any particular device, but rather to give the readers basic information on the measurement techniques to aid in the selection of the appropriate technique. When purchasing equipment readers should carefully review the literature provided by the manufacturers, discuss experience with the equipment with professional colleagues, and apply the time-honoured principle of *caveat emptor* (let the buyer beware).

2.2.1.1 Purposes of Flow Monitoring

There many reasons for flow monitoring, among the most common are:

(1) Real-time control (RTC) of the sewer system. Existing large sewers can be controlled by gates, e.g., to increase storage capacity and prevent overburdening of treatment plants (Curling *et al.*, 2003). RTC also can optimize treatment plant operation to ensure consent standards are met and to minimize the total pollutant load reaching the environment (Watt and Jeffries, 1996) or improve plant efficiency in order to provide capacity for future sewer extensions (Anon., 1996).

(2) Sewerage system operational considerations. Information about storage and discharge conditions can, for example, be the basis for optimizing cleaning and maintenance work (Weyand, 1996).

(3) In regional sewerage networks, flow monitoring can equitably allocate costs among communities.

(4) Compliance with regulatory requirements.

(5) Provide data for calibration and verification of numerical models (Baughen and Eadon, 1983).

(6) Identify inflow and infiltration (I/I) problems.

(7) Performance evaluations of pumps and hydraulic structures (e.g., overflow structures).

Items (1)–(4) typically require long-term, effectively permanent, monitoring, whereas items (5)–(7) generally require only temporary monitoring. On the basis of 10 years of sewer monitoring experience in Germany, Weyand (1996) made two very important, related observations: (1) it is important to start the planning of monitoring systems with the formulation of necessary demands; and (2) experience shows that the requirements of monitoring systems rise with their use. Thus, it is quite possible that sites that originally were established for a temporary study may become long-term sites, and careful selection of equipment and sites is necessary.

2.2.1.2 Equipment Selection Considerations

Huth (1998) prepared a list of considerations for selection of flow measurement equipment. Six of his eight issues are:

(1) Know the relative strengths and weaknesses of the available equipment.

(2) Buy the level of accuracy required for the application. For example, high accuracy is needed for RTC, cost allocation, and model calibration and verification; whereas lesser accuracy may be required for I/I studies, basic sewerage system operation, and performance evaluations of hydraulic structures. However, it must be remembered that requirements of monitoring systems rise with their use.

(3) Know your flow rate. Sanitary sewers may have fairly constant flows, whereas storm and combined sewers have wider flow ranges and require equipment that is accurate over a wide range of flow conditions. Curling *et al.* (2003) stressed the importance of having high accuracy over the full range of flows noting that varying accuracy will increase variation in modelling results and lead to poor understanding of problem sites, improper estimation of capacity, and improper allocation of capital improvement funds.

(4) Learn what is in the water. Debris may clog some equipment (flumes) and reduce the performance of other equipment (acoustic transducers) requiring frequent maintenance, also high particulate loads may affect the ability of sound waves to penetrate the flow.

(5) Location, location, location (discussed in detail in Section 2.2.1.3).

(6) Make sure there is power.

Similarly, Church *et al.* (1999) noted that selection of the most appropriate method for collection of accurate flow data that are representative of a particular site requires knowledge of the flow regime(s), range of flow rate and depth, rapidity of flow changes, channel geometry, and the capabilities and accuracies of the methods available for measuring flow.

2.2.1.3 Monitoring Locations

The importance of proper site selection cannot be overstated (Church *et al.*, 1999). Most of the flow measurement techniques described in this chapter work best at sites where fully developed, uniform, open channel flow not subject to backwater effects is present composing optimal hydraulic conditions. Fully developed, uniform open channel flow usually requires many diameters of straight, uniform, undisturbed pipe upstream and downstream of the measurement location. For example, Johnson (1995) notes that the British Standard 1042 recommends that upstream from the measurement point a straight length of pipe equal to 30 to 50 diameters is sufficient depending on the type of turbulence causing device, whereas downstream from the measurement point 5 diameters of straight pipe should be present. Shorter sections of straight pipe could affect flow measurement accuracy. The accuracy of some methods also may decrease due to backwater effects and transitions from open-channel to pressurized pipe-full flow.

Practical considerations may make it necessary to place a monitor at a location with nonoptimal hydraulic conditions. For example, important locations, such as overflows, bifurcations, and known flooding points, may require individual monitoring irrespective of hydraulic conditions (Baughen and Eadon, 1983). Borders between communities may require monitoring irrespective of hydraulic conditions for 'political reasons' in cost allocation. Accurate model calibration and verification may require monitoring of each subcatchment (Baughen and Eadon, 1983). Finally, local constraints such as accessibility, power supply, and nonhydraulic goals of monitoring may also necessitate using nonhydraulically optimal sites. For example, monitoring locations might be selected for ease of pollutant sampling regardless of hydraulic conditions, as was the case of combined sewer monitoring in the Chicago, USA, area reported by Waite *et al.* (2002).

Some of the techniques discussed in this chapter are better at measuring flow under nonhydraulically optimal conditions than others. Thus, once the monitoring locations are selected the following questions (after Church *et al.*, 1999) must be considered:

(1) Is the flow measuring technique applicable to the flow and channel characteristics at the site?

(2) Is the flow measuring technique capable of measuring the full range of flows?

(3) Will the flow measurements be of sufficient accuracy to meet the objectives of the study?

2.2.1.4 Characteristics of Ideal Sewer Flow Measurement Equipment

In order to deal with the complex hydraulic environment of sewer systems, Wenzel (1975) recommended that the ideal device for flow measurement should have the following characteristics:

(1) capability to operate under both open channel and full flow conditions;

(2) a known accuracy throughout the range of measurement;

(3) a minimum disturbance to the flow or reduction in pipe capacity;

(4) a minimum of field maintenance;

(5) compatibility with real-time remote data transmission;

(6) reasonable construction and installation costs.

Drake (1994) further suggested that the equipment must provide reliable and accurate level and/or flow measurements within dynamic conditions, withstand a corrosive environment, overcome turbulence, and resist entanglement with floating matter.

2.2.1.5 Quality Assurance and Quality Control

For any flow monitoring, but particularly for sewer flow, detailed quality assurance and quality control (QA/QC) programmes are necessary. Church *et al.* (1999) describe in detail the key components of a QA/QC programme for flow monitoring, and their main QA/QC components are summarized as follows:

(1) Frequent and routine site visits by trained/experienced personnel to maintain equipment and keep the site clean.

(2) Redundant methods for measuring flow.

(3) Technical training of project personnel. Weyand (1996) also stressed that it is necessary to have specially trained and qualified staff for operating and calibrating the sewer flow meters.

(4) Frequent review by project personnel of data collected. Weyand (1996) also noted that data quality must be continually checked to detect equipment malfunctions.

(5) Quality audits, in the form of periodic internal reviews.

(6) Quality audits, in the form of periodic external reviews.

Church *et al.* (1999) noted that frequent calibration of equipment is necessary because of the difficult monitoring environment, and that the difficulties of measuring in this environment result in a high probability of incomplete record, even when stations are well maintained and properly calibrated.

2.2.2 MANNING'S EQUATION

The simplest form of stage-discharge relation is obtained by assuming that Manning's equation is valid for the selected monitoring location. Using Manning's equation discharge, Q, is calculated as

$$Q = \frac{1}{n} A(h) R(h)^{2/3} S^{1/2} \tag{2.2.1}$$

where n is Manning's roughness coefficient, $A(h)$ is the cross-sectional area of flow, $R(h)$ is the hydraulic radius of the flow, h is the depth (or pressure head for full-pipe flow) of flow, and S is the energy slope of the flow. In this technique, h is measured using a pressure transducer or bubbler system, A and R are calculated as a function of h from the known conduit geometry, S is approximated as the pipe slope, and n is estimated from standard tables on the basis of pipe material and condition. Soroko (1973) noted that Manning's equation may be appropriate for discharge calculation in channels with a straight course of at least 61 m, preferably longer, the course being free of rapids, abrupt falls, and sudden contractions or expansions.

The primary advantage of this technique is that only a stage measurement device is needed to estimate flow. The primary disadvantages of this technique are that proper estimates of S and n are difficult to obtain. For steady, uniform flow in a channel as specified by Soroko (1973) the bed slope equals the energy slope, however, for unsteady, nonuniform flow common in storm and combined sewers the bed slope and energy slope diverge. Further, even in cases where the bed slope approximates the energy slope well, determination of the bed slope is difficult. Most often the bed slope is estimated from design plans, but this can be substantially different from the actual pipe slope. For example, Melching and Yen (1986) compared 'as built' measurements of pipe slopes between manholes with the slope indicated on the plans for 80 storm sewers in Tempe, (AZ, USA) and found a standard construction error of 0.0008 m/m. Given that slopes of these sewers ranged from 0.001 to 0.0055 m/m, the standard construction error represented a substantial portion of the design slope in this case. Even when the pipe slope between manholes at the measurement location has been measured in the field there may be inaccuracies in the estimated slope because differential settlement and/or sag of pipes in the measurement reach between manholes cause the measured slope to not be representative of the energy slope.

Determination of Manning's n from tables for sewer pipes is at best a 'guesstimate' (Soroko, 1973). Slime, debris, deposition, and decay of the pipes may cause Manning's n for a pipe to be substantially different from values in the standard tables. Further, in pipes Manning's n is a function of depth not a constant. Lanfear and Coll (1978) state that at depths of 5 to 70 % of pipe diameter, n is 20 to 30 % higher than the value for full pipe flow obtained from standard tables, failure to account for this phenomena will cause flows to be overestimated by more than 20 %. For improved estimates of n, Wright (1991) recommended that Camp's distribution of n as a function of depth (Figure 2.2.1) be used to adjust the value of Manning's n.

Wright (1991) presented the results of a field study for 22 sites in Grand Rapids (MI, USA) that illustrates the accuracy of the typical application of Manning's equation for estimation of flow in storm sewers. The pipes ranged in diameter from 45 to 243 cm, and in 5 of the 22 cases slopes were estimated from field measurements while the remaining slopes were based on design drawings. The actual flow rate was estimated using a hand-held electromagnetic velocity meter to measure the maximum

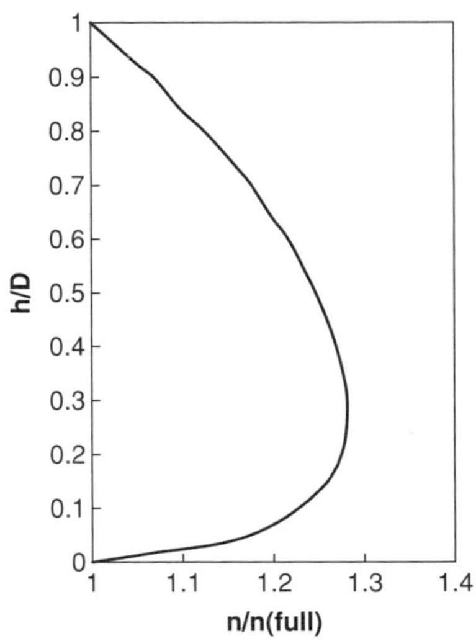

Figure 2.2.1 Camp's normalized distribution of Manning's n versus relative depth in a circular section

velocity and assuming that the mean velocity was 0.9 times the maximum velocity. At the 22 sites discharge was measured an average of nine times, and the measured discharge was used to calculate values of $S^{1/2}/n$ for each measurement. Average values of $S^{1/2}/n$ were determined for each site, and compared with the values of $S^{1/2}/n$ for each site estimated from field conditions including the variation of Manning's n with flow depth (as would be done in the typical application of Manning's equation). The mean percent error was 28.9 %. In 50 % of the cases the errors were greater than 25 %, and in 27 % of the cases errors were greater than 50 %. Wright (1991) also presented a case for sewers in Mobile, (AL, USA) that illustrated the even poorer results obtained with Manning's equation in pipes subject to surcharge and backwater. If a site is subject of surcharge and backwater, two stage gauges should be used, and the water-surface slope should be used to approximate the energy slope. This approach is rarely applied in practice.

Lanfear and Coll (1978) found that a 'fitted' Manning's equation, calibrated by a single flow measurement, provided good agreement with observed flow data and eliminated the need to measure slope. A single discharge measurement is used to calculate $S^{1/2}/n$, which then is applied to all other flows. This approach was illustrated for multiple flow measurements in three 122-cm diameter brick sewers with 'as built' slopes between 0.00044 and 0.0144 in Washington (DC, USA). Lanfear and Coll (1978) stated that if $S^{1/2}/n$ is determined for those flows of most concern, most of the error caused by variable Manning's n is eliminated. This implies that if a wide range of flows are of interest, the value of $S^{1/2}/n$ may need to be calibrated throughout the flow range. Marsalek (1973) stated that under conditions of unsteady, nonuniform flow in pipes, Manning's equation underestimates flows in the rising stage and overestimates flows in the falling stage. Finally, Alley (1977) reported that the accuracy of the Manning's equation technique is, at best, about 15 to 20 %.

2.2.3 FLUMES

Flumes have been used to measure open channel flows in small streams, irrigation canals, water and wastewater treatment plants, and sewers for more than 50 years. Flumes are flow-constriction structures that control the flow hydraulics such that flow is directly related to head (Church *et al.*, 1999). The most common type of flume constricts the flow such that critical flow results somewhere in the constricted section, which results in a unique relation between head and discharge as detailed later. These flumes are known as critical-flow flumes. Flumes work best at sites where the potential for surcharge, full-pipe pressurized flow, and backwater effects are expected to be negligible. Flume measurements are reliable for both uniform and nonuniform flow unless the sewer becomes surcharged (Parr *et al.*, 1981). Baughen and Eadon (1983) noted that flumes give misleading results if they are surcharged and this condition is not suspected.

The flow computation principle applied for critical-flow flumes may be derived as follows (Wenzel, 1975). The energy conservation equation is applied between a reference section 1 located immediately upstream of the flow constriction (flume) and section 2 is located in the constriction a distance L downstream from section (1) resulting in:

$$h_1 + \alpha_1 \frac{Q^2}{2gA_1^2} + z_1 = h_2 + \alpha_2 \frac{Q^2}{2gA_2^2} + z_2 + h_L \tag{2.2.2}$$

where α is the kinetic energy correction factor, g is the acceleration of gravity, z is the vertical distance from some datum, and h_L is the head loss between sections 1 and 2. In the application of Equation (2.2.2) the following assumptions are made: (1) steady flow; (2) hydrostatic pressure distribution at section 1; (3) small slope such that the flow depth h approximately equals the vertical component of depth; and (4) two- and three-dimensional effects are negligible or accounted for as coefficients or energy loss terms (Wenzel, 1975). Equation (2.2.2) can be solved for discharge if all other terms are measured or evaluated as follows:

$$Q = \left[\frac{2g(h_1 - h_2 + LS_0 - h_L)}{\frac{\alpha_2}{A_2^2} - \frac{\alpha_1}{A_1^2}} \right]^{1/2} \tag{2.2.3}$$

where S_0 is the bed slope. If open channel flow is present and A_2 is sufficiently small, critical flow will occur at some point in the constriction. If section 2 is defined as the point of critical flow the following relation is derived from the fact that the Froude number equals 1 for critical flow:

$$\frac{Q^2 B_2}{gA_2} = 1 \tag{2.2.4}$$

where B_2 is the width of the free surface at section 2. Substitution of Equation (2.2.4) into Equation (2.2.3) and using known relations between A, h, and B, the discharge can be implicitly determined by measuring only h_1 and evaluating h_L, since all other terms are known. The head loss can be determined from boundary layer theory (Wenzel, 1975), but typically the relation between flow and discharge for a flume is determined by laboratory ratings.

The Palmer–Bowlus flume was first proposed in the 1930s (Palmer and Bowlus, 1936), was extensively tested in the 1950s (Wells and Gotaas, 1958), and has become the most commonly used critical-flow flume in sewer systems. Palmer–Bowlus flumes have low head loss and can be installed in manholes where there is a standard, straight-through design, or they can be installed in the half section of the sewer conduit (Soroko, 1973). Palmer–Bowlus flumes can be permanently installed, or be portable devices which can be inserted in the downstream pipe of a manhole using a pneumatic seal (Baughen and Eadon, 1983). Wells and Gotaas (1958) extensive laboratory experiments on the Palmer–Bowlus flume indicated that accuracy within 3 % of the theoretical discharge is readily attainable at depths up to $0.9D$ (where D is the upstream pipe diameter) for flumes installed in circular conduits. However, Hunter *et al.* (1991) indicated that Palmer–Bowlus flumes typically are inaccurate at depths greater than $0.75D$.

Figures 2.2.2 and 2.2.3 show two standardized trapezoidal Palmer–Bowlus flume sections for which a rating table is presented in Ludwig and Parkhurst (1974). Ludwig and Parkhurst (1974) noted that it is believed that the typical trapezoidal sections offer advantages regarding flow range and the provision of more accurate measurements at low flow values. Figure 2.2.4 shows a standardized rectangular throat Palmer–Bowlus flume for which a rating table is presented in Ludwig and Parkhurst (1974). Ludwig and Parkhurst (1974) noted that a value of $D/10$ represents a desirable rise in the base of rectangular flumes installed within circular conduits. Standard Palmer–Bowlus flumes only have a stage measurement device in the approach section. To measure pressurized full-pipe flow, pressure should be measured at both sections 1 and 2 (approach and throat, respectively). Flumes with such two pressure sensor designs are known as Venturi flumes, which are discussed in the following paragraphs.

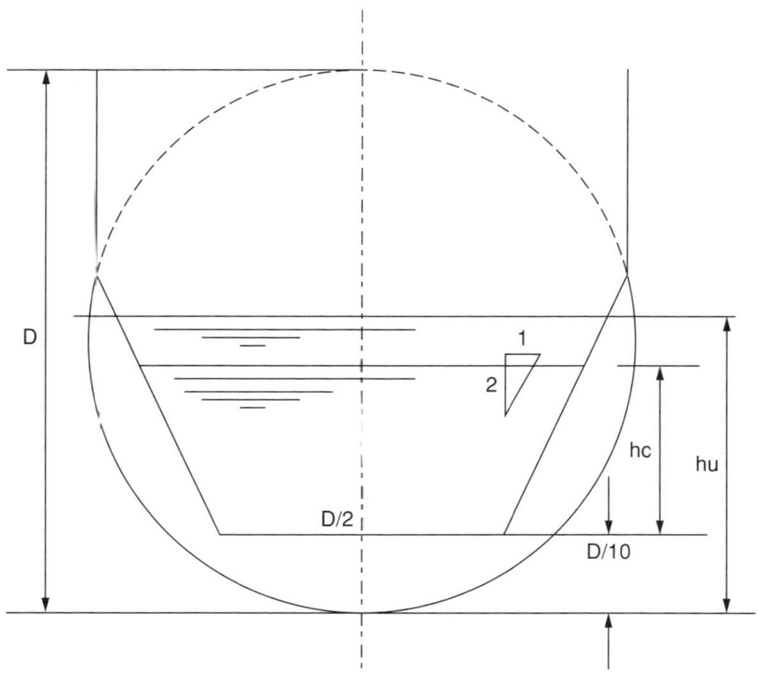

Figure 2.2.2 Standardized Palmer–Bowlus trapezoidal flume with a bottom width of one-half of the pipe diameter

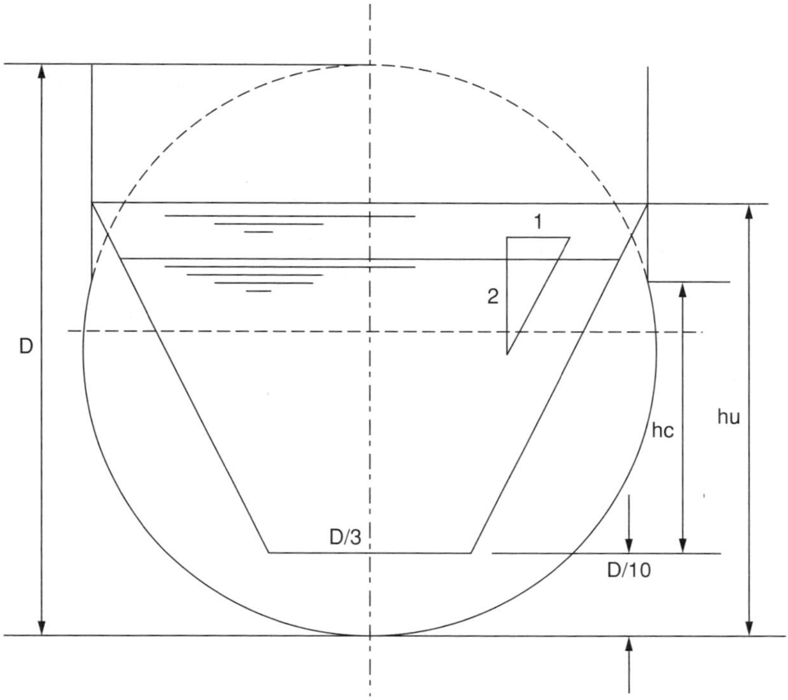

Figure 2.2.3 Standardized Palmer–Bowlus trapezoidal flume with a bottom width of one-third of the pipe diameter

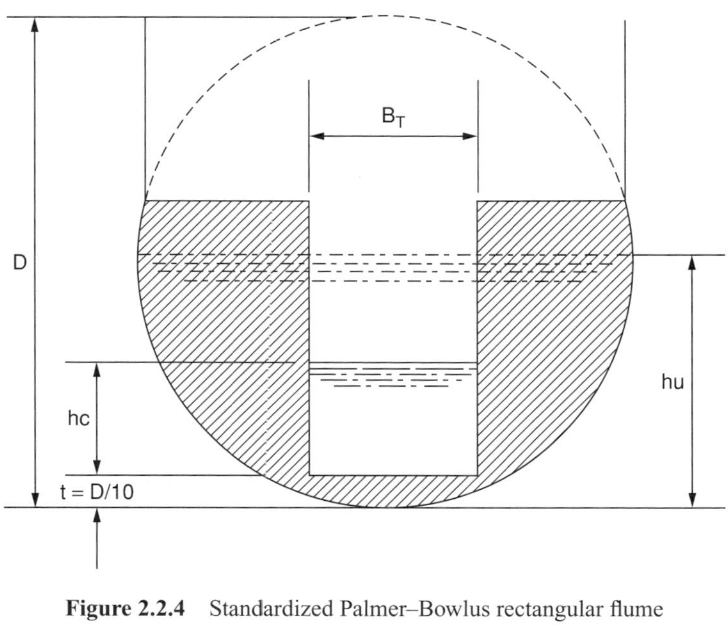

Figure 2.2.4 Standardized Palmer–Bowlus rectangular flume

Venturi flumes act as critical-flow flumes during free-surface flows and as Venturi meters during pressurized full-pipe flow. Equation (2.2.3) applies if the depth is replaced by the pressure head, p/γ (where p is the pressure and γ is the specific weight of the flowing fluid). For closed conduit flow Equation (2.2.3) becomes:

$$Q = \left[\frac{2g(H_1 - H_2 - h_L)}{\frac{\alpha_2}{A_2^2} - \frac{\alpha_1}{A_1^2}} \right]^{1/2} \tag{2.2.5}$$

where H is the piezometric head which equals $p/\gamma + z$. Assuming α_1 equals α_2 Equation (2.2.5) can be rewritten as follows forming the basic discharge equation for a Venturi meter:

$$Q = C_D \left[\frac{2gA_2^2 \Delta H}{1 - \left(\frac{A_2}{A_1}\right)^2} \right]^{1/2} \tag{2.2.6}$$

where $\Delta H = H_1 - H_2$, and C_D is a discharge coefficient which accounts for head loss and the nonuniform velocity distribution, α (Wenzel, 1975).

Diskin (1977) listed the following advantages of Venturi flumes:

(1) Ability to adapt the shape of the flume to the shape of the channel and the range of flows expected.

(2) The possibility of predicting the coefficients of discharge either by theoretical considerations or from the results of calibration tests.

(3) Relatively small head losses.

Venturi and critical-flow flumes also have the advantage of being self cleaning, i.e. deposits may often be reduced to a minimum and the danger of clogging the downstream channel is small (Hager, 1989), and, thus, maintenance costs are minimal. Diskin (1977) also noted that the disadvantage of Venturi flumes is the reduction in channel width also lowers the pipe capacity, this is especially true in comparison to the acoustic and electromagnetic flow meters discussed in the Sections 2.2.4–2.2.6. Another disadvantage of Venturi flumes is that the relation between head and flow breaks down in the transition zone from free-surface to pressurized flow. Such flows are difficult to measure with any device because the flow pulsates from free-surface to pressurized flow (Church *et al.*, 1999).

The varying shapes that Venturi flumes can take is documented in the literature. Kilpatrick and Kaehrle (1986) designed and calibrated a modified Palmer–Bowlus (MPB) type flume for both open channel and pressurized full-pipe flow. The modification involved the use of a longer flume with flatter side slopes and a greater floor thickness as well as adding a pressure sensor in the throat. The flume was designed so that its components could be passed through small manhole openings for assembly in a trunkline sewer (i.e. $D > 0.9$ m). The MPB was tested in the laboratory in 30.5- and 45.7-cm diameter pipes and in the field in a 122-cm diameter pipe. The field tests involved controlled tests with hydrant flow and storm data. For the hydrant flow, once each flow had stabilized, it was measured by both tracer dilution and acoustic meter; these discharges were in close agreement and also agreed closely with the MPB flume calibration curves. The storm data consisted of 12 dye-dilution measurements, which agreed well with the MPB and verified the laboratory conditions.

Wenzel (1975) developed a Venturi flume in which the constriction consisted of a cylindrical section with a radius greater than that of the pipe that could be fitted with a symmetrical (two-sided) or an asymmetrical (one-side) configuration. The theoretical rating developed from boundary layer theory was confirmed for both open channel and pressurized full-pipe flow in laboratory experiments in a 20 cm diameter, 43 m long acrylic circular pipe. Diskin (1977) presented the results of laboratory evaluations of Venturi flumes with rectangular, trapezoidal, and triangular throat shapes. Börzsönyi (1982) developed a Venturi flume involving a lateral constriction of the channel. In free, open-channel flow and in pressurized full-pipe flow the meter operates with accuracy better than ±5 %, and in submerged open-channel flow the accuracy is estimated at ±8 %. Hunter *et al.* (1991) proposed a Venturi flume with a truncated circular throat shape and transition slopes of 1:6 for the entrance and exit sections of the constriction. Eight 25- and 30-cm models were tested in two laboratories and it was found that this flume was capable of accurately measuring flows under conditions of free and submerged, forward and reverse, free-surface flow, and forward and reverse pressurized full-pipe flow.

Hager (1989) proposed and tested a substantially different critical-flow flume for use in circular conduits. Rather than contracting the sides or bottom of the conduit to cause critical flow, he proposed placing a cylinder in the centre of the pipe to cause critical flow. Hager (1989) noted that the major advantages of this flume include low cost, reading precision, insensitivity to submergence by backwater, and rapid installation in running water. He also noted that the disadvantage of this flume is that debris could get caught on the cylinder and clogging could result in sewers that carry appreciable debris loads. Laboratory experiments with this flume found that a diameter ratio of 0.3 between the pipe and the cylinder seemed optimal in terms of approaching flow stability and discharge capacity, the tailwater depth may be at least 80 % of the approaching flow depth (except for very low discharges), and the maximum relative deviation of discharge from the flume rating curve was always less than 3 %.

The accuracy of flow measurements is dependent on the accuracy of the construction and installation of the flumes in the pipe (i.e. level in a direction perpendicular to flow, no deformation during construction or installation, no leakage at approach section), and the measured geometry, slope, and friction of the flume surface (Church *et al.*, 1999). A well constructed, calibrated, and maintained flume may yield flows with accuracies of 2–3 %, however, when factoring in the error of the stage measurement device, the accuracy is about 5 % (Marsalek, 1973; Alley, 1977; Church *et al.*, 1999).

2.2.4 ELECTROMAGNETIC FLOW METERS

Electromagnetic flow measurement systems are based on Faraday's law, which states that the voltage induced in a conductor moving across a magnetic field is proportional to the average velocity of that conductor (Doney, 1999a). If the cross-section were rectangular with a uniform magnetic field, a true average velocity of the water in the section would be obtained by measuring the induced voltage, but for nonrectangular shapes and nonuniform magnetic fields the instrument must be calibrated (Newman, 1982). Electromagnetic flow measuring equipment measure pressurized pipe flows with high accuracy (0.5 % according to Soroko (1973)) and has been available for more than 30 years. Further, for full-pipe flows, Soroko (1973) reported that these flow meters have no straight run requirements.

When electromagnetic flow meters were first applied to partially filled pipes researchers were concerned that the presence of the free surface and the fact that gravity flow is usually much slower than pressurized pipe flow would result in low voltages that would be difficult to measure (Newman, 1982). However, Newman's (1982) practical experiments, supported by calibration tests at the British National Engineering Laboratories, found that accuracies on the order of $\pm 4\,\%$ could be achieved for typical variations in flow profiles and backwater effects in a pipe. Valentin (1981) also made an early application of an electromagnetic flow meter to free-surface flow in a pipe and found that for depths between $0.5D$ and $0.8D$ errors ranged from -4 to $-8\,\%$ and for depths between $0.8D$ and D the error decreases from -6 to $0\,\%$.

Doney (1999a) described a practical electromagnetic flow meter for use in partially filled pipes. This instrument uses three pairs of electrodes located at different flow heights in a flow tube to measure the induced voltage. The electronics unit of the instrument selects the optimally located pair as a function of flow depth and uses this pair to measure the induced voltage, which then is converted to the average velocity through a factory calibration. For full pipe flow the measurement accuracy is 1 %, while for free-surface flow the accuracy is 3–5 % depending on fill height. The instrument is accurate down to a flow depth of 10 %. The instrument will measure sub- or supercritical flows and function with pipe slopes up to 5 %. Finally, the meter requires straight runs of five pipe diameters upstream and three downstream.

Soroko (1973) and Doney (1999a) list the following advantages of electromagnetic flow meters: high accuracy, the ability to easily handle fluids with high solids content, extremely wide range including reverse flows, obstructionless flow path, minimal pressure loss, small straight pipe requirements, and low maintenance, i.e. it is unaffected by grease on electrodes and silt and debris deposited in the invert (Baughen and Eadon, 1983). The primary disadvantages of electromagnetic flow meters are high cost and difficult installation, especially in existing sewerage systems.

2.2.5 AREA–VELOCITY FLOW METERS

Area–velocity flow meters (AVFMs) have become the most commonly used devices for flow measurement in sewers. For example, the Milwaukee (WI, USA) Metropolitan Sewerage District (MMSD) had 158 temporary and semipermanent flow monitoring sites and 98 % of them had Doppler AVFMs (C. Schultz, MMSD, personal communication, 2005). AVFMs have acoustic/ultrasonic or electromagnetic components that are used to measure velocity at a point or throughout the profile (acoustic only), which in turn is used to estimate the average velocity of the flowing fluid. Depth also is measured and used to determine the flow area from the known sewer geometry. The flow rate is the product of area and average velocity. Most AVFM manufacturers report using these meters to measure submerged, surcharged, full-pipe, and reverse flow conditions (the velocity range of most devices is −1.5 to 6.1 m/s).

The acoustic velocity measurement devices send continuous ultrasonic signals and work as follows (Hughes *et al*., 1996):

> *When an ultrasonic beam is emitted into a fluid from an ultrasonic transducer, air bubbles and dirt particles in the flow cause the ultrasonic beam to be scattered. This scattering results in reflection of some acoustic energy in different directions, and the reflected energy may be picked up by a receiving transducer, either at the same position as the ultrasonic source or at some other position. If the fluid is in motion (relative to the source) and the flow is stable, then the reflecting particles will have approximately the same velocity as the moving fluid. The frequency of the reflected signal differs from the frequency of the originally transmitted signal owing to the Doppler effect. The frequency difference signal is known as the Doppler shift and is proportional to the velocity of the particles in the fluid at the point of reflection.*

The faster the object reflecting the sound waves moves in the water, the greater the phase shift, or change in tone the meter registers (Day, 1996). In the typical Doppler AVFM, a single ultrasonic beam directed on an angle into the flow from a single transducer is used. The Sontek Argonaut SW uses two beams at an angle to the flow, which increases the reliability and flow coverage of the instrument. Two types of ultrasonic beam have been used:

(1) A narrow beam that measures the velocity in a small volume of the flow, typically near the centre, an assumed velocity profile then is used to estimate the mean flow velocity.

(2) A wide beam that attempts to measure the total velocity of the flow and determine the average velocity directly.

The electromagnetic area–velocity meters operate similarly to the narrow-beam Doppler AVFMs.

2.2.5.1 Narrow-Beam Doppler Area–Velocity Flow Meters

In a fairly clean pipe, with sufficient up and downstream straight runs, and an axisymmetric velocity distribution, a fairly accurate average velocity will be rendered with the narrow-beam Doppler AVFM (Day, 1996). However, it is very important that an ideal or near-ideal velocity profile be present when utilizing a narrow-beam Doppler or electromagnetic AVFM (Johnson, 1995; Day, 1996). The theoretical velocity distribution used to compute the average velocity is based on uniform, turbulent flow with no backwater effects (Hunter *et al.*, 1991), and poor performance can be expected when these conditions are not present. In particular, backwater due to downstream flow restrictions is common in sewerage systems. Hughes *et al.* (1996) had the following comments on the narrow-beam approach. This approach is suitable for use in medium-sized sewers, but in larger sewers (>1 m diameter) the flow can have steep velocity gradients with depth and large variations in velocity across the section of the sewer flow. A single velocity reading derived from a volume near the centre of the sewer flow is not always sufficient to produce a good representation of the overall mean velocity of the flow. Previous studies using narrow-beam Doppler AVFMs have suggested that in large sewers a single velocity reading may only be accurate within $\pm 20\,\%$, even with sufficient depth (most meters need a flow depth of at least 7.5 cm to make a velocity measurement, and higher depths are preferred) and velocity of flow. However, under ideal conditions these systems can be accurate within a few per cent. The performance of narrow-band Doppler AVFMs can be improved through field calibration.

2.2.5.2 Wide-Beam Doppler Area–Velocity Flow Meters

Given the limitations of the narrow-beam Doppler AVFMs, most of the commercially available AVFMs in common use apply the wide-beam approach (including the ISCO, American Sigma and Automated Data Systems meters discussed in Section 2.2.7); these are referred to as Doppler AVFMs from this point forward. The primary concern for the wide-beam approach is whether the beam sufficiently samples the total flow to get a true average. For example, if the flow distribution is skewed and the ultrasonic beam is not of sufficient width, the true average may not be obtained. Watt and Jeffries (1996) also found that the concentration and composition of particulates in the flow have a marked effect on the velocity measured using Doppler AVFMs. They found that the fibrous waste from the paper making process has an absorbent effect on the ultrasound signal resulting in modified readings. They found similar results in the laboratory for sawdust. They reasoned that the ultrasound signal fails to penetrate the highly polluted flow and the received signal is from a zone very close to the sensor. The main problem with trying to simultaneously measure the entire flow at one time is 'range bias' wherein for deeper flows the difference in the strength of the return signal from particles close to the transducer and those far away causes a distortion in the return signal frequency spectrum that is difficult to resolve and typically results in the nearby, slower particles disproportionally affecting the measurement, which then is biased low (Metcalf and Edelhäuser, 1997).

2.2.5.3 Independent Evaluation of Doppler Area–Velocity Flow Meters

Watt and Jefferies (1996) reported the results of extensive field and laboratory evaluation of Buhler-Montec Flow Survey Package Doppler AVFMs in the UK. This involved laboratory and field checks of 6 monitors that had been deployed in the field and evaluation of 59 monitors that had been used for short-term field studies. The results of Watt and Jefferies (1996) are summarized in this section.

Field calibration of the depth sensors was carried out during each visit to the six monitoring locations. Zero drift was found to be frequent, and in two of the six cases it was greater than 5 cm, which was symptomatic of logger malfunction resulting in the rejection of all data from these meters. The zero offset drifted low in five of the cases, while three showed less than 0.5 cm drift in either direction. For all depths greater than 10 cm the maximum error of any calibration prior to adjustment was 16 % with an average error of −2 %. When the absolute error was considered, this latter figure became 3 % and when depths less than 25 cm were excluded the absolute error dropped to 2 %. Field inspections of depth measurement showed substantial need for calibration of the depth sensor.

Velocity measurements made by the six meters were evaluated in a 0.305 m wide rectangular flume both before and after installation. In the flume for depths greater than 10 cm the maximum velocity was 0.5 m/s. Although scatter was present in the velocity readings, no evidence of zero error (bias) was found. A large part of the scatter was attributed to turbulence in the flow resulting in fluctuations in the response of the ultrasonic equipment. A linear regression fit to the data indicated that, on average, the AVFM velocity was 5.6 % lower than the true mean velocity. The average absolute error ranged from 12 to 26 % at velocities less than 0.3 m/s and reduced rapidly at higher velocities. The data suggested that average errors no greater than 5 % in velocity should be expected at mean velocities greater than 0.5 m/s.

The final evaluation of Doppler AVFM usefulness involved a short-term flow survey in which 59 monitors were used for model verification. Five of the 59 monitors were found, after comparison with the model, to have yielded data that were so poor that they could not be used. In general, the flows were unbelievably high and three of the five were detected during comparisons with data from nearby sites at which flow discrepancies of up to 200 l/s were found. No specific reasons for the poor quality of data at these sites could be found, and the quality assurance checks were insufficient to prevent such data errors. Watt and Jefferies (1996) considered it disturbing that the data were rejected at the model verification stage and the fact that the majority of monitors giving poor data were found in this way raised questions as to validity of the data from other sites where no checking was possible. The accuracy of the verified model was, as a consequence, questioned in spite of a reasonable number of calibration readings taken.

At all sites, 35 % of the peak stage values during the verification events were greater than three times the highest calibration value and the corresponding value for velocity was 25 %. Thus, the field calibration of the Doppler AVFMs relied on recorded data from a small portion of the range of measurements to be made by the instruments. This may have contributed to the poor results at the five sites, but it does not explain why only these sites.

Watt and Jefferies (1996) quoted the UK Water Research Centre guide to short-term flow surveys in sewer systems as stating the following:

> *An accuracy in the region of ±10 % can be obtained with the measurement of flow when using a velocity and depth monitor provided that:*
>
> *(i) The instrument is fully serviceable and working in accordance with the specification.*
>
> *(ii) The site has suitable hydraulic characteristics.*
>
> *(iii) There is adequate flow for accurate measurement by the instrument.*
>
> *If an approximate velocity calibration has to be applied, the accuracy of the results will be reduced and an accuracy of ±10 % cannot be expected.*

Watt and Jefferies (1996) noted that in practice, the accuracy of the flows calculated will frequently be poorer than ±10 %, even though the UK Water Research Centre requirements are met. Nonetheless, Watt and Jefferies (1996) concluded that flow rates in small sewers without extreme velocities were found to be accurate within ±7 %.

2.2.5.4 Summary

AVFMs have many practical advantages including ease of installation, portability, and reasonable cost. Also, Day (1996) noted that the electronic packages for all of the Doppler AVFMs on the market adapt well to the sewer environment (moisture, corrosion, etc.), and that a unique feature of Doppler AVFMs that has been observed in the field is the ability to fire under mild fouling conditions and maintain reasonable accuracy. Doney (1999b) stated that manufacturers typically report accuracy as ±2 %. However, the literature search conducted here indicated that accuracies between 10 % and 30 % may be more likely without on-site calibration, and even after calibration a large range in accuracy for given events may result. Further, useful field calibration is difficult to achieve because the majority of checks are carried out for dry weather flows, while much of the interest in the results may be for storm flows (Watt and Jefferies, 1996). The accuracy of Doppler AVFMs relies on: the inherent accuracy of the equipment; its maintenance prior to installation; the quality of field site checks; site conditions; and a range of operational factors (Watt and Jefferies, 1996). Finally, Day (1996) noted that the performance of Doppler AVFMs is uncertain in large pipes (i.e. over 150 cm) simply from a lack of testing, however, for pipes under 91 cm, Doppler AVFMs have proven to be accurate *(given the previously stated caveats)* and repeatable.

2.2.6 ACOUSTIC DOPPLER PROFILER FLOW METERS

The acoustic Doppler profiler flow meter (ADFM) was exclusively developed by MGD Technologies Inc. (Complete details on the ADFM may be found at: http://www.mgdinc.com/.) The ADFM uses the Doppler shift principle to determine the velocity of flowing water, but instead of using a continuous ultrasonic signal as done with Doppler AVFMs, the ADFM emits short ultrasonic pulses called 'pings' into the water from each of four transducers that are angled into the flow (Figure 2.2.5). The ADFM range-gates the return signal from each ping allowing the ADFM to measure the velocity in many small volumes (called bins, cylinders about 4 cm in diameter and 5 cm long) that are regularly spaced throughout the water column. The velocity in each volume is measured independently, so that an independent velocity profile is obtained for each beam (Metcalf and Edelhäuser, 1997). There can be as many as 40 discrete bins in a 2 m deep flow. Narrow beams are used to minimize errors related to beam width. Since the acoustic pulses are very short and the velocity is measured in small bins, range bias is virtually eliminated

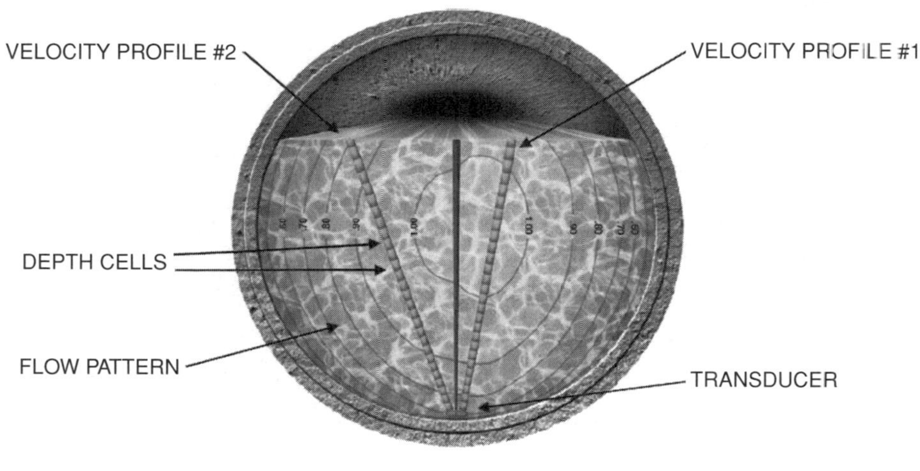

Figure 2.2.5 Schematic representation of the acoustic Doppler profiler flow meter (ADFM) measurement operation. (Provided by M. Metcalfe, MGD Technologies Inc.)

(Metcalf and Edelhäuser, 1997). Thus, the ADFM obtains an accurate measurement of the velocity distribution within a pipe or channel both vertically and transversely.

The velocity data from the profiles are entered into an algorithm to determine a mathematical description of the flow velocities throughout the entire flow cross-section. The algorithm fits a parametric model to the actual data. The result predicts flow velocity at every point throughout the flow, the velocity distribution then is integrated over the cross-sectional area to determine the discharge (Curling *et al.*, 2003). As hydraulic conditions change, the change will manifest itself in the distribution of velocity throughout the vertical. Thus, the ADFM does not require at site calibration.

Each time the ADFM pings, it collects velocity data. These data have a random error associated with them, which can be reduced by averaging many pings together into an ensemble. From sampling theory, the ensemble standard deviation is equal to the standard deviation of a velocity measurement for a single ping divided by the square root of the number of pings in an ensemble. Typically, the averaging of 400 pings takes only a few minutes (Metcalf and Edelhäuser, 1997). Thus, it is possible to get extremely precise estimates of velocity at a fairly rapid rate, avoiding the high degree of noise in the velocity data common for Doppler AVFMs (see Section 2.2.7).

The MGD Internet site (http://www.mgdinc.com/) provides the results of seven extensive laboratory and field tests of the ADFM. In each case the ADFM was placed in the centre of the channel or pipe and measured the flow without site calibration. The seven tests were as follows:

(1) Testing in rectangular flumes 1.22 and 3.66 m wide in the US Bureau of Reclamation Water Resources Research Laboratory.

(2) Field checking by Brown and Caldwell Inc. of measured flow in a 183 cm diameter concrete combined sewer interceptor in Onondaga County (NY, USA) using dye dilution.

(3) Field checking by Brown and Caldwell Inc. of measured flow in a 274 cm diameter concrete sewer line in Sacramento (CA, USA) using dye dilution.

(4) Laboratory testing in a 38 cm diameter PVC pipe done for the Pima County (AZ, USA) Wastewater Management Department.

(5) Field testing in the Salt River Project (AZ, USA) Arizona Canal in comparison with a rated broad-crested weir.

(6) Testing in a 76.2 cm diameter concrete test pipe in the US Department of Agriculture Water Conservation Laboratory.

(7) Testing in a 122 cm diameter concrete pipe and 0.91 and 2.13 m wide rectangular flumes at the Utah Water Research Laboratory of Utah State University.

Metcalf and Edelhäuser (Metcalf and Edelhäuser, 1997) also reported the results of tests in a 0.6 m wide rectangular flume at the Australian Water Technologies office in

Sydney, and in a 1 m wide rectangular flume at the River Hydraulics and Hydrology Section of the Civil Engineering Research Institute of the Hokkaido Development Bureau in Japan. Many of these tests were done over long periods involving a range of flows, and for all these tests the average error between the ADFM flow measurement and those of the laboratory rating or dye was less than 2 %.

The ADFM has been found to be highly accurate in a wide range of tests without at-site calibration. The disadvantage of the ADFM is moderately high cost (three to four times that of Doppler AVFMs). When choosing between AVFMs and the ADFM the potential cost of inaccurate measurements should be weighed against the extra cost of the ADFM. One case where high accuracy was sought was the Thames Tideway Study of the large combined sewers from the London, UK, area that drain into the Thames River for which 18 ADFMs were deployed and have provided high accuracy at sites with very complex hydraulics (Curling *et al.*, 2003).

2.2.7 COMPARISON OF FLOW MEASUREMENT TECHNIQUES

In 1995, the US Geological Survey (USGS) in cooperation with the Federal Highway Administration outfitted a 61 m length of straight, 137 cm diameter, 0.2 % slope, concrete storm-sewer pipe in Madison (WI, USA) with multiple instruments for the purpose of comparing these instruments. The details and results of this study are briefly summarized in Church *et al.* (1999). However, additional details of this field test were obtained as a written communication from D.W. Owens of the USGS Wisconsin Water Science Center (D.W. Owens, personal communication, 1998). Because this field test involved three of the previously discussed flow measurement techniques it is presented as a separate section.

Owens (D.W. Owens, personal communication, 1998) reported that the test site had the following characteristics that are typical of storm sewer locations where discharge monitoring may be desired:

(1) The concrete pipe sections had settled different depths creating pipe joints that acted as minor controls during lower flow conditions. As the water level increased, the smaller controls were drowned out.

(2) The flow conditions at the site change rapidly because of the small drainage area (77.7 ha), high amount of impervious surface, and intense summer rainstorms.

(3) Access to the pipe is limited creating a hazardous condition when the pipe is flowing.

(4) Standard discharge measurements are nearly impossible to collect because of the access and rapidly changing flow.

He also noted that the site was subject to relatively minor sediment loads.

The standard for judging the accuracy of the flows obtained from the various measurement techniques was a MPB flume (Kilpatrick and Kaehrle, 1986) that had been rated using 243 dye-dilution flow measurements over a flow range of 0.057 to 2.32 m^3/s (hereafter referred to as 'the measurement standard'). The average percentage error in the dye dilution discharge calculations was estimated as 4 % with a range from 1 to 14 % (D.W. Owens, personal communication, 1998). Fifty runoff events were monitored during a 6-month period and the resulting hydrographs and total storm runoff volumes obtained with the flow measurement techniques were compared with those obtained with the measurement standard. The flow measurement techniques evaluated included:

(1) Critical-flow flumes in the form of the theoretical rating for the MPB flume.

(2) Manning's equation applied at three locations in the study pipe.

(3) AVFMs–Automated Data Systems (ADS), ISCO 4250, and American Sigma 950 Doppler AVFMs and Marsh–McBirney Flow-Tote electromagnetic AVFM.

Data were collected at 1 min intervals for all meters except the ADS meter for which a 2.5 min interval was used. The meters were placed in series in the pipe.

The comparisons of the measured hydrographs revealed the following (D.W. Owens, personal communication, 1998):

(1) The hydrographs obtained from the AVFMs are noisier than hydrographs obtained with the measurement standard. Inspection of the data showed that this resulted from erratic velocity measurements.

(2) The AVFMs had periodic velocity dropouts wherein the velocity measurement dropped down to a value that was much lower than the previous and following measurements.

(3) At higher flows (>0.4 m^3/s), the Doppler AVFMs tended to underestimate the flow. At lower stages, the Doppler signal tended to work better. These results are indicative of the range bias for deeper flows that is common for the Doppler AVFMs.

(4) The electromagnetic AVFM tended to be the closest to the measurement standard. Furthermore, the electromagnetic velocity measurements displayed less noise than the Doppler measurements.

(5) The theoretical discharge for the MPB flume closely matched the measurement standard.

(6) The Manning equation technique produced mixed results based on monitoring location in the pipe.

Box plots were made of the percentage differences between the results of the various techniques/equipment and the measurement standard for the total storm runoff volume (Church *et al.*, 1999). Table 2.2.1 was prepared using the same data used to prepare Figure 7 in Church *et al.* (1999), which was provided by D.W. Owens of the USGS Wisconsin Water Science Center. The comparison of the total storm volumes in Table 2.2.1 yielded the following results:

(1) The electromagnetic AVFM yielded the best overall results with a median error of 0.4 % and an interquartile range of −9.4 to 4.4 %.

(2) The theoretical rating of the MPB flume also yielded good results with a median error of +10.8 % and an interquartile range of 2.7 to 17.9 %.

(3) All uncorrected Doppler AVFMs underestimated total storm volumes with median errors ranging from −6.6 to −28.8 % and mean errors ranging from −10.1 to −30.5 %. Again an indicator of the range bias for deeper flows.

(4) One of the Manning's equation sites was affected by backwater resulting in a median error of nearly 100 %. Another Manning's equation site was affected by drawdown resulting in 25 % of the storms having underestimates greater than 30 %. The final Manning's equation site was not affected by either backwater or drawdown and had a median error of 24.4 % and an interquartile range of −0.4 to 36.8 %.

It is difficult to derive general results from measurement comparisons at one site, but Church *et al.* (1999) raised two important conclusions from this study. The data clearly indicate the need to calibrate the flow measurement device using measurements obtained with an independent method. Further, although flow measurement techniques can be adjusted using verification data to minimize bias, the very large uncertainty in flow measurements exhibited by some of the flow measurement techniques is likely to remain after the adjustments.

2.2.8 CONCLUSIONS AND PERSPECTIVES

Many methods are available for measurement of flow in sewerage systems. Flumes have been available since the 1930s, and electromagnetic and acoustic methods for velocity measurement have been used since the 1970s and 1980s, respectively. During these long periods of use, manufacturers and users have fine-tuned the equipment so that reliable measurements may be obtained in real-time by telephone line or radio transmission. If real-time data are desired, users must pay special attention to the accessibility of the site to power and phone lines or radio transmission to a central station.

All the flow measurement equipment is capable of yielding accurate discharges for the appropriate hydraulic conditions (although the range of appropriate conditions for Manning's equation is quite limited). Flumes, electromagnetic flow meters and ADFMs have been found to yield high accuracy (within ±5 %) for a wide range of flow conditions. However, flumes and electromagnetic meters may be difficult

Table 2.2.1 Summary of storm volume errors in per cent relative to the rated modified Palmer–Bowlus flume for various flow measurement techniques/equipment applied in a 137 cm diameter in Madison (WI, USA). (Data provided by D.W. Owens, USGS Wisconsin Water Science Center)

Technique/Equipment	Mean	Median	Flow weight (mean)	Min.	25th Percentile	75th Percentile	Max.	No. of storms
Modified Palmer–Bowlus flume theoretical rating	10.1	10.8	10.2	−20.5	2.7	17.9	23.9	50
Electomagnetic AVFM/Marsh–McBirney Flow-Tote	−2.2	0.4	0.2	−30.1	−9.4	4.4	24.0	43
Doppler AVFM/American Sigma 950	−30.5	−28.8	−28.0	−58.6	−36.9	−26.0	−8.6	42
Doppler AVFM/ISCO 4250	−10.1	−6.6	−12.4	−27.8	−19.9	−4.4	11.8	33
Doppler AVFM/Automated Data Systems	−18.5	−19.0	−22.1	−38.8	−27.7	−14.0	−13.8	29
Manning's equation/Location 1	−11.6	−0.6	−18.5	−69.8	−30.5	8.0	16.8	50
Manning's equation/Location 2	18.8	24.4	8.4	−30.1	−0.4	36.8	64.8	48
Manning's equation/Location 3	99.5	99.3	86.5	34.0	73.0	121.7	155.2	48

to install at some locations. ADFMs are easier to install, but are more costly than acoustic Doppler area–velocity flow meters. Thus, users must consider site conditions, cost, use of the data, and desired accuracy when selecting the appropriate flow meter for the project at hand.

Probes for measuring dissolved oxygen concentration, conductivity and temperature in real-time are commonly used in treatment plants and stream systems. Their use in sewerage systems has been limited due to the possibility of damage by debris in the confined space of the sewer pipe and the difficulty to keep the probes clean in the harsh sewer environment. Other probes for measuring nutrients and other chemical constituents in real-time are in development. As these probes are improved, development of ways to use them in sewerage systems could be very valuable and is encouraged as a topic of future research and development.

REFERENCES

Alley, W.M. (1977) Guide for Collection, Analysis, and Use of Urban Stormwater Data: A Conference Report, Easton, MD, 28 Novembar–3 December 1976. ASCE.

Anon. (1996) *World Water Environ. Eng.*, **April**, 36.

Baughen, A.J. and Eadon, A.R. (1983) *J. Inst. Water Poll. Cont.*, **82**(1), 77–86.

Börzsönyi, A. (1982) *Advances in Hydrometry, IAHS Publ. No. 134*, 19–23.

Church, P.E., Granato, G.E. and Owens, D.W. (1999) Basic Requirements for Collecting, Documenting, and Reporting Precipitation and Stormwater-Flow Measurements. US Geological Survey Open-File Report 99–255.

Curling, T., Leafe, M., and Metcalfe, M. (2003) Avoiding the Pitfalls of Dynamic Hydraulic Conditions with Real-Time Data. Available online at http://www.mgdinc.com/pdfs/Thames%20 Tideway%20Hydraulic%20Results.pdf.

Day, T.J. (1996) *Water Eng. Manage.*, **143**(4), 22–24.

Diskin, M. (1977) *J. Irrig. Drain. Div., ASCE*, **102**(IR3), 383–387.

Doney, B. (1999a) *Water Eng. Manage.*, **146**(11), 32–34.

Doney, B. (1999b) *Water Eng. Manage.*, **146**(11), 11–12.

Drake, T. (1994) *Water Eng. Manage.*, **141**(12), 34.

Hager, W.H. (1989) *J. Irrig. Drain., ASCE*, **114**(3), 520–534.

Hughes, A.W., Longair, I.M., Ashley, R.M. and Kirby, K. (1996) *Water Sci. Technol.*, **33**(1), 1–12.

Hunter, R.M., Hunt, W.A., and Cunningham, A.B. (1991) *Water Environ. Technol.*, **3**(2), 47–51.

Huth, S. (1998) *Water Technol.*, **21**, 78–80.

Johnson, E.H. (1995) *Water SA*, **21**(2), 131–138.

Kilpatrick, F.A. and Kaehrle, W.R. (1986) *Trans. Res. Rec.*, **1073**, 1–9.

Lanfear, K.J. and Coll, J.J. (1978) *Water Sewage Works*, **125**(3), 68–69.

Ludwig, R.G. and Parkhurst, J.D. (1974) *J. WPCF*, **46**(12), 2764–2769.

Marsalek, J. (1973) Instrumentation for Field Studies of Urban Runoff. Research Program for the Abatement of Municipal Pollution Under the Provisions of the Canada-Ontario Agreement on Great Lakes Water Quality, Ontario Ministry of the Environment, Project 73-3-12.

Melching, C. S. and Yen, B.C. (1986) Slope influence on storm sewer risk. In *Stochastic and Risk Analysis in Hydraulic Engineering*, B. C. Yen, ed. Water Resources Publications, Littleton, CO, pp. 79–89.

Metcalf, M.A. and Edelhäuser, M. (1997) Development of a velocity profiling Doppler flow meter for use in the wastewater collection and treatment industry. Paper Presented at WEFTEC '97, available on-line at: http://www.mgdinc.com/pdfs/WEFTEC%20'97%20Velocity%20 Profiling%20for%20 Wastewater%20Collection%20and%20Treatment.pdf.

Newman, J.D. (1982) *Proc. Int. Symp. Hydrometeorology*, 15–26.

Palmer, H.K. and Bowlus, F.D. (1936) Adaptation of Venturi flumes to flow measurements in conduit. *Trans. ASCE*, **101**, 1195–1216.

Parr, A.D., Judkins, J.F., and Jones, T.E. (1981) *J. WPCF*, **53**(1), 113–118.

Soroko, O. (1973) Water and waste flow measurement. *TAPPI (Atlanta, GA) Engineering Conference*, Boston, MA, 9 October 1973 (Preprinted Proceedings), pp. 187–203.

Valentin, F. (1981) *Water Sci. Technol.*, **13**(8), 81–87.

Waite, A.M., Hornewer, N. and Johnson, G.P. (2002) Monitoring and Analysis of Combined Sewer Overflows, Riverside and Evanston, Illinois, 1997–99. *US Geological Survey Water-Resources Investigations Report 01-4121*.

Watt, I.A. and Jefferies, C. (1996) *Water Sci. Technol.*, **33**(1), 127–137.

Wells, E.A. and Gotaas, H.D. (1958) Design of Venturi flumes in circular conduits. *Trans. ASCE*, **123**, 749–771.

Wenzel Jr., H.G. (1975) *J. Hydr. Div., ASCE*, **101**(HY1), 115–133.

Weyand, M. (1996) *Water Sci. Technol.*, **33**(1), 257–265.

Wright, J.D. (1991) *Water Environ. Technol.*, **3**(9), 78–87.

2.3
Monitoring in Rural Areas

Ann van Griensven and **Véronique Vandenberghe**

Wastewater Quality Monitoring and Treatment Edited by P. Quevauviller, O. Thomas and A. van der Beken

2.3.1 INTRODUCTION

2.3.1.1 Monitoring for the European Union Water Framework Directive

Recently the European Union has approved the European Union Water Framework Directive (EU WFD). This directive claims that by the end of 2015 a 'good status of surface water' and a 'good status of groundwater' should be achieved (European Union, 2000). To make sure that the new water policy will succeed, a profound analysis of the actual and future state of the water is necessary. In this context, the evaluation of emissions into river water will be important.

To that end, the EU WFD provides several guidelines for monitoring the water bodies, leaving the practical implementation to the local governments. Since urban pollution has been strongly reduced in many western countries by collection and treatment of the urban wastewater, the remaining water quality problems require advanced management and optimisation techniques in an integrated manner. 'Integrated' is a term with many interpretations, but also a dangerous term to be used. Whereas 'integrated water management' at first referred to a holistic approach that linked the sewer–wastewater treatment plant–river systems, it soon became apparent that goals of good water quality were not reached, causing awareness that some other sources of pollution were involved. Indeed, after large investments to reduce urban pollution, managers were confronted with pollution from rural areas (Figure 2.3.1).

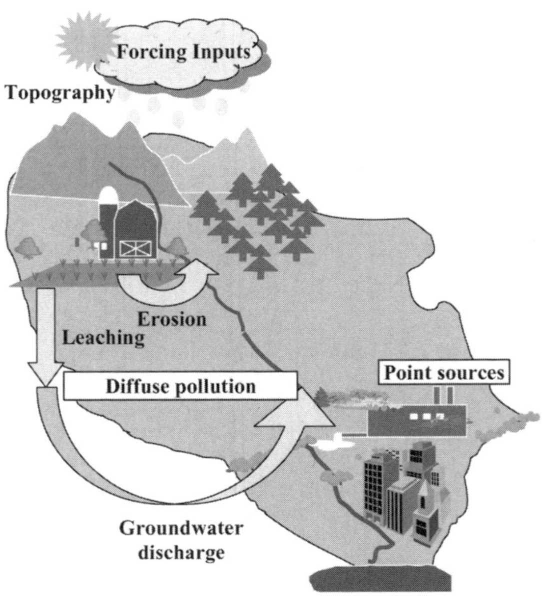

Figure 2.3.1 Sources of pollution in a river basin

2.3.1.2 Characterisation of Rural Areas and Pollution

Rural areas should not necessarily be considered as pollutant areas. Not-intensive grazing for instance has beneficial effects on erosion reduction and does not cause excessive nutrient loads to the receiving systems. In Europe, evolution towards more intensive practices took place during the past decades and has caused an increase of nutrient release into the environment (Poirot, 1999). Under the Common Agricultural Policy of the EU, the Gross Value Added (GVA) of the agricultural sector has raised sharply over the last 25 years. This was mainly due to increased investments giving in increase in the volume of production (Barthelemy and Vidal, 1999). The measures have generally led to a reduction of permanent grassland in favour of wheat, maize, the appearance of oilseed and protein crops and annual crops as fodder. Livestock production has also followed a trend to intensification, where small extensive holdings are replaced by modern and specialised ones. These 'nonland-bound' farms resulted in a considerable growth in the livestock sector (Boschma *et al.*, 1999). In particular, pig husbandry constitutes the most intensive type. The intensification in livestock production and crop culture has led to a high application of nutrients to agricultural land. Livestock manure is the second most important source in the EU. The Netherlands and Belgium had the highest input of nitrogen from manure per hectare coming mostly from pig production (Pau Val and Vidal, 1999). Within European soils, 115 million hectares suffer from water erosion and 42 million hectares from wind erosion (Montarella, 1999).

Most agricultural activities are considered to be nonpoint sources. This is not the case for the large 'nonland-bound' farms that are agricultural enterprises where a large number of animals are kept and raised in confined areas. The feed is generally brought to the animals, rather than the animals grazing or otherwise seeking feed in pastures, fields or rangeland. Such activities are treated in a similar manner to other industrial sources of pollution. Whereas point-source pollution can be measured by monitoring the discharge and the water quality, diffuse pollution sources are very difficult to monitor because the sources are distributed along the river.

2.3.1.3 Joint Use of Modelling and Monitoring

An integrated approach with regard to the nonpoint and diffuse pollution creates new challenges for monitoring and modelling, but it also promotes the interaction between these two. The water bodies are highly complex systems as they hold many unknowns and uncertainties due to the incomplete understanding of the processes, to scaling aspects and to the high variability of the variables in time and space. Consequently, it is not possible to develop one perfect model or to design an optimal monitoring network with present information and knowledge. An adaptive approach is therefore needed: besides linking available data and thereby improving the conceptual understanding of the water system, models may indicate errors and inadequacies in the monitoring network. Conversely, the model is revised and updated as new data become available. The effects of a pollution load into the river can be evaluated using models, especially for diffuse pollution, coming from rural areas, because complete monitoring of diffuse pollution input is impossible. Due to the characteristics of such a pollution that comes from land use practices, fertiliser and pesticide use, these are subject to different processes like runoff, leakage to groundwater, uptake by plants, conversion in the soil and absorption by soil particles. All these can be modelled, however, for several reasons those model outputs are uncertain (Beck, 1987). Model outcome uncertainties can become very large due to:

- input uncertainty;
- model uncertainty;
- uncertainty in the estimated model parameter values;
- mathematical uncertainty.

Therefore, estimating and calculating the diffuse pollution to a river can be subject to large input uncertainties, so in this chapter we will focus on monitoring with a view to making the input uncertainties of a model that calculates diffuse pollution towards a river smaller. To optimally allocate the efforts necessary to reduce those input uncertainties, it is useful to evaluate the sensitivity of the outputs, the water quality, to the different inputs needed for calculation of diffuse pollution.

In this study we focus on the diffuse pollution of nitrate in the water due to fertiliser use. With the use of an efficient Monte Carlo method based on Latin Hypercube sampling (McKay, 1988), the contribution to the uncertainty by each of the inputs is calculated. The methodology is applied to the Dender basin in Flanders, Belgium.

The following sections give a description of the river basin for the case study, the model environment and the applied methodology, which consists of a sensitivity and uncertainty analysis based on the studies performed by Vandenberghe *et al.* (Vandenberghe *et al.*, 2005).

2.3.2 A CASE STUDY

2.3.2.1 The Dender River in Flanders, Belgium

The catchment of the river Dender has a total area of 1384 km^2 and has an average discharge of 10 m^3/s at its mouth. Figure 2.3.2 shows how the Dender basin is situated in Flanders. As about 90 % of the flow results from storm runoff and the point sources make very little contribution, the flow of the river is very irregular with high peak discharges during intensive rain events and very low flows during dry periods (Bervoets *et al.*, 1989). The river Dender is heavily polluted. Part comes from point-pollution (e.g. industry) but also from diffuse sources of pollution originating mainly from agricultural activity. Although there is an unmistakable relation between intensive agricultural activity and the occurrence of high nutrient concentrations in

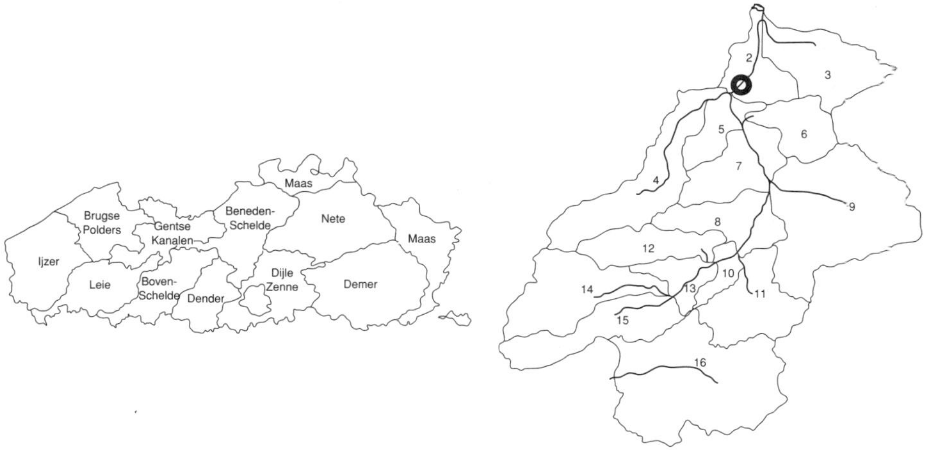

Figure 2.3.2 The River Dender basin in Flanders, Denderbelle (o) and the subbasins

the environment, few precise data are available about the contribution of agricultural activity to the total nutrient concentrations.

2.3.2.2 The Model Using ESWAT

ESWAT is an extension of SWAT (van Griensven and Bauwens, 2001), the Soil and Water Assessment Tool developed by the United States Department of Agriculture (Arnold *et al.*, 1996). ESWAT was developed to allow for an integral modelling of the water quantity and quality processes in river basins. The diffuse pollution sources are assessed by considering crop and soil processes. The crop simulations include growth and growth limitations, uptake of water and nutrients and several land management practices. The in-stream water quality model is based on QUAL2E (Brown and Barnwell, 1987). The spatial variability of the terrain strongly affects the nonpoint source pollution processes. GIS [Geographical Information System(s)] is used to account for the spatial variability. Based on soil type and land use a number of Hydrological Response Units (HRUs) can be defined. For each HRU, the ESWAT model simulates the processes involved in the land phase of the hydrological cycle, and computes runoff, sediment and chemical loading. Based on the areas of each HRU, the results are then summed for each subbasin.

Input information for each subbasin is grouped into categories for unique areas of landcover, soil and management within the subbasins. The main soil classes are sand, loamsand, silty loam and impervious areas. For landuse, five classes are important: impervious areas, forests, pasture, corn (maize and corn) and land for common agricultural use (crop culture, not corn). About 30 % of the landuse is pasture, while crop farming represents ca. 50 % of the landuse. To build the model, the total catchment was subdivided into 16 subbasins (Figure 2.3.2).

In terms of the nitrogen cycle, the three major forms in mineral soils are organic nitrogen associated with humus, mineral forms of nitrogen held by soil colloids and mineral forms of nitrogen in solution. Nitrogen may be added to the soil by fertiliser, manure or residue application, fixation by symbiotic or nonsymbiotic bacteria and rain. Nitrogen is removed from the soil by plant uptake, leaching, volatilisation, denitrification and erosion.

Nitrate is an anion and is not attracted to or sorbed by soil particles. Because retention of nitrate by soils is minimal, nitrate is very susceptible to leaching. In ESWAT the algorithms to calculate nitrate leaching simultaneously solve for loss of nitrate in surface runoff and lateral flow. Finally nitrate ends in the river.

A previous study for a nitrogen leaching model [implemented in the simulation model SWIM (soil water infiltration and move)] from arable land in large river basins (Krysanova and Haberlandt, 2001) showed that the relative importance of natural and anthropogenic factors affecting nitrogen leaching in the Saale river basin was as follows: (1) soil; (2) climate; (3) fertilisation rate; and (4) crop rotation. Reducing the uncertainty on inputs for soil and climate depends on better equipment to measure the different variables and proper use of sophisticated mathematical techniques to interpolate for places that are not measured. A lot of studies on that subject already exist (Sevruk, 1986). Until recently reducing the input uncertainty relating to fertilisation rate was not studied. In Flanders new legislation concerning fertilisation application was introduced in the late 1990s. Campaigns to list the fertiliser use were then started but it is known that a large amount of information is still wrong or missing. A lot of effort is still needed to complete the information. The evaluation and quantification of the impacts of land management practices on nitrogen wash-off to surface water is therefore very important.

In this study we focus on fertilisation rate and time of fertiliser application on the most important crops for the Dender river basin. For the application of fertiliser for the different land uses three application dates were assumed; 1 March, 1 April and 1 May. Also, operations such as planting and harvesting dates can be defined. Day and months were used to specify the planting and harvesting dates. Of course, those dates depend on the weather, the crop and the farmer, so assumptions had to be made concerning those dates. The output that was focused upon was the time that nitrate concentrations in the river Dender at Denderbelle (near the mouth) were higher than 3 mg/l.

The data needed for the model implemented in ESWAT were also very sparse and conversions had to be made to make the data useful for the model (Smets, 1999). Data on fertiliser and manure use were provided by The Flemish Institute for Land Use (Vlaamse Landmaatschappij, VLM). They provided data on the nutrient use and production for each municipality in Flanders. In SWAT, one has to specify for each subbasin the total amount of fertiliser and the detailed composition of the fertiliser. Some conversions of the supplied data had to be made so that they could be used in ESWAT. They consisted of recalculations of the application rates for each municipality to application rates per subbasin (Smets, 1999). Further, the same amount of fertiliser on all crops was assumed for this model. This is clearly different from practice but, at this stage, insufficient details are available to specify this more realistically.

First, a global sensitivity analysis is used to show what the most important factors are. Then the influence of uncertain data on the river nitrate concentrations time series is evaluated with an uncertainty analysis. For both the same Monte Carlo sampled inputs could be used. The sensitivity analysis focuses on the inputs to rank their importance while the uncertainty analysis assumes uncertain inputs and only considers and evaluates the outputs.

2.3.2.3 Sensitivity Analysis

The sensitivity analysis (SA) technique used here is based on a multilinear regression of the inputs on a specific output. A Monte Carlo technique, Latin Hypercube sampling, makes sure that the total range of inputs is covered. When the number of samples equals 4/3 times the number of inputs such a sampling is sufficient to perform a reliable SA (McKay, 1988). For the sampling of the inputs and the analysis of the outputs, a program was written to couple UNCSAM, the program used for the SA (Janssen *et al.*, 1992) with the management input files of ESWAT. We used the standardised regression coefficient (SRC) as an indication of the relative importance of the different inputs:

$$\mathrm{SRC}_i = \frac{\Delta y / S_y}{\Delta x_i / S_{x_i}}$$

with $\Delta y/\Delta x_i$ being the change in output due to a change in an input factor and S_y, S_{x_i} the standard deviation of, respectively, the output and the input. The input standard deviation S_{x_i} is specified by the user.

The ordering of importance of the input factors based on that statistic is as good as the associated model coefficient of determination R^2 of the whole multilinear regression. The closer R^2 is to 1, the better the results.

When the input variables are linearly related, the application of a linear regression can lead to an accuracy problem, the colinearity problem (Hocking, 1983). The variance inflation factor (VIF) is defined as:

$$VIF_i = [C_x]_{ii} = \left(1 - R_i^2\right)^{-1}$$

where $[C_x]_{ii}$ represent the diagonal elements of the covariance matrix relating y versus x and ${R_i}^2$ is the R^2 value that results from regressing y on only x_i.

For every subbasin the total amount of fertiliser and the time of planting and harvesting of crops have to be given to the model. A SA can now be performed to evaluate the influence of those inputs on the model results for nitrate in the river water. We evaluate the sensitivity of the model on the following result: the time that nitrate is higher than 3 mg/l. The fractions of mineral HNO_3, organic N, and NH_3-N in the fertilisers are considered to be known and fixed. Hence, we only analyse the total amount of fertiliser used (Table 2.3.1). As there are a lot of differences in

Table 2.3.1 Composition of the manure as input in SWAT

Chemical	Percentage of total fertiliser (100×kg/kg)
HNO_3	28.5
Mineral P	7.5
Organic N	28
Organic P	7.5
Ammonia	28.5

management practices between the different farmers and the time of planting and harvesting is different from year to year, the plant date and harvest date for the crops are also considered in the analysis. For a global SA we take the uniform distribution with standard deviation S_{x_i}. The ranges of the uniform distributions are given in Table 2.3.2. We assumed no correlation. To supply the information on those ranges a few farmers living in Maarkedal (situated in the Dender basin) were interviewed about their land management practices.

As the used SA technique is based on linear regression, two measures are calculated to see whether a linear regression is acceptable.

The first measure is the regression coefficient (RC) which was 0.845 for the whole multilinear regression. Because this value is close to 1 and the F-statistic showed that the regression is significant, a linear regression is adequate. The value of 0.845 means that there is a fraction of the output variance, 15.5 %, that is left unaccounted for.

The second measure is the VIF. The largest VIF for this analysis was 1.35. A VIF smaller than 5 means that the correlation between the inputs is small enough to allow application of a linear regression (Janssen *et al.*, 1992). The SRC is significant on the 10 % level for eight parameters. In Table 2.3.3 the parameters are ranked.

For river nitrate concentrations the amount of fertiliser used in the subbasins that are laying upstream are especially important. This SA also shows that it is more important to focus on the amount of fertiliser than on the management practices.

2.3.2.4 Uncertainty Analysis

The Flemish Institute for Land Use provided input data for the model. Due to unregistered manure and fertiliser use, it is very likely that those data are underestimated. The amount of fertiliser was the same for all crops, which is unrealistic and it is very

Table 2.3.2 Ranges for global sensitivity analysis of management practice inputs for nitrogen

Input	Uncertainty
Plant date for the crops	±1 month
Harvest date of the crops	±1 month
Amount of fertiliser applied per subbasin and per crop (kg/ha)	±25 %

Table 2.3.3 The eight most important inputs, their SRC and their sensitivity ranking

Input	SRC	Rank
Amount of fertilisation on pasture in subbasin 16	−0.303	1
Amount of fertilisation on farming land in subbasin 4	0.226	2
Growth date of pasture	−0.183	3
Plant date on farming land	0.171	4
Amount of fertilisation on corn in subbasin 5	−0.169	5
Amount of fertilisation on corn in subbasin 15	−0.165	6
Amount of fertilisation on pasture in subbasin 12	0.159	7
Amount of fertilisation on corn in subbasin 11	0.155	8

likely that the composition of the fertiliser used is not always the same as was assumed here. The uncertainty of the latter two generalisations is included in the uncertainty in the amount of fertiliser used on the crops. Such an uncertainty is propagated through the model and gives a final uncertainty on the model results of the water quality near the mouth of the river. An uncertainty analysis in which all of the uncertain sources are varied at the same time is performed to see the effects of the uncertainty of the inputs. For this analysis we calculate the uncertainty bands, i.e. the 5 % and 95 % percentiles for the results of the time series obtained after sampling the inputs as also done in the SA.

Figure 2.3.3 illustrates the contribution of an uncertainty of up to 25 % to the amount of fertiliser applied on the crops and an uncertainty of 1 month for the plant and harvest dates. The 95 % bound shows much higher peaks than the mean concentration–time series. This means that some peak values of nitrate in the river water at Denderbelle may not be predicted properly due to an underestimation of the amount of fertiliser used. Those peaks (e.g. days 156 and 260) are above levels of nitrate concentrations for basic water quality.

Only a few measurements were available for the calibration of the model using a multi-objective automated calibration method (van Griensven and Bauwens, 2003). As can be seen, some of the measuring points are available at time instants where the uncertainty on the inputs is not propagated. Fortunately, the model calibration was successful. The data around day 340 are within the uncertainty bound, but the ones on day 240 are clearly not predicted well.

The inputs we considered uncertain in this study are only a fraction of all the inputs that cause uncertainty, but it is shown that the inputs related to fertiliser use already give a large uncertainty in certain periods of the year (95 % percentile bounds differs up to ±50 % from the average nitrate predictions). Those are periods with rainfall and high flows. Knowing that there are more sources of uncertainty related to diffuse pollution, the input uncertainty is expected to be even higher. Therefore it is clear that care has to be taken during the gathering of data needed to run a dynamic process-based model for the prediction of effects of diffuse pollution.

This uncertainty analysis shows also some important results for future measurement campaigns. This study shows that we could obtain a better calibration for the

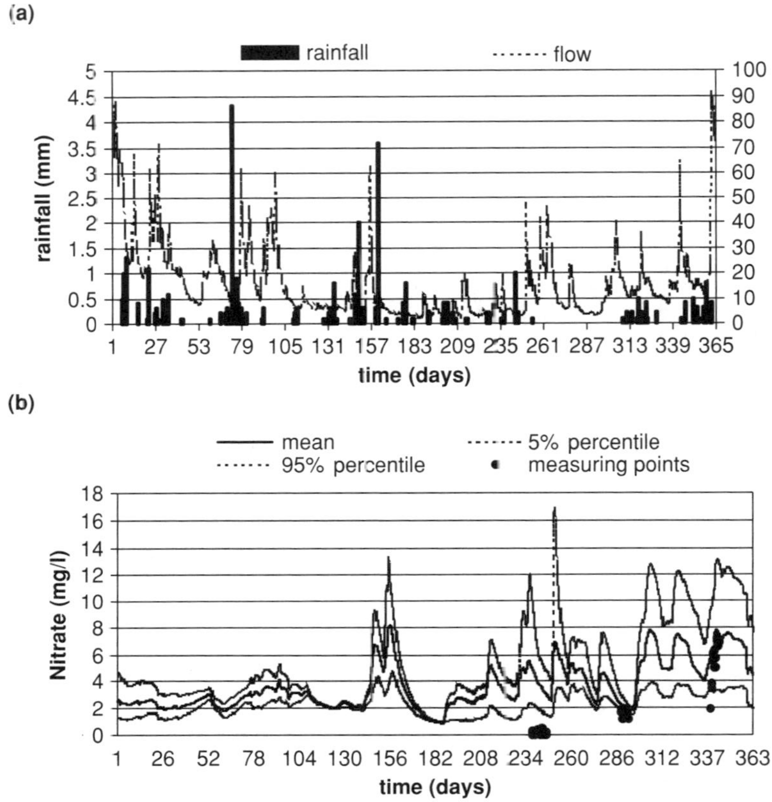

Figure 2.3.3 The simulated flow rate and rainfall measurements (a) and simulations of nitrate with the 5 % and 95 % confidence intervals (b) in the Dender river at Denderbelle for the year 1994

diffuse pollution part of the model with data that were taken during periods with rainfall and high flows, because the model output nitrate is more sensitive towards inputs of diffuse pollution in those periods. If you want to focus on calibrating the in-stream behaviour and point pollution then measurements during dry periods are needed as the model is then not sensitive towards input of diffuse pollution.

2.3.2.5 Discussion

The model output for nitrate in river water is sensitive towards eight inputs related to fertiliser use. These are the start date of the growth of pasture, plant date on farming land, the amount of fertiliser applied on pasture in subbasins 16 and 12, on farming land in subbasins 4 and on corn in subbasins 5, 11 and 15. The dates of planting and harvesting the crops appear not to be important. To reduce the output uncertainty it is best to focus on data about amounts of fertiliser applied on crops.

In the uncertainty analysis, the effects of uncertain input of fertiliser use and land management on nitrate concentrations in the river are shown. In certain periods of the year the uncertainty bounds are very wide. Because there are more sources of uncertainty, not considered in this study, it becomes clear that it is very important to gather accurate data to run a dynamic process-based model for the prediction of effects of diffuse pollution.

The uncertainty analysis is also of great use for experimental design. Measurements during dry periods can be used to better calibrate the model for point source pollution because the inputs of diffuse pollution are not important then. On the other hand, periods with rainfall and high flows are needed for the calibration of the model with diffuse pollution because the model output for nitrate is then very sensitive towards the inputs related to farmer's practices.

More detailed studies are needed to see the exact contribution of the input uncertainties to the output uncertainty. We can already conclude on the basis of this study that uncertainty analysis is an essential part of diffuse pollution modelling to evaluate and draw conclusions from model predictions.

2.3.3 AUTOMATED MONITORING

The Dender model clearly illustrates the high variability in water quantity and quality in rural areas. It is therefore hard to plan monitoring for certain preferred conditions (dry period versus rain event). In practice, it may not be feasible to catch these short events through manual sampling.

2.3.3.1 Automated Monitoring Stations

Automated Measuring Stations (AMSs) can be very helpful to capture specific dynamics through continuous monitoring or through controlled inducing of samplers by using the signals for the river level, for precipitation or for sediment concentrations (van Griensven *et al.*, 2002; Vandenberghe *et al.*, 2004). Figure 2.3.4 shows one of three AMSs that were placed on the Dender river in Belgium and that are connected through SMS (Short Message Service) communication or the internet to a central computer/database. In the station, the river water is pumped through a

Figure 2.3.4 AMS on the Dender river

Table 2.3.4 The sensors in the station

Variable	Method	Range	Compensation	Frequency
pH	Combined glass electrode	0–14	Temperature	Continuous
Dissolved oxygen (% sat)	Galvanic electrode	0–200	Temperature	Continuous
Redox potential (ORP) (mV)	Pt + Ag/AgCl electrode	−1000–1000	Temperature	Continuous
Turbidity (NTU)	Photo-electric meter	0–200	Temperature	Continuous
Conductivity (μS/cm)	Pt electrode	0–9999	Temperature	Continuous
Ammonium-N (ppm)	Ion-selective electrode	0.1–14000	Not automatic	Every 15 min
Nitrate-N (ppm)	Ion-selective electrode	0.1–14000	Not automatic	Every 15 min
Solar radiation (W/m^2)	Pyranometer	0–4000	No	Continuous
Precipitation	Tipping bucket		No	0.254 mm
Water level	Pressure electrode		No	Continuous
Temperature (°C)	Pt100	−10–120	—	Continuous

hydraulic loop. At the entrance of the loop, the turbidity is measured and a bypass to a sampling system is available. After filtration (100 μm), temperature, conductivity, pH, dissolved oxygen and redox potential are measured in the loop. Ammonia and nitrate can be measured with ion-selective electrodes in off-line reservoirs to which buffer solutions are added. Solar radiation, precipitation and water level are also measured *in situ*. Table 2.3.4 lists the characteristics of the sensors.

2.3.3.2 The Control of the Station – GSM Communication

In the case where only GSM (Global System of Mobile Communication) is available, SMS messages – containing the data and alarms – are automatically sent to a remote central computer (Figure 2.3.5). Daily, the data of the central computer are automatically backed up and imported into a relational database.

Remote control (through SMS) is limited to:

- the transmission of the latest data set (data and alarms);
- the start-up of a predefined sampling program.

2.3.3.3 The Control of the Station – Internet Communication

When an internet connection is available, the LabView interface enables remote interaction with the station. The interface optimises the follow up of the station by a front panel that visualises the hydraulic loop and displays the actual measurements

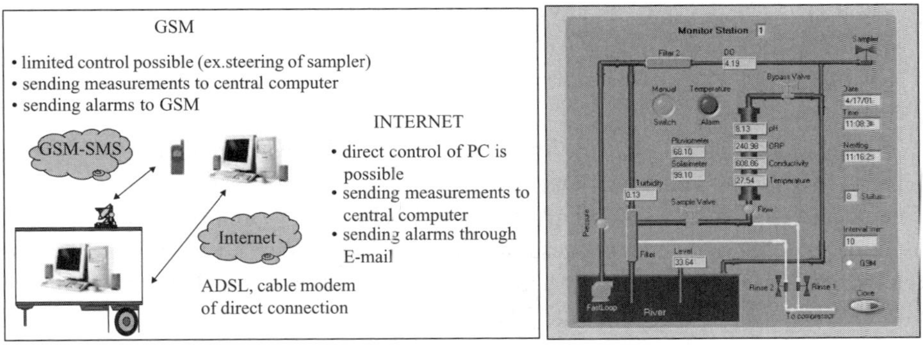

Figure 2.3.5 Communication of the automated and transportable on-line monitory system (ATOMS) and the LabView (National Instruments Software)

and alarms (Figure 2.3.5). A plot with the recent evolution of the measured water quality variables can also be displayed. The central database is hereby continuously updated.

Remote control of the operation of the system is facilitated by:

- the visualisation of alarms on the panels, in case of malfunctioning of the sensors (signal out of range), low flow or low pressure in the loop or low air temperature in the cabin;
- the sending of GSM messages and/or emails to selected team members;
- the on-line adaptation of control parameters (e.g. logging interval);
- the on-line control of the sampler;
- the availability of a WebCam.

2.3.3.4 Maintenance and Calibration

Automated maintenance consists of the rinsing of the filter and the electrodes by injection of air under pressure (after every logging). The station requires a weekly maintenance visit by two team members to clean the filters and the sensors and, if necessary, calibrate the sensors. A LabView program guides the calibration process. After the calibration, a log file with the new calibration parameters is logged and the measuring mode is started again.

2.3.3.5 Discussion

AMSs are useful tools for monitoring in rural areas under the conditions that the monitoring system coincides with good maintenance and quality control. In practice, this means that the system is not as fully automated as some might hope and that regular visits to the stations by employees should be foreseen. Also, a too complex 'black-box' concept of the system leads to a significant loss of data. The processing of the sensor signal to data should be transparent showing what can be done by using PC-based modules for the control of the station. A web-based communication enables remote control of the stations and the integration of the data into databases. This concept also allows for a full remote control of the station by authorised persons and a limited accessibility for data consultation by users through the web. A better spatial representation can be obtained by embedding the monitoring and the modelling in a GIS system (Vivoni and Richards, 2005).

2.3.4 CONCLUSIONS AND PERSPECTIVES

Monitoring in rural areas needs a different approach than in urban areas. The pollution in rural areas cannot be measured at certain points along the water body, but can only be estimated by making evaluations of the water quality together with information on what and how many polluting substances are applied in the area. Models, describing all processes on those substances before entering the water, can provide a means to evaluate the magnitude of pollution coming from diffuse pollution and to evaluate scenarios for diffuse pollution reduction. Specific data are needed to calibrate and build those models.

Therefore, the traditional cycle in water management should be inversed. Instead of starting from the data set to select an appropriate tool and hence use this tool for management, one should first define the problem, select a tool that can support this problem and then design an appropriate monitoring program to feed the tool. In that way, money is spent to generate primarily the information that is indeed needed. A closer cooperation between monitoring and modelling efforts will make sure that models for diffuse pollution can be used with sufficient reliability.

Automated monitoring can help to catch the high variability or short rain-driven events. Such tools can only provide reliable data provided that the monitoring system is transparent and follows quality control procedures with regard to maintenance and calibration. While a high level of automation may support such procedures, it still requires considerable manpower that should be foreseen in any monitoring budget.

REFERENCES

Arnold, J.G., Williams, J.R., Srinivasan, R. and King, K.W. (1996) *SWAT Manual*. USDA, Agricultural Research Service and Blackland Research Center, Texas.

Barthelemy, P.A. and Vidal, C. (1999) A dynamic European agricultural and agri-foodstuff sector. In: *Agriculture, Environment, Rural Development, Facts and Figures – A Challenge for Agriculture*. European Commission Report, Belgium.

Beck, M.B. (1987) *Water Resour. Res.*, **23**(8), 1393.

Bervoets, L., Schneiders, A. and Verheyen, R.F. (1989) Onderzoek naar de verspreiding en de typologie van ecologisch waardevolle waterlopen in het Vlaams gewest. Deel 1 - Het Denderbekken, Universitaire Instelling Antwerpen. In Dutch.

Boschma, M., Joaris, A. and Vidal, C. (1999) Concentrations of livestock production. In: *Agriculture, Environment, Rural Development, Facts and Figures – A Challenge for Agriculture*. European Commission Report, Belgium.

Brown, L.C. and Barnwell T.O. (1987) *The Enhanced Stream Water Quality Models QUAL2E and QUAL2E-UNCAS: Documentation and User Model*. EPA/600/3-87/007, USA.

Janssen, P.H.M., Heuberger, P.S.C. and Sanders, S. (1992) *Manual Uncsam 1.1, a Software Package for Sensitivity and Uncertainty Analysis*. Bilthoven, The Netherlands.

Krysanova, V. and Haberlandt, U. (2001) *Ecol. Modelling*, **150**, 255–275.

McKay, M.D. (1988) Sensitivity and uncertainty analysis using a statistical sample of input values. In: *Uncertainty Analysis*, Y. Ronen, ed. CRC Press, Inc., Boca Raton, FL, pp. 145–186.

Montarella, L. (1999) Soil at the interface between agriculture and environment. In: *Agriculture, Environment, Rural Development, Facts and Figures – A Challenge for Agriculture*. European Commission Report, Belgium.

Pau Val, M. and Vidal, C. (1999) Nitrogen in agriculture. In: *Agriculture, Environment, Rural Development, Facts and Figures – A Challenge for Agriculture*. European Commission Report, Belgium.

Poirot, M. (1999) Crop trends and environmental impacts. In: *Agriculture, Environment Rural Development, Facts and Figures – A Challenge for Agriculture*. European Commission Report, Belgium.

Sevruk, B. (1986) *Proceedings of the ETH, IAHS International Workshop on the Correction of Precipitation Measurements*, 1–3 April 1985. ETH Zürich, Züricher Geographische Schriften, Zürich, p. 23.

Smets, S. (1999) Modelling of nutrient losses in the Dender catchment using SWAT. Masters dissertation. Katholieke Universiteit Leuven –Vrije Universiteit, Brussels, Belgium.

Vandenberghe, V., van Griensven, A. and Bauwens, W. (2005) *Water Sci.Technol.*, **51**(3-4), 347–354.

Vandenberghe, V., Goethals, P., van Griensven, A., Meirlaen, J., De Pauw, N., Vanrolleghem, P.A. and Bauwens, W. (2004) *Environ. Monitor Assess.*, **108**, 85–98.

van Griensven, A. and Bauwens, W. (2001) *Water Sci. Technol.*, **43**(7), 321–328.

van Griensven, A. and Bauwens, W. (2003) *Water Resour. Res.*, **39**(10), 1348.

van Griensven, A., Vandenberghe, V. and Bauwens, W. (2002) *Proceedings of the International IWA Conference on Automation in Water Quality Monitoring*, 21–22, May 2002. Vienna, Austria.

Vivoni, E.R. and Richards, K.T. (2005) *J. Hydroinform.*, **7**(4), 235–250.

3.1
Elements of Modelling and Control of Urban Wastewater Treatment Systems

Olivier Potier and **Marie-Noëlle Pons**

Wastewater Quality Monitoring and Treatment Edited by P. Quevauviller, O. Thomas and A. van der Beken

3.1.1 INTRODUCTION

A wastewater treatment plant (WWTP) is an intricate system made of unit operations based on physical, biological and physico-chemical principles. Its aim is principally the removal of organic, nitrogen and phosphorus pollution. The basic processes are complex and the various arrangements of the unit operations which can be proposed lead to many possible configurations of WWTPs. It is difficult to describe in detail all of the processes here and only the basics of biological treatment by activated sludge will be examined. It is the most widespread for WWTPs of medium and large size. The interested reader will find more details in Henze *et al.* (Henze *et al.*, 2000). We focus our attention on the most important parameters for optimization and process control of pollution removal in large plants, where spatial distribution of substrate and nutrient in the reacting system plays a large role. In smaller plants, time scheduling can replace spatial gradients as in sequencing batch reactors for example. Whatever the case and in spite of the perturbations in terms of flow, composition and concentration experienced at the inlet of any WWTP, specifications on the discharged water should be kept within strict limits to avoid taxes and penalties. Different tools for monitoring and process control are also presented.

3.1.2 SHORT DESCRIPTION OF THE BIOLOGICAL PROCESS BY ACTIVATED SLUDGE

The biological step (often called secondary treatment) is an essential part of the WWTP. At the inlet of the plant, the water is usually pretreated to remove gross debris (grit removal) and can be further treated in a primary settler, which will eliminate a large part (usually 40–50 %) of the particulate pollution. In doing so, part of the biodegradable pollution is indeed removed, which might not always be a good idea: denitrification, one of the steps involved in nitrogen pollution removal, requires a certain balance between carbon and nitrogen and an external carbon source is often added in that step. This could be avoided (or at least limited) by direct injection into the biological reactor of unsettled wastewater. The principle of activated sludge is the intensification in a reactor of the principle of self-purification, which is naturally occurring in the environment, in presence of a much higher bacterial concentration than in rivers or lakes. The task of the secondary clarifier (Figure 3.1.1) is to separate the

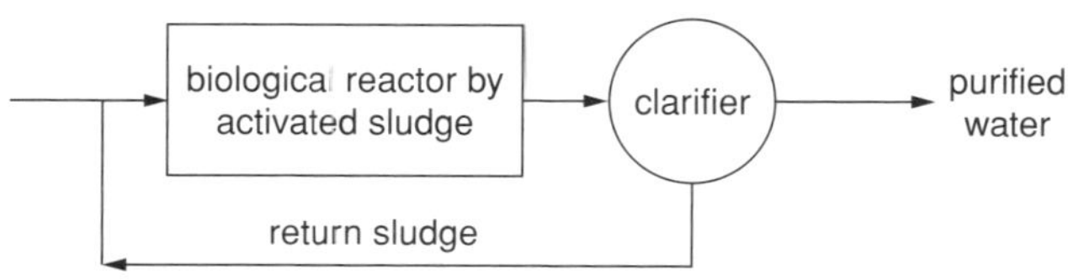

Figure 3.1.1 Schematic representation of an activated sludge system

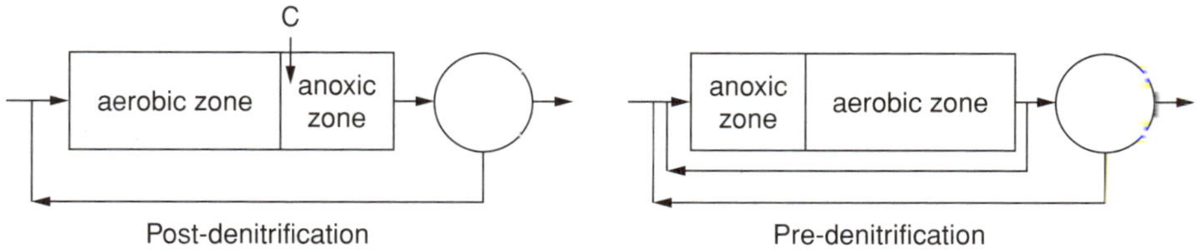

Figure 3.1.2 Schemes of different activated sludge reactors with anoxic zone

flocculated bacteria (sludge flocs) from the treated water. The sludge is returned to the inlet of the reactor and the purified water is polished in a tertiary stage (post-treatment of phosphorus, filtration, disinfection, etc.) and/or discharged.

In the presence of oxygen, carbon and a small amount of nitrogen (from ammonia and hydrolysed organic nitrogen) are metabolized by heterotrophic biomass and most of the nitrogen by autotrophic bacteria. The latter produced nitrates can be reduced by heterotrophs under anoxic conditions. As indicated previously, organic matter is needed for this reaction and therefore an addition of carbon (such as methanol) is often necessary. In the case of a pre-denitrification system, mixed liquor from the outlet of the reactor is recycled to the anoxic zone. Some of the most classical schemes are presented in Figure 3.1.2.

In order to ensure the best process efficiency, different parameters must be known and controlled: the main reactions of pollution removal and their kinetics; the spatial distribution of the substrates with respect to the micro-organisms and therefore the reactor hydrodynamics; the aeration capacity and therefore the oxygen transfer and the variability of the wastewater, in terms of composition, concentration and flow rate.

3.1.3 PROCESS PARAMETERS

3.1.3.1 Biokinetics

Many different compounds and micro-organisms are found in a biological wastewater system. In addition, the ecosystem is never at steady state. Therefore, an exact and complete kinetic model is out of reach. For many years the scientific community has tried to provide models of reasonable complexity, able to describe the main steps of activated sludge behaviour. The basic model is ASM1 (Activated Sludge Model n° 1), devoted to carbon and nitrogen removal (Henze *et al.*, 1987). Improved versions have been proposed, such as ASM2, which takes into account phosphorus removal, and ASM3 (IWA, 2000).

ASM1 is a good compromise between the description of the complex reality of biological reactions and the simplicity of a model. The identification of any model parameter should be possible theoretically (structural identifiability) and experimentally through experiments which can be run in the laboratory as well as on full-scale systems.

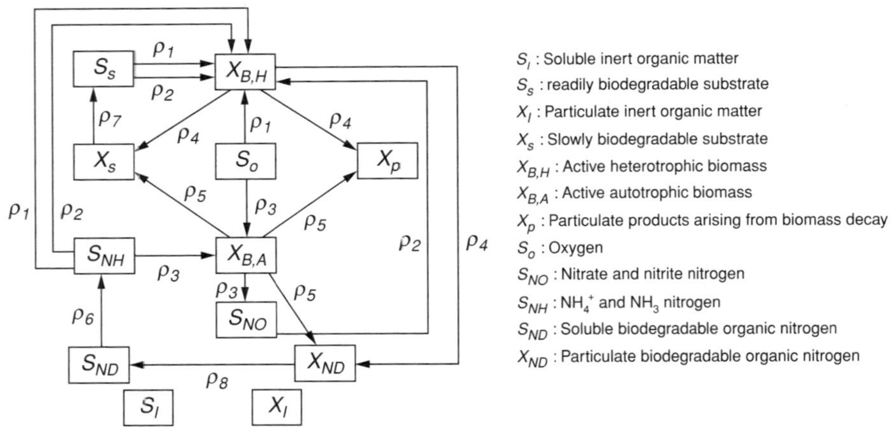

Figure 3.1.3 Schematic representation of the ASM1 kinetic pathways

As ASM1 is more particularly used, it will be described in some detail. In ASM1 (Figure 3.1.3), wastewater compounds are divided into different categories: inert (i.e. nonbiodegradable) versus biodegradable matter, particulate versus soluble. Particulate biodegradable matter should be hydrolysed to become readily biodegradable. The biomass is divided into two parts: heterotrophic and autotrophic.

Note that toxic events could trigger strong inhibition of bacteria. Inhibition terms can be added to the basic ASM1 model for specific purpose (industrial wastewater mainly). Autotrophs are deemed to be more sensitive to toxics than heterotrophs.

3.1.3.2 Oxygen Transfer

Influence of oxygen on pollution removal

Bacteria use oxygen for their respiration. In the ASM1 model, the oxygen concentration is considered to be a substrate:

- For the aerobic growth of heterotrophs, where readily biodegradable substrate is consumed:

$$\rho_1 = \mu_H \left(\frac{S_S}{K_S + S_S} \right) \left(\frac{S_O}{K_{O,H} + S_O} \right) X_{B,H}$$

where ρ_1 is the aerobic growth rate of heterotrophs, S_S the biodegradable soluble substrate concentration, S_O the oxygen concentration, $X_{B,H}$ the concentration of heterotrophs, K_S the heterotrophic half-saturation coefficient for S_S, $K_{O,H}$ the heterotrophic half-saturation/inhibition coefficient for oxygen and μ_H the maximum growth rate of heterotrophs.

- For the aerobic growth of autotrophs, where NH_4^+ and NH_3 nitrogen are transformed into nitrates:

$$\rho_3 = \mu_A \left(\frac{S_{NH}}{K_{NH} + S_{NH}}\right)\left(\frac{S_O}{K_{O,A} + S_O}\right) X_{B,A}$$

where ρ_3 is the aerobic growth rate of autotrophs, S_{NH} the ammonium concentration, $X_{B,A}$ the concentration of autotrophs, K_{NH} the autotrophic half-saturation coefficient for S_{NH}, $K_{O,A}$ the autotrophic half-saturation coefficient for oxygen and μ_A the maximum growth rate of heterotrophs.

For the anoxic growth of heterotrophs, a very low concentration of oxygen is required to avoid any inhibition:

$$\rho_2 = \mu_H \left(\frac{S_S}{K_S + S_S}\right)\left(\frac{K_{O,H}}{K_{O,H} + S_O}\right)\left(\frac{S_{NO}}{K_{NO} + S_{NO}}\right) \eta_g X_{B,H}$$

where ρ_2 is the anoxic growth rate of heterotrophs, S_{NO} the nitrate concentration, K_{NO} the heterotrophic half-saturation coefficient for S_{NO} and η_g the anoxic growth rate correction factor for heterotrophs.

Thus, the oxygen concentration has a great importance: it should be low in the anoxic stages and nonlimiting in the aerated zones. However, excessive oxygen supply should be penalized in terms of cost. Oxygen is provided by gas diffusers or surface aerators.

Oxygen transfer model

Generally, the gas–liquid transfer is modelled by means of the double film theory (Roustan *et al.*, 2003), according to which the gas–liquid interface is located between a gas film and a liquid film. For the oxygen–water system, the transfer resistance is found in the liquid film, due to the low solubility of oxygen in water. The oxygen flux is a function of the difference between the oxygen concentration at saturation (S_O^*) and the dissolved oxygen concentration in the reactor (S_O) and of the global coefficient of oxygen transfer ($k_L a'$). Experimental values of $k_L a'$ are generally between 2 h^{-1} and 10 h^{-1}. If it is assumed that the reactor can be modelled as a Continuous Perfectly Mixed Reactor (CPMR) (Figure 3.1.4), with a uniform oxygen concentration, the oxygen mass balance is written as:

$$QS_{OI} + k_L a'(S_O^* - S_O)V = QS_O + r_O V + V\frac{\mathrm{d}S_O}{\mathrm{d}t}$$

with S_{OI} the oxygen concentration at the inlet, r_O the oxygen consumption rate, V the reactor volume and Q the liquid flow rate.

The coefficient transfer is measured directly in the presence of sludge ($k_L a'$), or in clean water without sludge ($k_L a$) (Héduit and Racault, 1983a,b; ASCE, 1992;

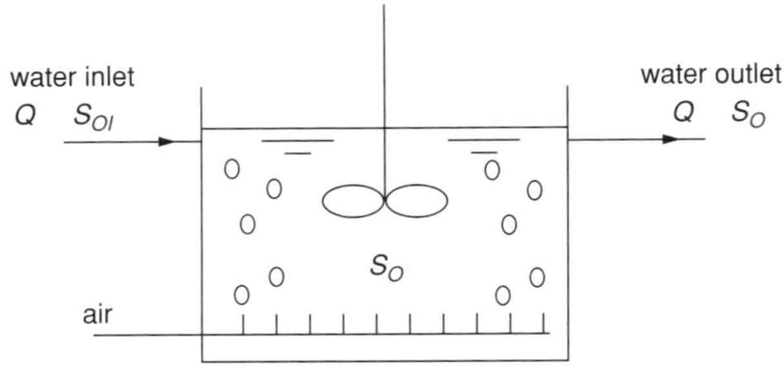

Figure 3.1.4 An aerated Continuous Perfectly Mixed Reactor

Roustan *et al.*, 2003). In this case, the so-called 'alpha' factor (α) must be taken into account (Boumansour and Vasel, 1996):

$$k_L a' = \alpha k_L a$$

Example of oxygen profile in a WWTP bioreactor

To illustrate the open loop behaviour of a biological reactor, with no aeration adjustment as a function of the oxygen demand, the oxygen profile was measured during 1 day in a 3300 m^3 channel reactor with a large aspect ratio. The reactor is 100 m long and 8 m wide and aerated by means of fine bubble diffusers located on its floor. The dissolved oxygen concentration was regularly measured in six locations along the reactor with a portable probe (WTW, Weilheim, Germany) (Figure 3.1.5). The

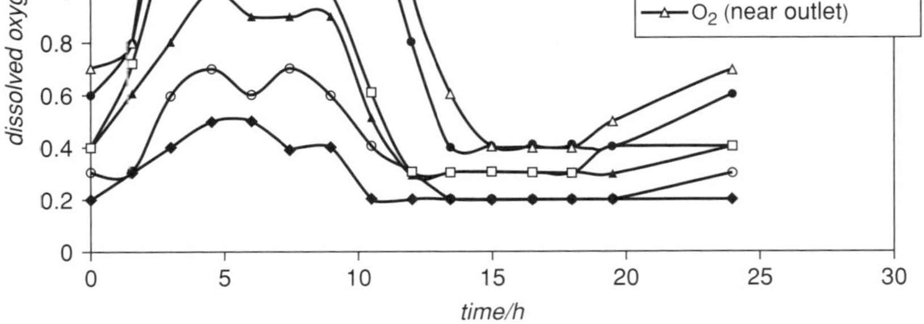

Figure 3.1.5 Variations of the dissolved oxygen concentration in different locations of an activated sludge channel reactor during 1 day

air flow rate was constant and equally distributed along the reactor, which was continuously fed by urban presettled wastewater. The dissolved oxygen concentration changes with the oxygen consumption, and therefore with biodegradable pollution concentration, which depends on time and space in the reactor.

In Figure 3.1.5, it can be seen that dissolved oxygen concentration is higher during the night, when pollution is lower. The concentration increases along the reactor as the oxygen consumption decreases due to a decrease in the biodegradable substrate availability. During the day, dissolved oxygen concentration remains very low, even near the reactor outlet, which indicates complete pollution removal is not achieved. Under such conditions oxygen limitation occurs. Better aeration with a larger air flow rate could alleviate such a limitation without increasing the reactor volume.

3.1.3.3 Hydrodynamics

In brief, two types of reactor shape are found: a compact, 'parallelepipedic' or 'cylindrical' design, often fitted with surface turbines for aeration; and an elongated design suitable for gas diffusion devices. Elongated reactors are often folded or built as 'race tracks', which avoids recirculation pumps (Figure 3.1.6). In this case they are generally called 'oxidation ditches' when the aerators are horizontal and 'carousels' when they are vertical. Many variations have been proposed by various manufacturers, such as sets of several concentric channels as in the Orbal™ system and OCO™ process, inclusion of anaerobic and anoxic zones equipped with mechanical mixing devices, or combination of spatial gradients along the tanks with alternating mode of operation, such as in the Biodenipho™ or Biodenitro™ process. Capacity, land availability, flow circulation, process type (carbon and/or nutrient removal) are some of the criteria for selection.

Hydrodynamics have a great importance in a process, because linked with kinetics, they affect pollution removal efficiency and the bacteria species selectivity. Usually the reactor behaviour is compared with one of two ideal types: the Continuous Perfectly Mixed Reactor (or CPMR) and the Plug Flow Reactor.

The CPMR is characterized by a uniform concentration of each component in all the volume of the reactor. This type of reactor can be found in small WWTPs, where the length is similar to the width.

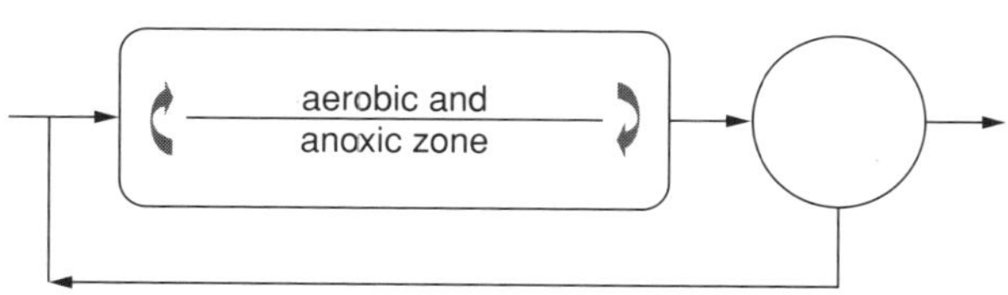

Figure 3.1.6 A 'race track' reactor

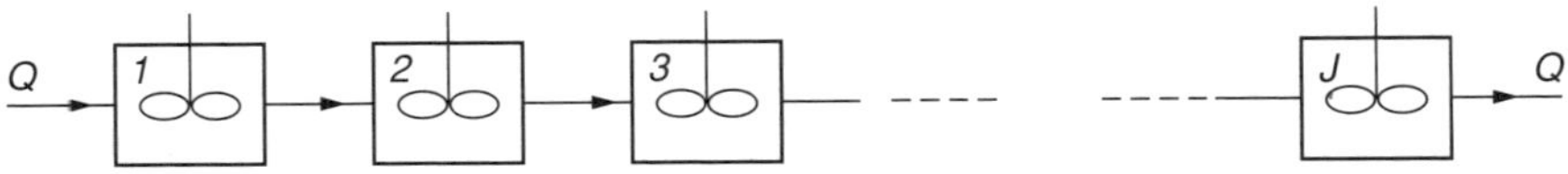

Figure 3.1.7 *J* CPMRs in series

The Plug Flow Reactor model is very different. It is composed of a succession of parallel volumes infinitiely small, perpendicular to the flow, with no transfer between them. These volumes move forward from the inlet to the outlet, at a velocity linearly related to the flow. There is a progressive change in concentrations. However, if the ideal Plug Flow Reactor model could be used for tubular or fixed-bed reactors in the chemical industry, it rarely represents in a satisfactory manner an aerated tank in a WWTP.

Models based on CPMRs in series (Figure 3.1.7) offer the best simple alternative to model full-scale plants and generally give a good agreement with experimental data. Theoretically, the number of reactors in series (J) can vary between 1 and infinity. In practice, J is determined by tracing experiments and takes values between 3 and 20. Although a series of J CPMRs is a discrete hydrodynamic model, it can model a continuous liquid system like a channel reactor.

Hydrodynamic characterization

A relatively simple method for the characterization of hydrodynamics is the Residence Time Distribution (RTD) method. Each molecule has is own residence time (r_t) in the reactor, which depends on the reactor hydrodynamics (Figure 3.1.8). The goal of the RTD method is to measure the different residence times based on statistics. A pulse of nonreactive tracer is injected at the inlet of the reactor. Different chemical substances are used, such as lithium chloride (detection by atomic absorption), rhodamine (detection by fluorescence sensor) and radioactive elements. The tracer is dissolved in the mixed liquor in the reactor and behaves as the liquid phase. At the reactor outlet, the tracer concentration is measured to calculate the RTD (Villermaux, 1993; Levenspiel, 1999).

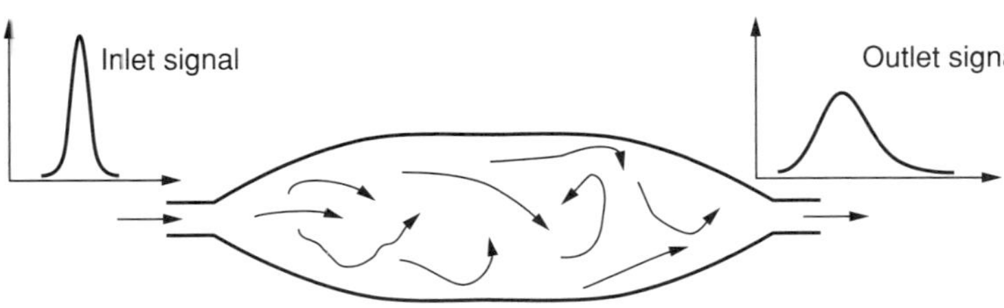

Figure 3.1.8 Inert tracing of a reactor

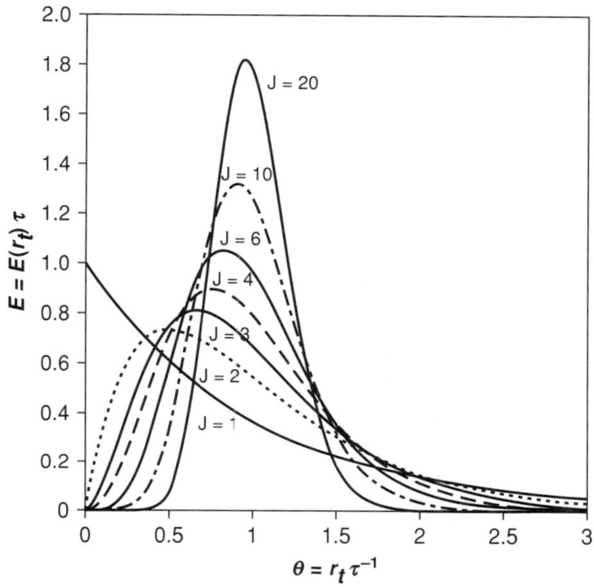

Figure 3.1.9 Theoretical RTD tracings of different sets of CPMRs in series

For CPMRs in series, the RTD is a function of J and of the space time $\tau = V/Q$:

$$E(r_t) = \left(\frac{J}{\tau}\right)^J \frac{r_t{}^{J-1} exp\,(-Jr_t/\tau)}{(J-1)!}$$

In Figure 3.1.9, theoretical RTD tracings for series of J CPMRs ($J = 1$–20) are plotted. The parameters are normalized by the space time τ.

Influence of hydrodynamics on pollution removal

In biological wastewater treatment, kinetics are a function of the biodegradable substrate concentration (S_S); the larger the concentration, the larger the reaction rates. In this case, it can be demonstrated that better pollution removal efficiency is obtained with CPMRs in series than with a single CPMR. The larger the J, the better the efficiency. Therefore, between two reactors with the same volume, the better one is the longest.

Moreover, in activated sludge, filamentous bacteria, which constitute the backbone of activated sludge flocs, could overgrow, which creates a problem called filamentous bulking. This problem has many causes, but it has been noticed that the hydrodynamics of a CPMR favours this phenomenon (Chudoba *et al.*, 1973). Conversely, a reactor with a high aspect ratio, behaving as a series of CPMRs favours a more 'normal', i.e. well balanced, biomass. It is a problem of selectivity.

Computer Fluid Dynamics

Computer Fluid Dynamics (CFD) is a recent tool used to analyse in detail the flow characteristics in a number of systems, including chemical and biological reactors (Ranade, 2002). However, many hurdles remain, especially in the field of wastewater treatment. Assumptions concerning limit and initial conditions, turbulence model, etc., should be made. Activated sludge processes are multiphase systems but the liquid and 'solid' (biomass) phases are generally considered as a single homogenous liquid in which the bubbles (gas phase) are in motion. Two approaches are generally utilized to simulate this gas–liquid system. The Euler approach is used in both cases for the liquid phase. The gas phase can be treated by a Eulerian approach (Euler–Euler) or a Lagrangian approach (Euler–Lagrange). There is a third method, Volume of Fluid (VOF), but this is generally only used for small systems with few bubbles.

In spite of the increase in computer speed and the possible parallelization of some calculations, the simulation time remains very long and it is still difficult to introduce mass transfer and kinetics in this type of simulation.

3.1.3.4 Wastewater Variability

Different types of variability

Wastewater characteristics change with time, not only in terms of flow rate, but also in terms of composition and concentration. Wastewater variability depends on human and industrial activities and on weather conditions, especially in combined sewer networks where sewage is mixed with run-off water from roofs, pavements, etc. Several scales of dry-weather variability are recognized: daily, weekly and seasonally disturbances affect the wastewater characteristics.

Example of variability

Figure 3.1.10 illustrates the variability of chemical oxygen demand (COD) at the inlet of the wastewater treatment system of a 2000 inhabitants' community in France, under summer dry weather conditions. A 24-h period is clearly visible. Week days (Monday through Friday) present a similar pattern, where morning, lunchtime and evening activities induce COD peaks. In weekend days pollution is higher as inhabitants tend to remain at home and are not going to work in the nearby large city. In large urban centres the trend will be the opposite, with less pollution during weekends than week days.

Flow rate variability and its influence on hydrodynamics

Reactor hydrodynamics are affected by the diurnal flow rate variations. Figure 3.1.11 presents the example of the flow characteristics at the inlet of a 350 000

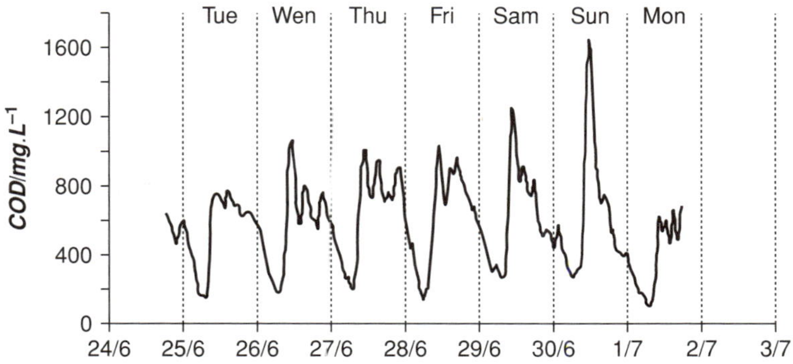

Figure 3.1.10 Example of COD variations in a 2000 inhabitants' community

person-equivalent plant. The effect of rain can also be seen in the middle of the week. Part of the incoming wastewater was bypassed and directly discharged to the river, which explained the limitation at 6500 m^3/h.

RTDs were determined in the channel reactors previously described under different flow conditions and they show that the tanks can be modelled by CPMRs in series. The hydrodynamic behaviour is modified by the flow rate and the number of CPMRs (Figure 3.1.12), J, changes with the space-time τ, and therefore with the water flow rate Q (Potier *et al.*, 2005):

$$J = \frac{L^2}{2\tau D} + 1 \quad \text{and} \quad \tau = \frac{V}{Q} \qquad \text{where } L \text{ is the reactor length.}$$

$$\text{or} \quad J = \frac{K}{\tau} + 1 \quad \text{with} \quad K = \frac{L^2}{2\ D}$$

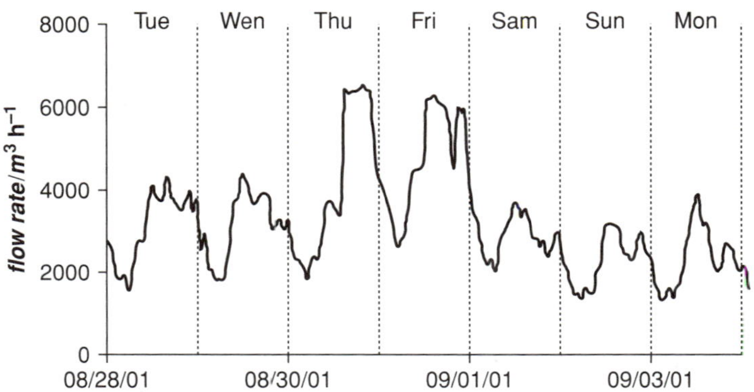

Figure 3.1.11 Flow rate variations, at the inlet of a 350 000 person-equivalent WWTP

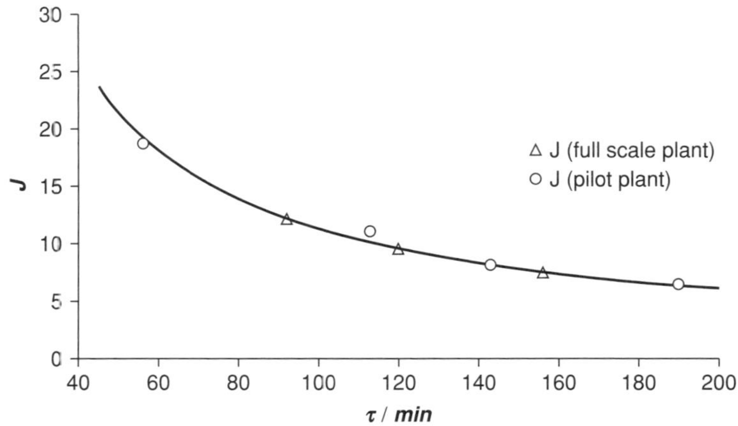

Figure 3.1.12 Number of CPMRs (J) versus liquid space-time (τ) for a full-scale plant and a bench-scale plant

Often, hydrodynamics are considered as a fixed parameter. In order to facilitate the modelling task, it is convenient to work with a constant number of CPMRs. By introducing the concept of CPMRs in series with back-mixing (Figure 3.1.13), a compromise is reached.

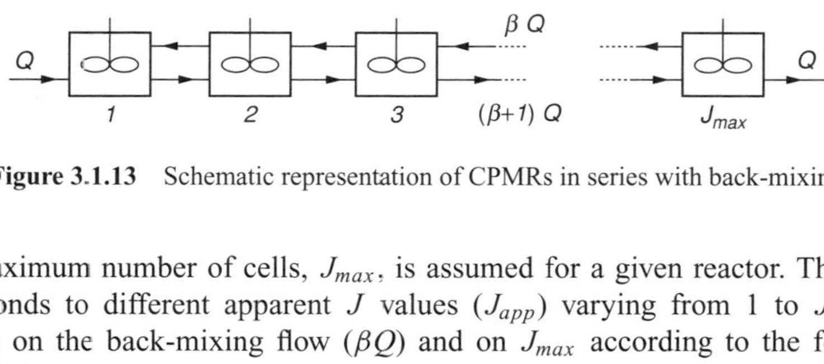

Figure 3.1.13 Schematic representation of CPMRs in series with back-mixing

A maximum number of cells, J_{max}, is assumed for a given reactor. The model corresponds to different apparent J values (J_{app}) varying from 1 to J_{max}, depending on the back-mixing flow (βQ) and on J_{max} according to the following relationship:

$$\beta = \frac{1}{2}\left\{(J_{max} - 1) - \left[1 + J_{max}^2\left(1 - \frac{2}{J_{app}}\right)\right]^{1/2}\right\}$$

which is valid for J_{app} larger than 2.5.

3.1.3.5 Mass Balance

The full mass balance enables finally to bring together the different aspects: hydrodynamics, kinetics and mass transfer. It is the basis of the global model used for the understanding of the system, its simulation, its optimization and even its automation. The first stage is to identify the hydrodynamic model, as described previously. For a CPMR, a mass balance equation is then written for each component (substrates, metabolites). For CPMRs in series, a mass balance equation is necessary for each component in each CPMR.

3.1.4 SENSORS

Many different sensors are found on WWTPs. They give information about treatment efficiency and they are necessary to monitor, control and optimize the processes (Vanrolleghem and Lee, 2003; Degrémont, 2005). There are three types of sensors: in-line sensors situated directly in the process; on-line sensors, based on automated sampling and conditioning of the sample; and off-line devices, in plant laboratories, which require human operators. In any case in-line and on-line sensors will require careful maintenance, including automated cleaning sequences, and calibration.

3.1.4.1 In-line Sensors

In-line sensors are mostly devoted to physical parameters: flow rate [water, gases (air, methane from sludge digesters, etc.), sludge, reagents such as polymers for sludge conditioning, precipitants such as ion chloride for phosphorus removal], level (liquid, sludge blanket), pressure, temperature, electrical power, suspended solids, turbidity, etc. A few chemical sensors are also available such as pH, redox, dissolved oxygen, conductivity, ammonia (with an ion-selective probe). More recently devices based on UV-visible spectrophotometry have been proposed as surrogate measurements for COD, which requires a 2 h digestion. These systems operate at a fixed wavelength (254 nm in general) or collect spectra in the range 200–600 nm (Spectro::lyser, Scan Messtechnik GmbH, Vienna, Austria). Fluorescence sensors (BioView, Delta Light & Optics, Denmark) and infrared technology (Steyer *et al.*, 2002) offer also new prospects for *in-situ* wastewater quality monitoring based on spectroscopy (Pons *et al.*, 2004).

3.1.4.2 On-line Sensors

On-line sensors have been proposed for nitrate, ammonia, phosphate, short-term biological oxygen demand (BOD) (to evaluate the oxygen demand and control the aeration rate), toxicity (based on bacterial respiration) (Vanrolleghem *et al.*, 1994) and sludge volume index (to detect settling problems such as filamentous bulking) (Vanderhasselt *et al.*, 1999).

3.1.5 INTRODUCTION TO THE CONTROL METHODS OF A WASTEWATER TREATMENT PLANT BY ACTIVATED SLUDGE

The aims of control systems are to maintain the concentration and the flux of pollutants below the limits fixed by the environmental norms and to reduce the operating costs. WWTPs are often designed so as to operate near their limits in order to minimize the investment costs. Therefore they become more sensitive to inlet perturbations. It is even more the case for nutrient removal systems which are more easily disturbed by an ammonia / carbon inbalance than carbon removal plants.

In some cases treatment capacity increases up to 25 % can be obtained by carefully designed control strategies, without increasing the reactor volume. Within Europe, the level of implementation of instrumentation, control and automation systems varies depending on the country (Jeppsson *et al.*, 2002; Olsson *et al.*, 2005).

WWTPs are very complex to control because of the composition and the time variability of the biomass and the wastewater. As shown previously, their models need many parameters, are nonstationary and strongly nonlinear. Many control strategies, presenting different degrees of sophistication, have been proposed but they are often difficult to test and validate at full-scale. For this reason a benchmarking procedure, initiated at the European level in COST Actions (Jeppsson and Pons, 2004) and further developed in an IWA Task Group, has been proposed (http://www.benchmarkWWTP.org). In many cases, the control techniques used in WWTPs are simple and pragmatic.

The open loop combined with some time scheduling is the simple 'control' system. For example, for the aeration of the activated sludge, the air flow rate can be set to a lower value during the night than during the day. Because of disturbances, this basic control method gives limited results.

When a valve or a pump manipulation is triggered by a measurement provided by some sensor situated after the controlled process, the system works in closed loop. In feedback control the measured variable is compared with a set point (Figure 3.1.14). A control law, usually of the PID (proportional-integral-derivative)

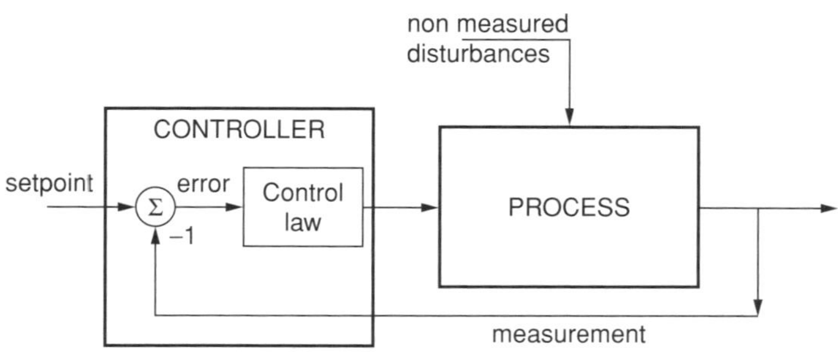

Figure 3.1.14 Schematic representation of a basic feedback control loop

or PI (proportional-integral) type (Corriou, 2004), transforms the resulting error into information for the actuator (situated before the controlled process), which has an action on the process. Three types of actuator operations can be found: on/off, continuous or discrete. When possible feedforward control, which causes the system to react before the perturbations could have effects on the plant, should be implemented: for example, the inlet flow rate variations can be used to predict the variability of the incoming load.

In some small plants a unique reactor is used with alternated periods of aeration and anoxia. The phase durations are deduced from the measurement and the control of the dissolved oxygen. Better results are obtained, if this information is combined with a nitrate or a redox potential sensor (Chachuat *et al.*, 2005; Fikar *et al.*, 2005).

In larger reactors, the anoxic reaction and the aeration are taking place in different zones. In the aeration tank, the controlled variable can be the dissolved oxygen and the actuator the valve controlling the air flow (Olsson *et al.*, 2005). To illustrate our purpose, a schematic representation of a 600 000 person equivalent WWTP control system is shown in Figure 3.1.15. There are three lines and each line is divided into two parts. Each part is controlled by a cascade of two PIDs. Sensor redundancy is provided by two dissolved oxygen (DO) probes.

Nitrate concentration can be controlled by the addition of an external carbon source, in order to keep the correct ratio between nitrate and carbon during denitrification or by adjustment of the internal recycle flow in a pre-denitrification scenario (Gernaey and Jørgensen 2004).

The control loops can be independent but in general interactions between them exist and make the life of the control engineer difficult. They can be organized in a hierarchical control system. The basic control loops are taken care of at the lowest level, close to the process. Their setpoints are defined at a higher level. In the event

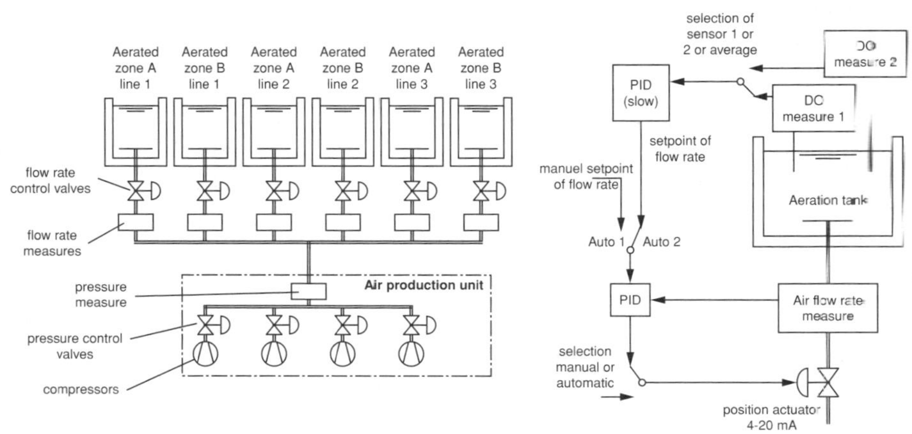

Figure 3.1.15 Schematic representation of a closed loop aeration control in a 600 000 person equivalent WWTP in France (Courtesy of Degrémont)

of a problem (such as a sensor or actuator fault), the automated control system could be stopped and the WWTP be controlled manually, from the upper level.

3.1.6 CONCLUSION AND PERSPECTIVES

New approaches of control are proposed, but must be more widely tested in WWTPs. Different techniques are available such as fuzzy control, internal model control, which requires biological and hydrodynamical models, and adaptive control, which permits on-line identification of the model parameters. However, managers are often reluctant to implement such sophisticated control strategies, as they are under the constant pressure of achieving stricter quality limits on the discharged water, while minimizing operation cost. The availability of a plant-scale dynamic model representing in sufficient detail the behaviour of WWTPs and that can be used to 'benchmark' control strategies (Jeppsson and Pons, 2004) could help the modernization of plants from the control point of view.

ACKNOWLEDGEMENT

The authors wish to thank Degrémont and particularly Eric Garcin, Françoise Petitpain-Perrin, Jean-Pierre Hazard and Didier Perrin.

REFERENCES

ASCE (1992) *Standard Measurement of Oxygen Transfer in Clean Water.* American Society of Civil Engineers, Reston, VA.

Boumansour, B.E. and Vasel, J.L. (1996) *Tribune de l'eau*, **5–6,** 31–40.

Chachuat, B., Roche, N. and Latifi, M.A. (2005) *Chem. Engin. Proc.*, **44**, 591–604.

Chudoba, J., Ottava, V. and Madera, V. (1973. *Water Res.*, **7**, 1163–1182.

Corriou, J.P. (2004). *Process Control.* Springer, Berlin.

Degrémont (2005) *Mémento technique de l'eau*, 10th Edn. Lavoisier SAS.

Fikar, M., Chachuat, B. and Latifi, M.A. (2005) *Control Engin. Pract.*, **13**, 853–861.

Gernaey, K. and Jørgensen, S.B. (2004) *Control Engin. Pract.*, **12**, 357–373.

Héduit, A. and Racault, Y. (1983a) *Water Res.*, **17**, 97–103.

Héduit, A. and Racault, Y. (1983b) *Water Res.*, **17**, 289–297.

Henze, M., Grady, C., Gujer, W., Marais, G. and Matsuo, T. (1987) Activated sludge model no.1. *IAWPRC* Task Group Report. IWA, London.

Henze, M., Harremöes, P., La Cour Jansen, J. and Arvin, E. (2000) *Wastewater Treatment. Biological and Chemical Processes*, 3rd Ed. Springer, Berlin.

IWA (2000) Task group on mathematical modelling for design and operation of biological wastewater treatment, Activated sludge models ASM1, ASM2, ASM2D and ASM3. Scientific and Technical Report no. 9. IWA, London.

Jeppsson, U., Alex, J., Pons, M.N., Spanjers, H. and Vanrolleghem, P.A. (2002) *Water Sci. Technol.*, **45** (4–5), 485–494.

Jeppsson, U. and Pons, M.N. (2004) *Control Engin. Pract.*, **12**, 299–304.

Langergraber, G., Fleischmann, N. and Hofstädter, F. (2003) *Water Sci. Technol.*, **47** (2), 63–71.

Levenspiel, O. (1999) *Chemical Reaction Engineering*, 3rd Edn. John Wiley & Sons, Inc., New York.

Olsson, G., Nielsen, M.K., Yuan, Z.,. Lynggaard A. and Steyer, J.P. (2005) Instrumentation, control and automation in wastewater systems. Scientific and Technical Report no. 15. IWA Publishing, London.

Pons, M.N. (1992) Physical and chemical sensors – actuators. In: *Bioprocess Monitoring and Control*, M.N. Pons, ed. Hanser Publishers, New York.

Pons, M.N., Le Bonté, S. and Potier, O. (2004) *J. Biotechnol.*, **113**, 211–230.

Potier, O., Leclerc, J.P. and Pons, M.N. (2005) *Water Res.*, **39**, 4454–4462.

Ranade, V.V. (2002) *Computational Flow Modelling for Chemical Reactor Engineering. Process systems engineering.* Academic Press, San Diego, CA.

Roustan, M., Wild, G., Héduit, A., Capela, S. and Gillot, S. (2003) *Transfert gaz-liquide dans les procédés de traitement des eaux et des effluents gazeux*. Tec & Doc Editions, Paris.

Steyer, J.P., Bouvier, J.C., Conte, T., Gras, P., Harmand, J. and Delgenes, J.P. (2002) *Water Sci. Technol.*, **45** (10), 133–138.

Vanderhasselt, A., Aspegren, H., Vanrolleghem, P.A. and Verstraete, W. (1999) *Water SA*, **25**, 453–458.

Vanrolleghem, P.A., Kong, Z., Rombouts, G. and Verstraete, W. (1994) *J. Chem. Technol. Biotechnol.*, **59**, 321–333.

Vanrolleghem, P.A. and Lee, D.S. (2003) *Water Sci. Technol.*, **47**(2), 1–34.

Villermaux J. (1993) *Génie de la réaction chimique. Lavoisier, Paris.*

3.2
Treatability Evaluation

Gianni Andreottola and **Paola Foladori**

3.2.1 INTRODUCTION

'To know treatability is to know the fate of contaminants in WWTPs'.

The pollutants introduced into the sewerage collecting system and reaching municipal wastewater treatment plants (WWTPs) derive principally from human activities

Wastewater Quality Monitoring and Treatment Edited by P. Quevauviller, O. Thomas and A. van der Beken

and in particular from domestic sources, industrial districts and urban run-off rainwater. A very large amount of different organic and inorganic compounds, estimated as several thousand, has been detected in raw wastewater. The treatability of these compounds in the conventional WWTPs can differ significantly depending on each considered contaminant. The importance of knowing the treatability of the different kinds of pollutants present in municipal wastewater is related to the prediction of the fate of these contaminants in WWTPs before the discharge in the receiving water bodies. The following principal categories of contaminants in municipal raw wastewater can be distinguished:

- *Organic compounds as aggregate parameters*. The whole amount of organic matter is generally measured as aggregate organic parameters, such as chemical oxygen demand (COD), total organic carbon (TOC), or biological oxygen demand (BOD) in the case of the measurement of only biodegradable compounds. Aggregate organic constituents are comprised of a number of individual compounds that cannot be distinguished separately. Eventually the fractionation of COD can be performed with the aim to discriminate biodegradable and nonbiodegradable fractions of organic matter:
- *Organic micropollutants*. The determination of these organic compounds is done as individual parameters; some of them are associated with a potential toxic risk to health and the environment.
- *Nutrients, such as nitrogen (N) and phosphorus (P)*. Among the inorganic nonmetallic compounds, N and P in their different ionic or organic forms, represent the most important pollutants and are also, in most cases, the major nutrients of importance.
- *Metallic compounds*. Some, including cadmium, chromium, copper, mercury, nickel, lead and zinc, are characterized by a potentially toxic action.

The effectiveness of the removal of these categories in WWTPs depends on the plant configuration and not all WWTPs are able to remove all the pollutants present in the influent wastewater.

Most WWTPs designed or upgraded in the last decades to European level are characterized by primary and secondary treatment (adopting activated sludge or biofilm configurations) able to achieve complete removal of biodegradable COD in influent wastewater. Furthermore, plants located in areas sensitive to eutrophication reach high efficiency in nitrification, denitrification and P removal, as directed by the European Directive promulgated in 1991 (91/271/CEE) that imposed more restrictive effluent limits for the discharge of treated wastewater in the receiving water bodies (see Chapter 1.1). In particular, the effluent concentration limit for total nitrogen is equal to 15 or 10 mg/l for a population equivalent (PE) lower or higher than 100 000, respectively. Analogously in the same Directive, the effluent limit for phosphorus is 2 and 1 mg/l for plant capacity below or above 100 000 PE, respectively.

Plants currently guaranteeing to meet the discharge limits for COD, biological oxygen demand for 5 days (BOD_5) and total suspended solids (TSS), could not meet the limits for N and P as imposed by 91/271/CEE for sensitive areas, requiring further upgrading.

Discharge limits are indicated also for other constituents, such as metals or organic micropollutants; due to their wide heterogeneity and their different treatability not all the WWTPs are suitable for the complete removal of these contaminants, but many of them can be removed only partially. For example, organic micropollutants can be biodegraded only in part, but often are removed physically from water and accumulated in excess sludge, transferring the pollution problem from water to sludge. This occurs also in the case of metals.

For evaluating the wastewater treatability, two key aspects have to be considered: the composition of the influent wastewater; and the treatment capacity in the WWTPs. In particular, the treatment capacity is related to the physico-chemical processes performed in the plant and the biodegradation capacity of activated sludge or biofilm processes in the secondary treatment. The wastewater composition in combination with the plant treatment capacity constitutes the basis of the 'treatability' concept. The knowledge of these aspects is fundamental in order to evaluate the entity of pollutants removal in the plant and to predict the quality of the treated effluents aimed to respect the imposed limits and to reduce the impact in receiving water bodies.

In the following paragraphs the fate through WWTPs of the categories of pollutants cited above are described and the repartition of contaminants in sludge or effluent water is indicated. In particular, the influence of the various treatment processes (physico-chemical primary treatment, biological secondary treatment and eventually tertiary treatment) is considered for each category of contaminants.

3.2.2 ORGANIC COMPOUNDS AS AGGREGATE PARAMETERS

The quantification of the total organic matter in wastewater and its characterization is of primary importance for the correct design, management and optimization of a WWTP. Carbonaceous substrates are generally quantified by using aggregate parameters such as BOD_5 or COD, but only the analysis of COD is able to represent the whole amount of organic matter, while BOD_5 is representative of the biodegradable fraction only.

As far as the BOD_5 parameter is concerned, it has been widely applied in the field of receiving water bodies and for wastewater characterization. Due to the 5-day duration of the BOD test (BOD_5), the measurement of oxygen consumption (index of biodegradability) is relative to 5 days and therefore very different from wastewater retention time in WWTPs where the biodegradation occurs. The problems related to the interpretation of the BOD_5 test for the measurement of biodegradable compounds in wastewater and its use in the design and management of treatment processes gave increasing interest to new characterization proposals. In particular, interest in biodegradability characterization has been increased from the simulation models for the activated sludge process that do not use traditional parameters [for example, the Activated Sludge Model, from ASM No. 1 (Henze *et al.*, 1987) to ASM No. 3 (Gujer *et al.*, 1999)]. In the literature, proposals for the characterization of the biodegradability of carbonaceous substrates are available, especially based on respirometry (Henze, 1992; Spanjers and Vanrolleghem, 1995; Orhon *et al.*, 1997; Spanjers *et al.*, 1999). Respirometry is defined as the measurement and the interpretation of the rate of oxygen consumption (oxygen uptake rate, OUR) by activated sludge or wastewater under different load conditions. The consumption of oxygen is due to two different factors:

(1) *Endogenous respiration* (OUR_{endo}) measured for a biomass in the absence of external substrate and due to cellular maintenance and oxidation of dead cells.

(2) *Exogenous respiration* (OUR_{exo}) measured during the oxidation of biodegradable COD present in wastewater added to a biomass.

The quantification of biodegradable COD in wastewater can be assessed through respirometric tests carried out on activated sludge after the addition of an adequate amount of wastewater. The dynamics of OUR_{exo} are monitored for a period of about 10–20 h and the data are interpreted as described in more detail in Section 3.2.2.2.

Alternatively, in the absence of respirometric measurements, a rapid estimation of COD fractions (less precise than the results obtainable by respirometry) can be done in existing WWTPs, according to an easy calculation based on BOD_5 and COD analyses in influent and effluent wastewater, as indicated in Section 3.2.2.3.

3.2.2.1 Fractions of Total COD in Wastewater and their Treatability

While some organic compounds are easily biodegradable in WWTPs, others are persistent and refractory and they are found in the treated effluents or in the excess sludge. The complete fractionation of COD in raw wastewater is shown schematically in Figure 3.2.1, in which symbols are adopted according to ASM models. The total COD concentration is subdivided into two biodegradable and nonbiodegradable fractions and into an active biomass fraction. A soluble part (S) and a particulate part (X) are distinguished for both biodegradable COD (indicated by subscript S) and nonbiodegradable COD (indicated by subscript I).

In COD fractionation the following terms are introduced and defined:

(1) *Total COD*: determined experimentally by chemical analysis without any pretreatment of the wastewater (APHA, AWWA and WPCF, 1998).

(2) *Soluble COD* (S): determined experimentally by means of the chemical analysis of COD after a pretreatment of wastewater with coagulation, flocculation

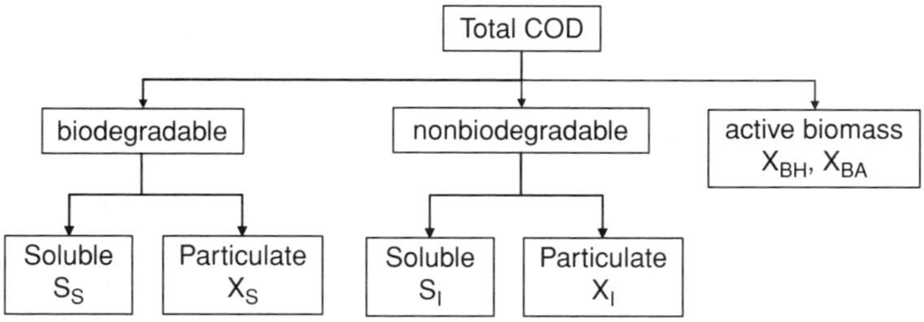

Figure 3.2.1 Scheme of total COD fractionation in wastewater

and 0.45-μm-filtration, according to the procedure proposed by Mamais *et al.* (Mamais *et al.*, 1993). Alternatively, the determination of soluble COD can be carried out by the direct filtration of wastewater at 0.1 μm, in order to minimize the occurrence of colloidal solids. The results obtained from the two kinds of measurements are similar with a difference of about 1 % (Roeleveld and van Loosdrecht, 2002);

(3) *Particulate COD* (X): determined as the difference between total COD and soluble COD.

(4) *Soluble biodegradable* COD (S_S): made up of simple molecules ready to be assimilated through the cellular membrane (readily biodegradable COD) or easy to be hydrolysed (rapidly hydrolysable COD); it can be measured by respirometry.

(5) *Particulate biodegradable COD* (X_S): made up of suspended and colloidal solids and compounds with high molecular weight that require enzymatic hydrolysis before being metabolized. It is also called 'slowly biodegradable COD' and can be measured by respirometry; the biodegradation rate of X_S is about 10 times smaller than the rate of S_S.

(6) *Soluble inert COD* (S_I): made up of dissolved nonbiodegradable molecules. It is calculated as the difference between S and S_S.

(7) *Particulate inert COD* (X_I): made up of nonbiodegradable compounds, both in suspended and colloidal forms. It is calculated as the difference between X and X_S.

(8) *Heterotrophic and autotrophic active biomass* (X_{BH} and X_{BA}, respectively): made up of the cellular active biomass present in wastewater and represents an inoculum for the biological process in the WWTP. The value of X_{BH} can be quantified by respirometry, while the amount of X_{BA} is often neglected in the COD fractionation.

The total COD is given by:

$$\text{total COD} = S_S + X_S + S_I + X_I + X_{BH} + X_{BA}$$

Table 3.2.1 Fractionation of COD in raw and presettled wastewater (percentages are referred to total COD)

Fraction	Raw wastewater (%)	Presettled wastewater (%)
S_S	10–30	20–40
X_S	40–60	30–50
S_I	5–10	5–15
X_I	10–20	7–15

Typical percentages of the COD fractions for raw wastewater and presettled wastewater (after primary sedimentation) are indicated in Table 3.2.1.

This fractionation allows understanding of the composition of organic matter in wastewater and to predict its fate during treatment in WWTPs. The fate of each individual fraction is:

- S_S is rapidly biodegraded in the biological stage of the WWTP, requiring a short time (generally less than 1–2 h).
- S_I is transferred in the effluent without any modification, being not biodegradable and not settleable; for its reduction a tertiary treatment is eventually required.
- X_S is mostly biodegraded during the biological treatment and eventually part is transferred in primary or secondary sludge. The amount of X_S discharged in the final effluent is negligible.
- X_I is transferred in primary and secondary sludge, without any significant modification, being nonbiodegradable.
- X_{BH} is an inoculum in the biological process in WWTP (and subjected to growth and death) and it is separated with the primary and secondary sludge.

3.2.2.2 Respirometric Approach for COD Fractionation

Many authors have proposed methods based on respirometry for the assessment of the COD fractions in wastewater (Ekama *et al.*, 1986; Kappeler and Gujer, 1992). In depth contributions about wastewater characterization have been published by Henze (Henze, 1992) and Vanrolleghem *et al.* (Vanrolleghem *et al.*, 1999). Furthermore, methods have been proposed to obtain the complete fractionation of COD in wastewater and other kinetic parameters by modelling the respirometric data acquired during a single batch respirometric test. This opportunity requires however the availability and the implementation of a simulation model and the extraction of accurate data requires specific competences (Spanjers *et al.*, 1999).

In this section an approach is described for the complete fractionation of COD based on the measurement of OUR and without the need of modelling. This 10-step procedure is summarized in Table 3.2.2.

Table 3.2.2 Synthesis of the respirometric approach for the complete fractionation of total COD

Step	Parameter	Method
1	Total COD	Lab. analysis (APHA, AWWA and WPCF 1998)
2	Soluble COD (S)	Lab. analysis of soluble COD
3	Particulate COD (X)	As difference of 1 and 2: $X = \text{total COD} - S$
4	Biodegradable COD ($S_S + X_S$)	Respirometry
5	Soluble biodegradable COD (S_S)	Respirometry
6	Particulate biodegradable COD (X_S)	As difference of 4 and 5
7	Heterotrophic active biomass (X_{BH})	Respirometry
8	Autotrophic active biomass (X_{BA})	Considered as negligible
9	Soluble nonbiodegradable COD (S_I)	As difference of 2 and 5: $S_I = S - S_S$
10	Particulate nonbiodegradable COD (X_I)	As difference of 3, 6 and 7: $X_I = X - X_S - X_{BH}$

The biodegradable COD, subdivided into the readily (S_S) and slowly (X_S) biodegradable fractions, can be quantified by using respirometric tests, while the remaining inert fractions, X_I and S_I, are calculated as the difference of known values. Also the content of heterotrophic active biomass (X_{BH}) can be measured by respirometry. The proposed respirometric methods and the laboratory instrumentation used for tests are described below.

Description of instrumentation for respirometric tests

The OUR tests were carried out using a series of closed respirometers. A closed respirometer is made up of a temperature controlled 2 l reactor. Aeration and mixing are guaranteed by compressed air and magnetic stirrer. The revolution speed of the magnetic stirrer must avoid spontaneous reoxygenation of the mixed liquor. Dissolved oxygen was monitored by an oxymeter (OXI 340, WTW GmbH, Germany) connected to a data acquisition system. A scheme of the instrumentation used is shown in Figure 3.2.2. OUR is measured during programmed phases without aeration.

Evaluation of biodegradable COD by respirometry (step 4 of Table 3.2.2)

For the estimation of biodegradable COD ($S_S + X_S$) the respirometric test is carried out with 1–1.5 l of activated sludge in which an adequate amount of wastewater (about 0.5 l) and allylthiourea (ATU) are added. An example of the dynamics of OUR(t) versus time (respirogram) obtained after the addition of raw wastewater is shown in Figure 3.2.3.

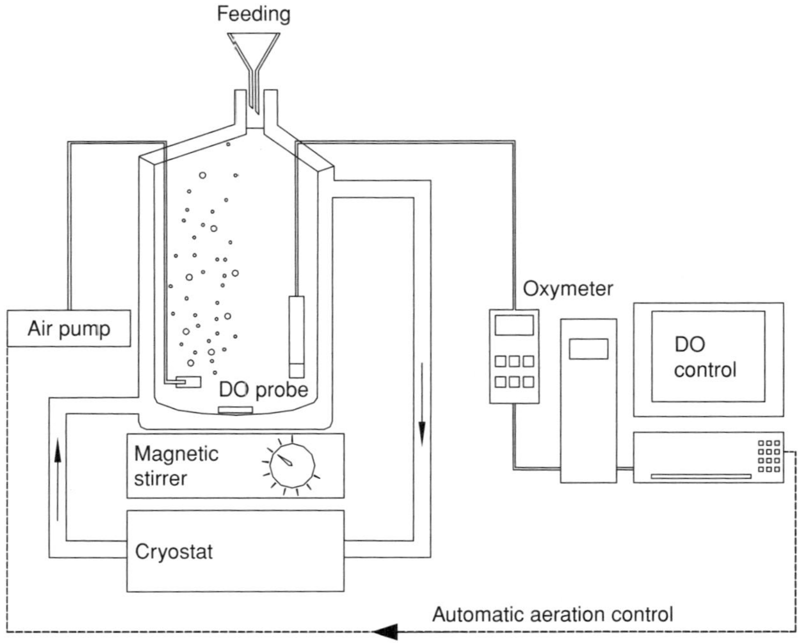

Figure 3.2.2 Scheme of the instrumentation utilized for the respirometric runs

At the beginning of the test the higher OUR values are due to the oxidation of readily biodegradable substrates, while successively, after the complete depletion of S_S, a gradual decrease of OUR is observed due to the consumption of slowly biodegradable compounds limited by hydrolysis. When all the biodegradable substrates are completely oxidized, the OUR values reach the endogenous respiration.

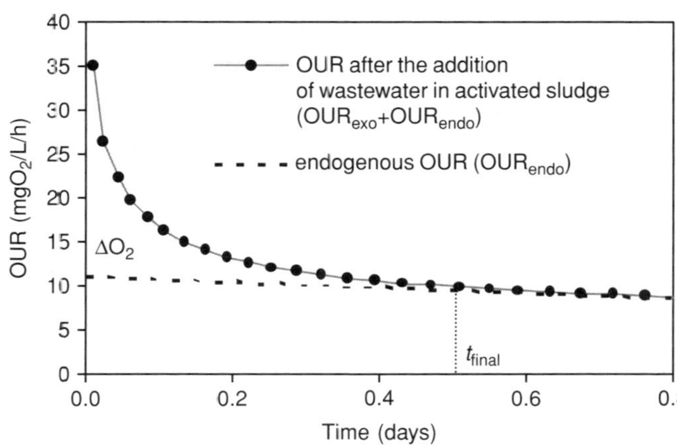

Figure 3.2.3 Respirogram obtained for activated sludge after the addition of municipal raw wastewater. The contributions from both OUR_{endo} and OUR_{exo} are indicated

The area between OUR_{exo} and OUR_{endo} (ΔO_2) represents the total oxygen consumed for the oxidation of biodegradable COD present in the added wastewater. The conversion from oxygen into the equivalent amount of COD is calculated by applying the following expression, in which the contribution of the biomass yield is subtracted (Ekama *et al*., 1986):

$$\Delta O_2 = \int_0^{t_{final}} OUR_{exo}(t)\,dt \qquad (mg\,O_2/l)$$

$$S_S + X_S = \frac{1}{1 - Y_H} \cdot \frac{V_{ww} + V_{as}}{V_{ww}} \int_0^{t_{final}} OUR_{exo}(t)\,dt \qquad (mg\,COD/l)$$

where V_{as} is activated sludge volume (l), V_{ww} is wastewater volume (l), Y_H is the yield coefficient, assumed equal to 0.67 mg COD/mg COD and t_{final} is the time corresponding to the complete oxidation of biodegradable COD in wastewater.

Evaluation of soluble biodegradable COD by respirometry (step 5 of Table 3.2.2)

A method for the estimation of S_S has been proposed by Xu and Hultman (Xu and Hultman, 1996), who put forward a method based on a calibration curve between a readily biodegradable substrate having a known COD (acetic acid or sodium acetate) and the oxygen demand for its removal. S_S in wastewater can be assessed from the measurement of the oxygen consumption and the conversion into COD by using the calibration curve. In particular, this technique allows the assessment of S_S concentration through a so-called 'single-OUR' method, because only an oxygen depletion curve is necessary and therefore the time required for the test is very short (Ziglio *et al*., 2001).

Calculation of particulate biodegradable COD (step 6 of Table 3.2.2)

Knowing the value of S_S the concentration of X_S in wastewater is obtained immediately as:

$$X_S = \left(\frac{1}{1 - Y_H} \cdot \frac{V_{ww} + V_{as}}{V_{ww}} \cdot \int_0^{t_{final}} OUR_{exo}(t)\,dt \right) - S_S \qquad (mg\,COD/l)$$

Evaluation of heterotrophic active biomass by respirometry (step 7 of Table 3.2.2)

For evaluating X_{BH} in wastewater the respirometric test has to be carried out only in the presence of wastewater, without any addition of activated sludge, according

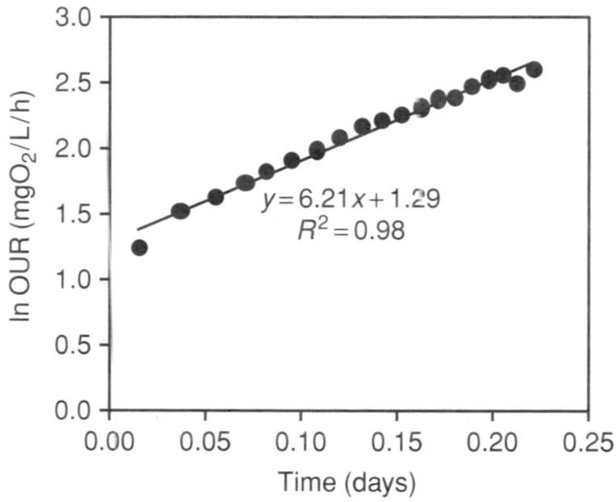

Figure 3.2.4 Results of the respirometric test for the estimation of X_{BH} in wastewater. The y-intercept is 1.29 mg O_2 1L/h and the OUR slope is $\mu_{H,max} - b_H = 6.21/\text{day}$ [where $\mu_{H,max}$ is the specific maximum growth rate (/day) and b_H is the decay rate (/day)]. $X_{BH} = 27.1$ mg COD/l

to the method proposed by Kappeler and Gujer (Kappeler and Gujer, 1992). At the beginning of the test the ratio S_0/X_0 (substrate/biomass) must be higher than 4 in order to reproduce the optimal organic load for nonlimiting bacterial growth. The value of X_{BH} is derived easily from the OUR dynamic during the exponential growth phase. By plotting ln OUR values versus time (Figure 3.2.4) the linear interpolation of the data allows to calculate the slope ($\mu_{H,max} - b_H$) and the y-intercept on the vertical axis. The specific decay rate (b_H) is assumed equal to 0.24 day^{-1}.

Finally the active heterotrophic biomass in wastewater is obtained by the following relationship:

$$X_{BH} = \frac{e^{(y-\text{intercept})} \cdot 24}{\frac{1-Y_H}{Y_H} \cdot (\text{slope} + b_H)} \qquad (\text{mg COD/l})$$

where Y_H is the yield coefficient for heterotrophic biomass, assumed equal to 0.67 mg COD/mg COD.

Calculation of inert COD (steps 9 and 10 of Table 3.2.2)

Finally, after the experimental determination of S_S, X_S and X_{BH}, the two remaining inert fractions of COD can be calculated immediately as difference. In particular the value of S_I is obtained as the difference from the soluble COD in wastewater and the value of S_S:

$$S_I = S - S_S \qquad (\text{mg COD/l})$$

Analogously, the value of X_I is obtained subtracting the biodegradable fraction X_S and the active biomass X_{BH} from the particulate COD:

$$X_I = X - X_S - X_{BH} \qquad (\text{mg COD/l})$$

With this 10-step procedure the whole fractionation of COD in wastewater indicated in Figure 3.2.1 is obtained.

3.2.2.3 COD Fractionation from Data of Conventional Analytical Monitoring in WWTPs

Alternatively to the respirometric approach, COD fractionation can be obtained through simple calculations by using data acquired during the conventional monitoring of WWTPs. The need to measure data in influent and effluent wastewater is the main limitation of this procedure, that is applicable only in the case of existing and fully monitored plants. In particular the following analytical parameters are required: COD, soluble COD and BOD_5 in influent wastewater and soluble COD in effluent wastewater, collected after secondary treatment. The 11-step procedure is summarized in Table 3.2.3.

Table 3.2.3 Synthesis of the approach for COD fractionation by using data from conventional analytical monitoring

No.	Parameter	Method
1	Total COD	Lab. analysis (APHA, AWWA and WPCF, 1998)
2	Soluble COD (S)	Lab. analysis of soluble COD
3	Particulate COD (X)	As difference of 1 and 2: $X = \text{total COD} - S$
4	BOD_5	Lab. analysis (APHA, AWWA and WPCF, 1998)
5	Biodegradable COD (S_S+X_S)	Conversion of the BOD_5 value
6	Soluble non biodegradable (S_I)	Lab. analysis of soluble COD in the final effluent after treatment
7	Soluble biodegradable COD (S_S)	As difference of 2 and 6: $S_S = S - S_I$
8	Particulate biodegradable COD (X_S)	As difference of 5 and 7
9	Particulate nonbiodegradable COD (X_I)	As difference of 3 and 8: $X_I = X - X_S$
10	Heterotrophic active biomass (X_{BH})	Considered as negligible
11	Autotrophic active biomass (X_{BA})	Considered as negligible

Calculation of biodegradable COD (step 5 of Table 3.2.3)

The biodegradable COD, equal to $S_S + X_S$, is obtained from the conversion of the BOD_5 value. This conversion has been recently evaluated in depth by Weijers (Weijers, 1999) and Roeleveld and van Loosdrecht (Roeleveld and van Loosdrecht, 2002), with the aim to apply a simplified procedure for the advanced fractionation of COD in numerous WWTPs in The Netherlands.

Firstly, the BOD_5 value is converted to the corresponding $BOD\infty$ value (that is, the oxygen consumption for $t = \infty$) introducing the first-order kinetic constent, k_{BOD}:

$$BOD\infty = \frac{BOD_5}{1 - e^{-5k_{BOD}}} \qquad (mg\ O_2/l)$$

By assuming a typical value for k_{BOD} equal to 0.23 day^{-1} at 20 °C (STOWA, 1996; Weijers, 1999; Metcalf and Eddy, 2003), this conversion gives:

$$BOD\infty = \frac{BOD_5}{0.68}$$

Secondly, the value of $BOD\infty$ is converted to an equivalent value of biodegradable COD by using a correction factor f_{BOD}, ranging from 0.1 to 0.2, with a typical value equal to 0.15 (Roeleveld and van Loosdrecht, 2002):

$$S_S + X_S = \frac{BOD\infty}{1 - f_{BOD}} \qquad (mg\ COD/l)$$

Calculation of soluble nonbiodegradable COD (step 6 of Table 3.2.3)

For low-loaded plants the soluble nonbiodegradable COD in influent wastewaster can be determined from the soluble COD in the treated effluent. The soluble inert fraction in influent wastewater can be considered as conservative in the WWTP, due to the fact that this fraction cannot be removed being not biodegradable and not settleable. Furthermore, the soluble COD concentration in the effluent from low-loaded WWTPs is made up only of nonbiodegradable compounds (in fact in the effluent $S_S = 0$). Therefore, by measuring the concentration of soluble COD in the effluent, the value of S_I in influent wastewater is known. This correlation suffers from an approximation: some authors have highlighted that the biological process in WWTP contributes to an additional production of soluble inert COD that is then discharged in the effluent. Therefore the value measured in the effluent could be higher than the value of soluble inert COD in the original raw wastewater (Orhon *et al.*, 1989; Sollfrank *et al.*, 1992).

Calculation of the remaing fractions (steps 7, 8 and 9 of Table 3.2.3)

After the estimation of $S_S + X_S$ and S_I as explained above, all the other fractions of COD can be calculated, as summarized in Table 3.2.3. In particular, the value of S_S is obtained by subtracting the known value of S_I from the soluble COD:

$$S_S = S - S_I \qquad (\text{mg COD/l})$$

Analogously, the value of X_S is obtained as the difference of two known values:

$$X_S = (S_S + X_S) - S_S \qquad (\text{mg COD/l})$$

Finally, the value of X_I is obtained as the difference between the particulate COD and the biodegradable fraction X_S:

$$X_I = X - X_S \qquad (\text{mg COD/l})$$

In this procedure both X_{BH} and X_{BA} are neglected.

3.2.2.4 A Case Study at Regional Level

COD fractionation was evaluated on a wide set of influent wastewaters of about 70 full-scale municipal WWTPs located in the Province of Trento (Italy). Capacities up to 150 000 PE have been considered, comprising also some cases smaller than 1000 PE. In this large number of plants, cases with discharges of industrial wastewater and plants located in tourist sites have been also considered.

For each plant the COD fractionation was calculated on the basis of the conventional data acquired during the routine monitoring of the plants according to the procedure described in the Section 3.2.2.3. The average results of COD fractionation, calculated for all 70 WWTPs, are summarized in Table 3.2.4.

The size of the municipalities and the consequent capacity of the WWTP affects the COD fractionation of raw wastewater, as shown in Figure 3.2.5. In particular, wastewaters of larger plants are characterized by a lower percentage of readily

Table 3.2.4 Results of COD fractionation of influent wastewater of 70 WWTPs (percentages are referred to total COD)

	S_S	S_I	X_S	X_I
Average (%)	32.1	3.1	46.9	17.9
Maximum (%)	50.3	14.0	66.9	33.1
Minimum (%)	12.2	1.2	27.6	8.3
Standard deviation (%)	9.2	1.7	8.6	5.2

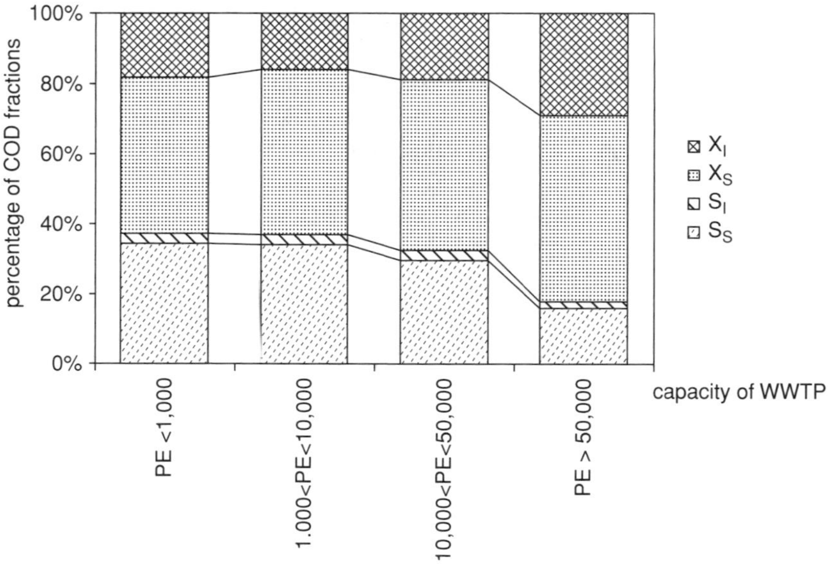

Figure 3.2.5 Comparison between the COD fractionation for WWTPs with different capacities (expressed as population equivalent, PE)

biodegradable COD (S_S) and a higher percentage of particulate inert COD (X_I) with respect to smaller plants. The reason is that large WWTPs are generally associated with a longer retention time of wastewater in sewerage where the biodegradation process starts, causing the consequent reduction of the readiy biodegradable fraction in raw wastewater. Furthermore, the increase in the percentage of inert COD and biorefractory compounds in municipal wastewater is generally attributed to discharges from industrial activities, that are higher in the larger WWTPs with industrial catchments with respect to the smaller WWTPs mostly located in rural regions.

With regards to the absolute value of COD concentration in the influent wastewater, its value ranges typically from 250 to 800 mg/l (Metcalf and Eddy, 2003). Three categories of influent COD concentration are considered and the results are indicated in Figure 3.2.6. COD fractionation is quite independent from the absolute COD concentration in wastewater. In fact, in all three categories of Figure 3.2.6 the percentages of COD fractions are very similar demostrating a negligible influence of the concentration.

3.2.3 ORGANIC MICROPOLLUTANTS

In the previous paragraph, organic compounds have been considered in aggregated parameters as COD. The main part of COD or BOD_5 is made up of organic matter of biological origin, but the measurement of COD comprises also organic micropollutants characterized especially by synthetic origin. Due to the potential health and environmental risks of these latter compounds, they have to be identified as

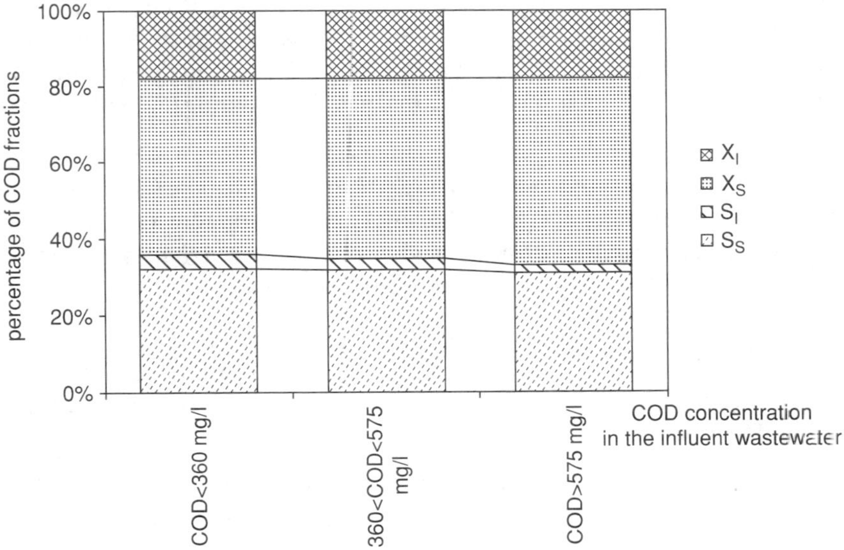

Figure 3.2.6 Comparison between the COD fractionation for WWTPs with different concentrations in the influent wastewater (expressed as mg COD/l)

individual parameters. The main limitation in the measurement of individual parameters is the large number of organic pollutants, which may enter the WWTP. More than 100 organic micropollutants can be found in influent and effluent municipal raw wastewater (Paxéus, 1996).

3.2.3.1 Categories of Organic Micropollutants

Organic micropollutants cover a wide range of substances and products. The main characteristics of some categories and the fate in WWTPs, that may be significantly different depending on the type of organic compounds, are summarized as follows:

(1) *Polycyclic aromatic hydrocarbons (PAHs)*. PAHs are produced from incomplete combustion of organic substances and are characterized by low biodegradability. During the treatment in WWTPs they are separated from water and concentrated in sludge; in fact, due to their low biodegradability complete oxidation is not possible and only physical removal from water to sludge is obtained.

(2) *Polychlorinated biphenyls (PCBs)*. PCBs are used for lubricants, emollients, impregnating agents or produced by combustion. Analogously to PAHs, PCBs in wastewater are separated from water in WWTPs and are transferred into sludge, due to the mechanism of absorption on solids and sludge.

(3) *Polychlorinated dibenzo-p-dioxins and dibenzofurans (PCDDs and PCDFs).* PCDDs and PCDFs are produced by thermal processes or incomplete combustion and released into the atmosphere and reach WWTPs due to deposition and run-off. Analogously to PCBs the degradation in WWTPs during biological treatment is negligible.

(4) *Plasticizing agents.* Among them, Di-(2-ethyhexyl)phthalate (DEHP) is used as emollient, antifoaming agent or emulsifier and it is found widely in municipal raw wastewater. Because of its lipophilic properties it is removed from water and concentrated in excess sludge produced in secondary treatment.

(5) *Surfactants and detergent residues.* Surfactants are used in washing and cleaning products and are always present in municipal raw wastewater. Linear alkylbenzene sulfonates (LASs) are commonly used in detergents and are biodegradable in aerobic processes.

(6) *Pharmaceutical products.* Medical substances are considered as micropollutants, due to the fact that they are developed with the intention of performing biological effects.

With regards to the quantitative measurements of organic micropollutants, their regular monitoring requires a high number of parameters to be analysed with complex and expensive laboratory procedures. The analytical measurement for the quantification of PAHs, PCBs, PCDDs and PCDFs in wastewater or sludge is very expensive and it is the reason for the scarce availability of these data in wastewater. Recently, the importance of these contaminants in urban wastewater and sludge has significantly diminished and there may be little practical or environmental benefit gained from adopting limits controlling PAHs, PCBs or PCDD/PCDFs (European Union, 2001).

Alternatively, the relatively easy analysis of adsorbed organic halogens (AOX) as an indicator of priority micropollutants can be applied for screening the wastewater quality. AOX is a mass parameter that includes many substances, released mainly from industrial processes, which are adsorbable and contain halogens (usually chlorine but also fluoride, bromine and iodine) that can envelop organic substances. These substances are often not easily degradable in WWTPs and highly toxic in the environment. Moreover, some AOX are known to have endocrine effects.

3.2.3.2 Treatability of Organic Micropollutants

The first three categories (PAHs, PCBs and PCDDs/PCDFs) are present in municipal wastewater as a consequence of the diffuse atmospheric deposition and the urban run-off during rainfall events. All these organic pollutants are characterized by hydrophobic properties, a low treatability and a high resistance in WWTPs.

The degradation of PCB, PCDD or PCDF in the environment is reviewed by Sinkkonen and Paasivirta (Sinkkonen and Paasivirta, 2000); because of the long half-lives it is not possible to biodegrade these persistent organic pollutants in WWTPs.

With regards to aromatic hydrocarbons, the complete biodegradation of three- and four-ringed PAHs (anthracene, phenanthrene, pyrene) occurs under both aerobic and anoxic conditions, in times between 12 h and 80 h operating with pseudomonad strains (McNally *et al.*, 1998). Probably in full-scale WWTPs, under real operating conditions limiting the kinetic rates, the required times might be longer, reducing the efficiency of PAH biodegradation during the biological process. Nevertheless, the concentrations of PAHs in final effluents are efficiently reduced in WWTPs because of their accumulation in sludge solids. In fact, the mechanisms of PAH removal in WWTPs are: (1) through adsorption on sludge in the case of the higher molecular weight PAHs; and (2) through biodegradation and/or volatilization in the case of the lower molecular weight PAHs (Samara *et al.*, 1995; Manoli and Samara, 1999).

Plasticizing agents, surfactants and detergent residues, mainly produced by domestic activities, can be biodegraded in WWTPs during the secondary biological treatment. In particular aerobic processes, such as activated sludge, are suitable for detergent removal, while in anaerobic treatment the removal of detergents and DEHP is not achieved. Removal of anionic surfactants, LASs, in WWTPs ranging from 70 to 99 % is referred to by Holt *et al.* (Holt *et al.*, 1998) and Prats *et al.* (Prats *et al.*, 1997). The maximum degradation of surfactants takes place in the aeration tank, while an amount of nonbiodegraded surfactants, around 16 % of the influent mass, is found in sludge (Field *et al.*, 1995), indicating that the mechanism of surfactant removal is mainly biodegradation and then sorption on sludge. Thanks to their easy treatability, low concentrations of these organic compounds are generally measured in treated effluents discharged in surface waters. However, residues remain in the final treated effluents due to the large amount of surfactants present in the influent wastewater.

The biodegradability of pharmaceutical compounds or residues of medicines can differ significantly and therefore their treatability through WWTPs varies considerably depending on the type of compound. Some pharmaceutical substances, such as antibiotics, are persistent during treatment in WWTPs. In contrast, the concentration of estrogens (natural and synthetic) can be reduced in conventional WWTPs. In fact the load of estrogenic activity of the wastewater is reduced by about 90 % in WWTPs, while the estrogenic activities found in sludge are negligible, being 3 % (Körner *et al.*, 2000). However, residues remain after the treatment and are discharged with the effluent in surface waters; this discharge can impact on the estrogenic activity in aquatic life. A list of the fate of medical compounds during the sewage treatment is referred to in Halling-Sørensen *et al.* (Halling-Sørensen *et al.*, 1998).

Summing up the treatability of organic micropollutants in WWTPs, the main mechanisms involved in their removal are: (1) during the biological secondary treatment and especially under aerobic conditions, some organic compounds, such as surfactants, can be biodegraded to a certain extent; (2) other organic micropollutants, characterized by low biodegradability and tendency for adsorption on particulate matter, are separated and concentrated in primary or secondary sludge through sorption mechanisms depending on the properties of the organic species. As a consequence the effluents discharged in receiving water bodies show a significant reduction in the micropollutants concentration, while sludge may contain most of the organic contamination present in the influent raw wastewater. The fate and the effective treatability of organic micropollutants is not easy to predict, because of the very large number of different organic species that may be present in municipal raw wastewater and the complex aspects of the physico-chemical sorption mechanisms onto sludge solids.

3.2.4 NUTRIENTS: NITROGEN AND PHOSPHORUS

The removal of nutrients, N and P, from wastewaters is of primary importance, especially to reduce eutrophication of sensitive water courses.

The main process for the removal of P from wastewaters is chemical precipitation, while the biological process, carried out in anaerobic–aerobic systems, is currently applied only in a few cases at full-scale. Efficient P removal is achieved by using common precipitants such as aluminium sulfate (alum) or ferric chloride. The chemical precipitation allows the simultaneous enhancement of the removal of potentially toxic micropollutants, for example increasing the transfer and accumulation of metals to sludge.

With regards to N removal, the processes of nitrification and denitrification theoretically allows the complete removal of N from wastewater, depending on an adequate N/COD ratio in wastewater. From a practical point of view the removal efficiency performed in WWTPs depends in general on the required discharge limits imposed by the law. Most of the incoming N can be treated and removed in conventional low-loaded WWTPs designed for nitrification/denitrification and the effluent N concentration is mainly due to: (1) residues of ionic forms, NH_4, NO_3 or NO_2; (2) the N content in suspended solids effluent from final sedimentation; and (3) the organic N in soluble nonbiodegradable form. For the estimation of the latter, the fractionation of N in the influent wastewater is required as discussed in the next section.

3.2.4.1 Fractions of Nitrogen and their Treatability

The fractionation of N has generated increasing interest when nutrient removal is a priority in WWTPs.

Most N in urban wastewater is represented by ammonia, while the remaining part is organic nitrogen, nitrite and nitrate often being negligible. Organic nitrogen can be subdivided into biodegradable, nonbiodegradable, soluble and particulate fractions, exactly in the same manner as for COD (Section 3.2.2.1). The scheme of total Kjeldhal nitrogen (TKN) fractionation (comprehensive for ammonia and organic nitrogen) is indicated in Figure 3.2.7. The symbols used in the scheme are those adopted in ASM models.

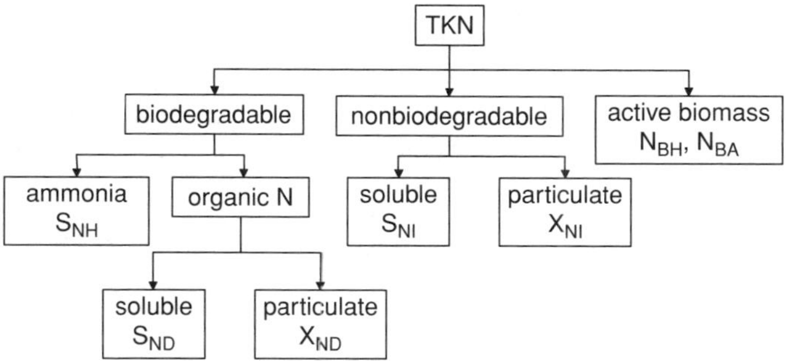

Figure 3.2.7 Scheme of TKN fractionation in wastewater

On the basis of this fractionation the TKN concentration in wastewater is made up of the following:

$$\mathrm{TKN} = \mathrm{S_{NH}} + \mathrm{S_{ND}} + \mathrm{X_{ND}} + \mathrm{S_{NI}} + \mathrm{X_{NI}} + \mathrm{N_{BH}} + \mathrm{N_{BA}}$$

The terms and the fate in WWTPs of these seven fractions are:

- *Ammonia (S_{NH})*: determined by conventional chemical analysis (APHA, AWWA and WPCF, 1998); it is nitrified in low-loaded WWTPs and when the volume of the biological reactor is sufficient the complete oxidation of ammonia can be achieved.
- *Soluble biodegradable N (S_{ND})*: it is the content of nitrogen in the rapidly biodegradable COD; it is rapidly hydrolysed into ammonia.
- *Particulate biodegradable N (X_{ND})*: its hydrolysis requires a longer time, analogously to the hydrolysis of the slowly biodegradable COD, $\mathrm{X_S}$.
- *Soluble inert N (S_{NI})*: it cannot be removed in WWTPs, neither by biological process nor through sedimentation; this behaviour is analogous to that of S_I; the $\mathrm{S_{NI}}$ concentration will be found in the final effluent after secondary treatment without any modification.
- *Particulate inert N (X_{NI})*: it is removed from wastewater and transferred into sludge together with the particulate inert fraction of COD, $\mathrm{X_I}$.
- *N in biomass (N_{BH}, N_{BA})*: it is the content of nitrogen in the microbial biomass, both hetrotrophic and autotrophic.

In order to obtain the fractionation of TKN, the application of respirometric tests is quite difficult and further reseach is needed to obtain suitable procedures. Except for ammonia, determined by chemical analysis, the fractions of organic N are calculated from the fractionation of organic matter (COD), when the latter is available. This procedure is based on the assumption of a specific N/COD ratio for each COD fraction, as indicated in Table 3.2.5. In practical applications the value of $\mathrm{N_{BA}}$ is neglected.

Table 3.2.5 N/COD ratios and calculations of the single fractions of TKN

Calculation	Symbol of N/COD ratio	Values of N/COD ratios (g N/g COD)	
		Typical value	Range
$S_{ND} = i_{NSS} \cdot S_S$	i_{NSS}	0.02	—
$X_{ND} = i_{NXS} \cdot X_S$	i_{NXS}	0.04	0.02–0.06
$S_{NI} = i_{NSI} \cdot S_I$	i_{NSI}	0.01	0.01–0.02
$X_{NI} = i_{NXI} \cdot X_I$	i_{NXI}	0.03	0.01–0.06
$N_{BH} = i_{XB} \cdot X_{BH}$	i_{XB}	0.086	—

3.2.5 METALLIC COMPOUNDS

The concentration of metals in raw wastewater can differ significantly depending on the domestic, commercial or industrial activities collected by the sewerage. The main interest is in metals characterized by potential toxic impact on health or the environment, such as Cd, Cr , Cu, Hg, Ni, Pb and Zn. The load of these components at the inlet of a WWTP can be several times greater in industrial sites than in residential areas far from industrial activities. Urban run-off during storm events is also a source of metals and other pollutants, and contributes to the total influent load into a WWTP.

3.2.5.1 Treatability of Metallic Compounds

Metals in raw wastewater are removed in WWTPs through two different mechanisms:

- *Primary sedimentation*: metals are separated as insoluble precipitates or adsorbed on settled particulate matter and then extracted with primary sludge. In contrast the removal of metals in soluble form is negligible.
- *Secondary treatment*: during the biological process metals are integrated into activated sludge or biofilm (adsorbed on flocs or in extracellular polymers). They are removed at the same efficiency as the sludge solids in the secondary settler and extracted together with the excess sludge.

Some values for metal removal in primary and secondary treatments are summarized in Table 3.2.6 (European Union, 2001).

Similar patterns of removal percentages are observed in primary and secondary treatments. Lower removal is observed in both cases for Ni due to its high solubility that limits the presence of Ni in the particulate matter and sludge. In contrast Pb, one

Table 3.2.6 Percentage of metals removed in WWTPs, calculated with respect to the concentration in the influent raw wastewater

Metal	Removal in primary treatment (%)	Removal in primary + secondary treatment (%)
Ni	24	40
Cd	40	65–75
Cr	40	75–80
Zn	50	70–80
Cu	50	75–80
Hg	55	70–80
Pb	55	70–80

of the least soluble metals, shows higher removal in both the primary and secondary stages. For the majority of metals a significant percentage of the influent load, up to 70–80 %, is transferred into primary and secondary sludge. As a consequence the concentration of metals in dry sludge (measured as TSS) reaches levels of several thousand mg/kg TSS, about 1000 times higher than the concentration of metals in raw wastewater.

In synthesis, the majority of metals entering the WWTPs with the raw wastewater is transferred to the sludge extracted from primary and secondary treatments. Depending on the metal solubility, a smaller amount, ranging from 20 to 40 % (60 % only for Ni), is however discharged in water bodies with the final effluent. With regards to the fate of sludge separated by settlers, the stabilization processes through aerobic or mesophilic anaerobic digestion cause the biological reduction of the volatile solids (30–50 %) and the specific metal content increases, metals being conserved during stabilization. Due to the presence of metals the final disposal of sludge may be problematic especially in the case of accumulation in soils interfering with the long-term sustainable use of sludge on land.

For the prediction of the removal of metals from raw wastewater and the partitioning into final effluent and sludge, mechanistic approaches have been proposed (Monteith *et al.*, 1993). On the basis of influent wastewater characterization (flow rate, metal concentrations) and the layout of the WWTP (unit volumes, operational conditions) the metal concentration in primary sludge, secondary sludge and final effluent can be predicted. The calculation is performed on the basis of mass balances by considering the main chemical and physical mechanisms (precipitation of soluble metals into a settleable form, sorption onto settleable solids, surface volatilization). In the model the mass of primary and biological sludge produced by primary sedimentation and secondary treatment is calculated and partitioning coefficients are introduced in the model for the estimation of the metal concentrations in the soluble and solid phases. A similar approach can be applied also for estimating the fate of organic contaminants instead of metals in WWTP. Modelling can be performed both under steady-state or dynamic conditions.

3.2.6 FINAL CONSIDERATIONS

WWTPs are effective in the reduction of most pollutants present in wastewater (such as organic matter, nutrients, potentially toxic elements or some micropollutants), before the discharge of the treated effluents in surface waters. In WWTPs several biological and physico-chemical processes can be implemented, but the main pathways for pollutants removal are: (1) the biological oxidation by activated sludge or biofilm systems; or (2) the accumulation of contaminants in excess sludge.

In this chapter the main categories of pollutants present in influent wastewater and their fate in WWTPs has been discussed. The assessment of the treatability of a specific wastewater in WWTPs is strictly dependent on the fate of contaminants in the treatment stages. The amount of pollutants removed in conventional WWTPs or passing into the effluent has been indicated depending on the category of pollutants, separated into organic compounds, organic micropollutants, nutrients and metallic compounds. These main categories were identified in order to make an aggregation of the large number of individual pollutants; a much longer and detailed report would be required for the explanation of the fate of each single element. Therefore the present description is not exhaustive for understanding the fate of each single compound; the objective of this chapter is to explain the main pathways in WWTPs for macro-categories of pollutants.

The wastewater characterization can be investigated more or less in depth depending on the particular needs in management of WWTPs, the requirement for discharge, and the practicalities of operators that make the measurements. The increasing detail in characterization and control of effluent wastewater from WWTPs coupled with the more stringent limits for discharge in receiving water bodies, necessitates a more complex and sophisticated monitoring. This causes considerable additional effort and expense to obtain a high degree of knowledge about the type and the concentrations of pollutants and micropollutants in influent and effluent wastewaters.

With regards to COD fractionation the routine measurement of all the parameters indicated in Section 3.2.2.1, according to the respirometric approach (described in Section 3.2.2.2), is extremely time-consuming because of the time required for the respirometric tests and the time need for data elaboration. Therefore, COD fractionation could be done only occasionally in a WWTP and the percentages obtained can be assumed as typical for a specific wastewater. Of course a periodic validation of fractionation is required. Alternatively, the simplified procedure described in Section 3.2.2.3 can be applied, which is more approximate but is advantageously fast to use. A more detailed characterization, performed by using the respirometric approach, could be required in order to observe daily, weekly or seasonal variation or fluctuation occurring in the COD fractions. In the case of industrial sources, shock loadings are by their nature difficult to predict.

In the case of N fractionation, the calculation described in Section 3.2.4.1 can be easily done thanks to its dependence on COD fractionation.

In general a good characterization of COD and N in the influent wastewater is very important to understand the fate of these components in WWTPs and to predict the quality of the effluent wastewater before discharge in receiving water bodies.

With regards to metals or nutrients, they are routinely measured in wastewater and sludge and an extensive knowledge about these components is usually available in WWTP management. The measurement is done often routinely in influent and effluent wastewater due to the relative ease of the analysis and the moderate expense involved.

In contrast, organic micropollutants, such as PAHs, PCBs, PCDD/PCDFs or pharmaceuticals, are rarely monitored because of the high cost of analysis and the need for specialized laboratories and, sometimes, the lack of unified and standardized methodologies. Furthermore, the limitation in the evaluation of the fate of organic micropollutants and potentially toxic elements is mainly related to the lack of studies on mass balance in WWTPs and with regards to partitioning in water and sludge. Further research is needed to improve knowledge in this field.

REFERENCES

APHA, AWWA and WPCF (1998) *Standard Methods for the Examination of Water and Wastewater*. American Public Health Association, American Water Works Association and Water Environment Federation, Washington DC, USA.

Ekama, G.A., Dold, P.L. and Marais, G.v.R. (1986) *Water Sci. Technol.*, **18**(6), 91–114.

European Union (2001) *Pollutants in Urban Wastewater and Sewage Sludge*. Office for Official Publications of the European Communities, Luxembourg.

Field, J.A., Field, T.M., Poiger, T., Siegrist, H. and Giger, W. (1995) *Water Res.*, **29**(5), 1301–1307.

Gujer, W., Henze, M., Mino, T. and van Loosdrecht, M.C.M. (1999) *Water Sci. Technol.*, **39**(1), 183–193.

Halling-Sørensen, B., Nors Nielsen, S., Lanzky, P.F., Ingerslev, F., Holten Lützhøft, H.C. and Jørgensen, S.E. (1998) *Chemosphere*, **36**(2), 357–393.

Henze, M. (1992) *Water Sci. Technol.*, **25**(6), 1–15.

Henze, M., Grady J., C.P.L., Gujer, W., Marais, G.v.R. and Matsuo, T. (1987) Activated Sludge Model No. 1. IAWQ Scientific and Technical Report No. 1, London, UK.

Holt, M.S., Fox, K.K., Burford, M., Daniel, M. and Buckland, H. (1998) *Sci. Total Environ.*, **210/211**, 255–269.

Kappeler, J. and Gujer, W. (1992) *Water Sci. Technol.*, **25**(6), 125–139.

Körner, W., Bolz, U., Süßmuth, W., Hiller, G., Schuller, W., Volker, H. and Hagenmaier, H. (2000) *Chemosphere*, **40**, 1131–1142.

Mamais, D., Jenkins, D. and Pitt, P. (1993) *Water Res.*, **27**, 195–197.

Manoli, E. and Samara, C. (1999) *J. Environ. Qual.*, **28**(1), 176–186.

McNally, D.L., Mihelcic, J.R. and Lueking, D.R. (1998) *Environ. Sci. Technol.*, **32**, 2633–2639.

Metcalf and Eddy. (2003) *Wastewater Engineering. Treatment and Reuse,*. 4th Edn. McGraw-Hill, New York.

Monteith, H.D., Bell, J.P., Thompson, D.J., Kemp, J., Yendt, C.M., Melcer, H., (1993) *Water Environ. Res.*, **65**(2), 129–137.

Orhon, D., Artan, N. and Cimsit, Y. (1989) *Water Sci. Technol.*, **21**(4–5), 339–350.
Orhon, D., Ateş, E., Sözen, S. and Ubay Çokgör, E. (1997) *Environ. Pollut.*, **95**(2), 191–204.
Paxéus, N. (1996) *Water Res.*, **30**(5), 1115–1122.
Prats, D., Ruiz, F., Vázquez, B. and Rodriguez-Pastor, M. (1997) *Water Res.*, **31**(8), 1925–1930.
Roeleveld, P.J. and van Loosdrecht, M.C.M. (2002) *Water Sci. Technol.*, **45**(6), 77–87.
Samara, C., Lintelmann, J. and Kettrup, A. (1995) *Toxicol. Environ. Chem.*, **48**(1–2), 89–102.
Sinkkonen, S. and Paasivirta, J. (2000) *Chemosphere*, **40**, 943–949.
Sollfrank, U., Kappeler, J. and Gujer, W. (1992) *Water Sci. Technol.*, **25**(6), 33–41.
Spanjers, H, Takács, I. and Brouwer, H. (1999) *Water Sci. Technol.*, **39**(4), 137–145.
Spanjers, H. and Vanrolleghem, P. (1995) *Water Sci. Technol.*, **31**(2), 105–114.
STOWA. (1996) Methoden voor influentkarakterisering (in Dutch). STOWA Report 96–08, STOWA, Utrecht, The Netherlands.
Vanrolleghem, P.A., Spanjers, H., Petersen, B., Ginestet, P. and Takacs, I. (1999) *Water Sci. Technol.*, **39**(1), 195 – 215.
Weijers, S.R. (1999) *Water Sci. Technol.*, **39**(4), 177–184.
Xu, S. and Hultman, B. (1996) *Water Sci. Technol.*, **33**(12), 89–98.
Ziglio, G., Andreottola, G., Foladori, P. and Ragazzi, M. (2001) *Water Sci. Technol.*, **43**(11), 119–126.

3.3 Toxicity Evaluation

Martijn Devisscher, Chris Thoeye, Greet De Gueldre and **Boudewijn Van De Steene**

3.3.1 INTRODUCTION

Under the Urban Wastewater Treatment Directive 91/271/EEC, the quality of effluents has been based on the monitoring of global chemical parameters, such as BOD (biological oxygen demand), COD (chemical oxygen demand) or TSS (total suspended solids). Wastewaters from various origins may contain compounds, toxic to the aquatic ecosystem, or even to the biocommunity responsible for the treatment of the wastewater. These toxic effects are insufficiently expressed in the currently practiced measurements.

Wastewater Quality Monitoring and Treatment Edited by P. Quevauviller, O. Thomas and A. van der Beken

Although some countries impose toxicity tests on effluents, there is currently no general European legal framework that systematically prescribes toxicity tests on effluents. Nevertheless, it is expected that the role of toxicity tests will become more important in the near future. Indeed, the European Union Water Framework Directive 2000/60/EC places more emphasis on the reduction of discharges of toxic elements, and the Integrated Pollution Prevention and Control Directive (96/61/EC), coming into effect by October 2007, is based on a permit system requiring the use of best available technology (BAT). In this, toxicity measurements may play an important role.

This chapter presents an overview of the common toxicity detection methods in use today. The discussion is limited to 'conventional' toxicity tests. In recent years, there has been increased concern over the release of pharmaceutically active compounds, personal care products and endocrine disrupting compounds into the environment. These compounds occur in low concentrations in the environment and are unlikely to cause acute toxicity. Highly sensitive bioassays have been developed to screen wastewater effluents on their (anti-)estrogenicity, (anti-)androgenicity, mutagenicity and cytotoxicity. Developments in these fields are extensive, evolve fast and deserve separate chapters in their own right.

However, we have limited the discussion to tests that are most relevant to the operation of wastewater treatment plants (WWTPs): the detection of toxic influents that can disturb the treatment process, and of toxic compounds in the effluent, which may be an indication of diminished treatment efficiency.

3.3.2 NEED FOR TOXICITY MEASUREMENTS

Toxic compounds are present in wastewater from various sources. In many countries in Europe, industrial plants are connected to the sewer. Industrial wastewaters can contain large amounts of toxic material, such as heavy metals, or synthetic chemicals and their waste products. These pollutants can even be present after conventional wastewater treatment (Paxeus, 1996).

Also purely domestic wastewater can contain toxic elements. Domestic discharges can contribute toxins from consumer products (e.g. cleaning products) or liquid wastes. Urban run-off may contain leachates or organic pollutants deposited from the atmosphere onto paved surfaces. In combined sewer systems this run-off is also introduced into the sewer system. Other known sources of potentially toxic compounds include commercial premises such as health establishments, small manufacturing industries or catering/hotel enterprises. It is obvious that also illegal discharges to the sewer represent a potential source of toxicity.

Chemical analyses alone are insufficient for assessing the toxicity of a wastewater. In the first place, the toxic compounds may be unknown. Indeed, the composition of wastewater is traditionally expressed in nonspecific terms such as BOD, COD or TOC (total organic carbon). These rather general measures reflect the general poor knowledge of the exact composition of wastewaters. Even if an exact composition of the wastewater is known, it is impossible to have a comprehensive overview on all compounds that are effectively present in the wastewater upon arrival at the treatment plant or in the environment. Several transformations may occur and create additional toxic content. Physico-chemical transformations may be occur, e.g. under the influence of sunlight UV, and toxic metabolites may originate via biodegradation, for example during storage in cesspits, during sewer transport or in activated sludge treatment.

In addition to the presence of unknown compounds, the (eco)toxicity of the known components may not be well documented. Although databases of such data exist (e.g. ECOTOX:http://www.epa.gov/ecotox/), important gaps remain. The lack of this kind of information on thousands of chemicals on the market today has been acknowledged by the European Union, and has prompted the REACH (Registration, Evaluation, Authorisation and Restrictions of Chemicals) proposal (CEC, 2001). The goal of this proposal is to secure data on and regulate some 30 000 chemicals produced in excess of 1 ton for which there is limited information with regard to toxicity and environmental effects. These data will expand the knowledge on toxic effects of pure compounds.

However, even when all toxic components in a wastewater have been identified, and detailed ecotoxicity information would be available for each of these components, an additional difficulty is the assessment of the effect of complex mixtures. Interaction of the compounds with each other, with the wastewater matrix or with the environment may result in synergistic or antagonistic effects, the matrix may render certain compounds biologically unavailable or may even increase toxicity (Hernando *et al.*, 2005).

A more direct measure of toxicity consists of submitting the whole complex mixture to a toxicity test. Although interactions with the final environment are not modelled precisely, it is a measure of the resultant toxicity of the complex wastewater mixture, integrating the combined effect of known and unknown toxic components and their interactions with the wastewater matrix. This type of testing is known in the USA as WET (whole effluent testing; US EPA, 1994). and in the UK as DTA (direct toxicity assessment; Tinsley *et al.*, 2004).

3.3.3 INFLUENT VS EFFLUENT TOXICITY OF WASTEWATER

The first major distinction to be made is whether the wastewater is monitored *before* or *after* treatment. We will refer to these techniques as influent toxicity monitoring and effluent toxicity monitoring, respectively.

This distinction is different because both the goal and requirements, and therefore the adopted methods differ whether the wastewater is monitored before or after treatment.

3.3.3.1 Influent Toxicity Evaluation

These tests have the intention to protect the biological wastewater treatment process from the effect of toxic influents. Although Annex 1 of the Urban Wastewater Treatment Directive already states that 'Industrial wastewater entering collecting systems and urban wastewater treatment plants shall be subject to such pretreatment as is required in order to . . . ensure that the operation of the wastewater treatment plant and the treatment of sludge are not impeded', these tests are not commonly imposed by regulators. The tests used are sometimes referred to as upset early warning devices (UEWDs; Love and Bott, 2000). The sensitivity of these tests should be representative for the biocommunity of the wastewater treatment process. This sensitivity can differ greatly from that of the receiving ecosystem.

3.3.3.2 Effluent Toxicity Evaluation

The purpose of effluent toxicity evaluation is to assess the effect of a certain wastewater on the receiving waters. The methods used are essentially the same as those used for ecotoxicity testing of pure compounds. Effluent toxicity tests are imposed by some discharge consents and have been extensively studied and standardized. The conventional approach is the use of bioassays. In these tests, the biological response of a certain bioindicator species is monitored in response to the wastewater to be tested. These bioassays can be further subdivided according to the species involved, the duration (acute/chronic toxicity test) or to the effect on the indicator organism (mortality, reproduction, motility). The requirements of these tests are a high sensitivity and representativity for the receiving ecosystem.

Although the distinction between influent and effluent toxicity is clear, it is evident that there is a strong link between the two. The effluent of an industrial treatment plant may be part of the influent to a municipal plant, and highly toxic substances in the influent may inhibit the treatment process in such an amount, that the toxic compounds break through to the effluent to cause effluent toxicity.

3.3.4 UNITS

Central to (eco)toxicity evaluation is a dose–effect relationship. Since bioavailability of a compound introduced in wastewater differs greatly for each individual compound, test species and wastewater matrix, the exact dose imposed on the test organism is difficult to quantify. Therefore, in aquatic toxicity testing, a concentration–effect relationship is considered, relating the concentration in the wastewater to the effect on the test organism. This relationship becomes evident in

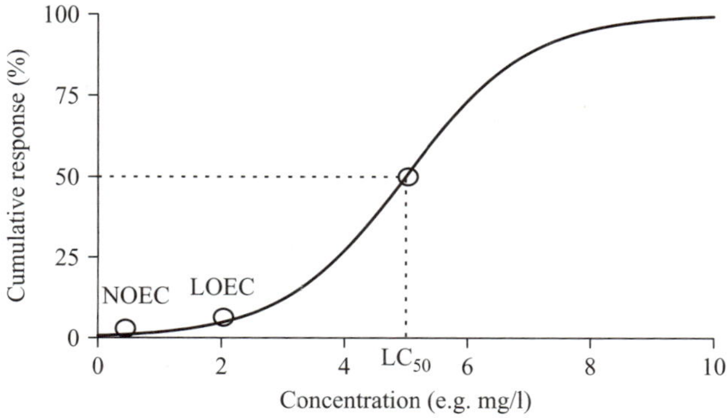

Figure 3.3.1 Sigmoidal response curve. (Adapted from Connell *et al.*, 1999 with permission from Blackwell Publishing)

the commonly used units for ecotoxicity:

- EC_{50}: The concentration at which 50% of the effect is observed.
- LOEC: Lowest observable effect concentration, i.e. the lowest concentration at which an effect can be observed.
- NOEC: No observable effect concentration, i.e. the highest concentration at which no effect can be observed.

The term concentration, in the context of whole effluent testing, refers to dilution series of the original wastewater, ranging from 0 to 100 % of the wastewater.

These measures are graphically represented in Figure. 3.3.1.

(Eco)toxicity is determined by studying quantifiable effects. The effects studied are specific to each toxicity test. A commonly observed effect is mortality (lethal effect). In this case, the term LC_{50} is used rather than EC_{50}. This determines the concentration at which 50 % mortality is observed. Another commonly used measure is IC_{50} which is the concentration at which 50 % inhibition of a certain activity (e.g. light emission) is observed.

There is no such thing as a EC_{50} of a certain compound. Toxicity is a measurement of an effect to a certain organism or community of organisms. It is therefore important that the test method is specified together with the EC values.

3.3.5 SOURCES OF TOXICITY

Influents of industrial WWTPs may contain a large variety of toxic compounds. It is practically impossible, and certainly beyond the scope of this chapter, to give a comprehensive overview of possible toxicity sources, given the variety of industrial processes, and hence waste products existing today.

Toxic compounds may originate in industrial plants directly, or by biodegradation of production or waste chemicals. These may break through the industrial wastewater treatment process because of plant upsets or reduced treatment efficiency, or because these compounds are simply left untouched by the treatment process. These compounds subsequently represent a cause of toxicity to receiving waters, or when the industry is connected to the sewer, to the receiving municipal WWTP.

Toxicity may also originate from domestic sources. In the first place, essentially all chemicals on the market today are potential sources of toxicity. Examples are cleaning products, personal care products, pharmaceuticals, or biocides, available on the market today. Several commercial sources can be identified to contribute to wastewater toxicity. For example, small manufacturing industries with metal/vehicle related industries, health establishments and hotel/catering enterprises are important sources of contamination of urban wastewater with potentially toxic elements. In combined sewer systems, storm water flows can contribute toxins from leachates, paved surface wash-off containing residues from tyre and brake-lining wear, or heavy metals from potable water ducts, painted surfaces or roofing materials (Thornton *et al.*, 2001).

Toxicity furthermore can originate in the treatment process itself. Firstly, a poor breakdown of conventional pollutants (e.g. BOD and nitrogen) can have an adverse effect on toxicity reduction, since toxic components that are otherwise decomposed by the normal carbon degradation pathways also suffer from treatment deficiencies. Sometimes, selected effluents from e.g. the food industry, may be added as a carbon source to enhance nitrogen removal. It is important to screen these streams for potentially toxic by-products before introduction into the treatment process. Also, chemical additives used in wastewater and sludge treatment such as coagulants, flocculant aids, or disinfectants or chemicals for phosphorus precipitation, when not dosed in an adequate manner, can form serious threats to the health of the biocommunity and the receiving ecosystem.

When persistent toxicity is observed, identification of the source is necessary. Toxicity tests on strategic locations in the wastewater transport system can be used to track down the source of toxicity (Geenens and Thoeye, 1998). Toxicity tests are also extensively used in toxicity identification evaluation (TIE) procedures (US EPA, 1991). These procedures are intended to identify the sources of toxicity, by testing toxicity on parts of the sample, that have undergone laboratory manipulations. These manipulations include for example pH adjustment, addition of chelating agents such as EDTA, or addition of reductants. The difference in toxicity observed before and after these manipulations can yield clues regarding the sources of toxicity. For example, disappearing toxicity after pH decrease indicates the presence of a pH dependent toxicant (a well-known example is ammonia, with the undissociated form being the main toxic agent).

3.3.6 TOXICITY TESTING

3.3.6.1 Influent Toxicity

The goal of influent toxicity testing is the protection of the biocommunity of the wastewater treatment system against toxic influents.

Of all biological treatment processes, suspended growth activated sludge is the most widespread. In these processes, the wastewater is brought into close contact with a concentrated suspension of micro-organisms, which degrade the pollutants by various biochemical pathways. After treatment, these micro-organisms are separated from the treated effluent, and are reused. Despite recent developments in membrane-based separation, the existing patrimonium and the majority of newly built plants still perform this separation process by gravitational settling. An overview of the conventional activated sludge treatment process is given in Figure. 3.3.2.

Toxic shocks can be expressed in various ways. They can result in inhibition or inactivation of certain micro-organisms that perform the biological degradation of the pollutants. This in turn results in a reduced treatment efficiency of the plant, and possibly violations of effluent consents.

A more serious effect is the possible complete loss of viability of the organisms. Although rare, examples exist of total loss of viable biomass in the treatment plant. Treatment plants may take weeks to recover from such an event, and restoring treatment capacity is very costly, since it involves disposing large volumes of intoxicated sludge, and re-seeding the system with micro-organisms.

Loss of treatment capacity can also be the result of deflocculation resulting in sludge washout (Geenens and Thoeye, 1998). Deflocculation is the breakup of flocs of micro-organisms into smaller fragments. As these have a larger specific surface area, they settle more slowly, and cannot be removed by gravitational settling.

It is important that the sensitivity of the test be representative for the treatment plant organisms. The sensitivity of the biomass to toxic substances is inevitably lower than that of the receiving environment. Indeed, otherwise there would be no breakdown of these substances in the plant. For this reason, tests designed for effluent toxicity testing are likely to be too sensitive for application to influents, and will give rise to false alarms (Gutiérrez *et al.*, 2002).

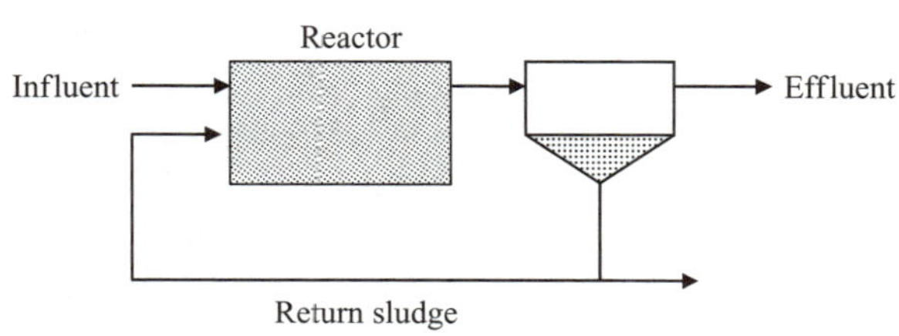

Figure 3.3.2 Conventional activated sludge treatment

Activated sludge is a complex ecosystem of hundreds of micro-organisms. As it is the case for effluent toxicity testing, a single test species is insufficient to fully assess toxicity to the biocommunity. Test batteries form a potential solution to this, provided that the tests, in addition to being representative, yield complementary results (Ren and Frymier, 2004).

An additional difficulty is the fact that adaptation mechanisms can reduce the sensitivity of the sludge community to certain compounds. For example, phenolics, cyanides and thiocyanates are known to be toxic for biological treatment systems (Blum and Speece, 1991). Grau and Da-Rin (Grau and Da-Rin, 1997) reported serious municipal plant upsets as a response to phenol concentrations in the influent. Nevertheless, certain wastewaters, such as those from cokes plants, contain high amounts of these components, and are adequately treated by activated sludge plants.

Another important aspect of early warning systems is the need for short-term testing, and preferably on-line instruments. It is obvious that an influent for a plant with a hydraulic residence time of 24 h should not be monitored using e.g. a 21-day reproductivity test, if the goal is to protect the plant from toxic shocks. Longer term testing of influents does occur to evaluate treatability of a wastewater before introduction, or for confirmation of the results of on-line testing.

An extensive overview on influent toxicity detection methods has been given in Love and Bott (Love and Bott, 2000), and an update in Ren (Ren, 2004). We will restrict the discussion to the most commonly used methods: bacterial luminescence, nitrification inhibition and respirometry.

Bacterial luminescence

The principle of bioluminescence toxicity detection is discussed in more detail for effluent testing (see below). The method has been applied to raw influents for a long period of time, and a lot of data has been accumulated that can be used as reference data. However, there are significant disadvantages of using this test for assesssing toxicity to wastewater treatment bacteria. *Vibrio fischeri*, the standard organism at the basis of the commonly used bioluminescence tests, is a marine bacterium, and therefore the relevance to the activated sludge community is at the very least questionable. Furthermore, because of its marine origins, the salinity of the test solution needs to be adapted. This manipulation diverts the measurement conditions from the environmental conditions in the treatment plant.

Several adaptations have been proposed to address these disadvantages. For example, Hoffmann and Christofi (Hoffmann and Christofi, 2001) proposed a method where a population of the luminescent marine bacterium was incorporated into a sludge testing matrix. Other authors (Kelly *et al.*, 1999; Ren and Frymier, 2003) have transferred the *lux* operon of *V. fischeri* (i.e. a group of genes coding for the bioluminescence) into a bacteria isolated from activated sludge.

These measures improve the representativity of the methods. An extensive overview of current developments in bacterial luminescence methods is given in Philp *et al.* (Philp *et al.*, 2004).

Nitrification inhibition

Conventional biological nitrogen removal involves a two-step process. The first step, nitrification, comprises the oxidation of reduced nitrogen compounds to nitrite and eventually to nitrate. Nitrification occurs in two major steps: the oxidation to nitrite, mediated by a group of bacteria called the ammonia oxidising bacteria (AOB, usually represented by the species Nitrosomonas), and the subsequent oxidation of nitrite to nitrate by nitrite oxidising bacteria of which Nitrobacter is the most well-known example.

The second step of biological nitrogen removal is denitrification, in which the oxidised nitrate forms are used as an electron acceptor, resulting in N_2 gas, which dissipates into the atmosphere, and finally removes nitrogen from the water.

It is well-known that nitrifying bacteria are the most sensitive to toxic substances among the activated sludge consortium (Blum and Speece, 1991). A survey performed by Jönsson (Jönsson, 2001) revealed that of 75 interrogated nitrifying wastewater treatment plants, 48 have experienced nitrification problems. Of these 48, approximately 20 % attributed the problems to an industrial discharge.

Follow-up of nitrification can be done using several methods. Since the first nitrification step is known to be most sensitive to toxic substances (Blum and Speece, 1991), most methods monitor either the first step (ammonia to nitrite) or the whole nitrification process (ammonia to nitrate).

Some methods use pure cultures of Nitrosomonas and Nitrobacter. However, it should be realised that the traditional role of these species as 'key' nitrifiers is currently being criticised (Blackall, 2000), therefore the test may not be in line with the species actually performing nitrification in the treatment plant. Alternatively, enriched cultures from nitrifying WWTPs can be used (Gernaey *et al.*, 1997).

Some methods monitor directly the consumption of ammonia, and/or the production of nitrite or nitrate (Hayes *et al.*, 1998). Since each completely oxidised ammonia molecule yields a proton production of two protons per ammonia molecule, nitrification can be measured by titrimetry. This is the monitoring of added quantities of (in this case) base needed to keep the pH in a reactor at a constant level (Gernaey *et al.*, 1998). A third method is based on observing the oxygen consumption associated with ammonia oxidation. The main difficulty in this approach, is separating oxygen demand of nitrification from background oxygen consumption (originating for example from heterotrophic respiration). This can be done by comparing the oxygen consumption before and after the addition of allylthiourea (ATU), a known specific inhibitor of nitrification (Gernaey *et al.*, 1997). The difference of the two represents the oxygen consumption of nitrification alone.

Nitrification inhibition methods have the advantage of their high sensitivity and their relevance to the biological nutrient removal process. However, they do not yield information about toxicity to heterotrophic bacteria, and obviously their relevance is lost in nonnitrifying plants. In addition, in some plant layouts, nitrifying organisms are exposed to wastewater after BOD removal. In these situations, the biodegradation of certain toxic compounds by the heterotrophic biomass during BOD removal is not taken into account, and toxicity may be overestimated.

Respirometry

One of the most widely used influent toxicity detection techniques is respirometry. The significance of respirometry in activated sludge systems is largely recognised in the literature and its uses exceed toxicity detection alone (Bixio *et al.*, 2000; Copp *et al.*, 2002).

Respirometry monitors the oxygen uptake rate of activated sludge with one or more oxygen sensors placed in a test reactor. Toxicity is measured by the inhibition of the oxygen uptake rate following the addition of a test substance. The oxygen uptake rate of activated sludge is directly coupled with energy metabolism of the activated sludge micro-organisms. In an indirect way, respiration rate is also indicative for growth and reproduction, since a decreased growth eventually results in less energy needs. Therefore, respirometric experiments can be designed to detect toxic effects on both energy metabolism and growth/reproduction of the biomass.

In influent monitoring, a small biomass sample is subjected to the influent under a loading rate that is typically higher than that of the actual plant. During breakdown, various parameters such as oxygen uptake rate are monitored and compared with the response to a reference influent. The experiment is usually performed in a short time span, and can therefore be automated and included in the on-line supervision and control systems of the plant.

All respirometry-based methods in some way refer the measurements to a reference influent, known not to be toxic to the biomass. By careful selection of this reference influent and the evaluation method, it is possible to estimate toxicity both to the heterotrophic and the autotrophic community (Kong *et al.*, 1996). In this way, additional information on nitrification inhibition can help an early detection of toxic episodes.

Respirometric measurement methods differ in the way oxygen uptake rate is monitored. The main difficulty is separating oxygen supply (aeration) from oxygen uptake. Some methods separate aeration and oxygen decay in time by subsequent aeration and decay (possibly in repeating cycles, e.g. de Bel *et al.*, 1996), others separate them in space by cycling a biomass between an aerated and an unaerated vessel (Spanjers, 1993), while other methods use mathematical methods to separate the two, simultaneously occurring, processes (Vanrolleghem, 1994).

Another important distinction is in the biomass used for the respirometric measurements. Some respirometers grow an internal biomass, independent from the plant which is protected by the device, while others sample the plant's activated sludge for the toxicity test. The advantage of the first type of respirometers is their independence from plant performance, and their adaptation to the reference influent, yielding a fast and well-defined response. The advantage of the second type is obvious: since the activated sludge itself is sampled for every measurement, the test species and the activated sludge itself become identical, ensuring maximal representativeness.

Several commercial devices are available, both for laboratory use and for on-line application. Although reliability of the on-line instruments has been criticised, successful full-scale applications exist (Devisscher *et al.*, 2001), provided a thorough maintenance and control scheme is implemented and respected.

3.3.6.2 Effluent Toxicity

Effluent toxicity tests attempt to quantify the toxic effect of the effluent on the receiving ecosystem. Bioassays consist of monitoring a quantifiable effect on an indicator organism. These tests have been used for this purpose for a long time, and extensive documentation, toxicity data and standard procedures are available.

It is impossible to represent an entire ecosystem by one specific indicator species. Therefore, in order to have meaningful results, a battery of bioassays representing locally relevant species from all trophic levels is considered a prerequisite. It is important to realise that, even with these precautions, considerable differences may exist between the predicted effect and the actual *in-situ* effect of the studied effluent to the receiving water (La Point and Waller, 2000).

An overview of effluent toxicity measurements can be found in Farré and Barceló (Farré and Barceló, 2003). These authors classified the toxicity detection methods according to the test species used. The same classification is used here.

Fish bioassays

Traditionally used species include rainbow trout (*Onchorhynchus mykiss*) and the fathead minnow (*Pimephales promelas*). A routinely used test is the 96-h lethality assay (European Commission, 1992a). In this test, fish are exposed to a dilution series of the wastewater for 96 h. Mortality is recorded at 24-h intervals, and used to calculate the LC_{50}.

Three types of lethality test can be used:

- Static test: no flow of the test solution occurs.
- Semi-static test: test with regular batch-wise renewal of the test solution.
- Flow-through test: the water is renewed constantly in the test chamber.

Other species have been proposed, and besides lethality, other fish bioassays are based on larval growth, larval survival and adenosine triphosphate (ATP) measurements. Recent developments are ongoing to replace fish tests by direct measurements on cultured cells.

Fish bioassays are quite laborious. They require specialised equipment and staff.

Invertebrate bioassays

Popular species for invertebrate toxicity testing include Daphnia and Ceriodaphnia. The 48-h immobilisation test (European Commission, 1992b) is widely used. In this test, young daphnids are exposed to a dilution series of the wastewater. Immobilisation is recorded at 24 and 48 h and the data are then used for calculating the EC_{50}.

Other tests exist, such as the 21-day reproduction test, and many other invertebrates have been proposed, such as mayflies (*Baetis* spp.) amphipods (e.g. *Gammarus lacustris*) or stoneflies (*Pteronarcys* spp.).

Several invertebrate bioassays are being marketed in user-friendly kits.

Plant and algae bioassays

Several bioassays based on plants exist, but are seldom used. A typical algae indicator species is *Selenastrum capricornutum*. In the algal growth inhibition test (European Commission, 1992c), the exponentially growing test species are incubated in the test solution for 72 h and cell density is measured every 24 h. The quantified effect is the inhibition of growth relative to a control culture.

Bacterial bioassays

A widespread toxicity test is based on the luminescence inhibition of luminescent bacteria, such as *V. fischeri* or *Photobacterium phosphoreum*. The bioluminescence reaction involves the oxidation of a long chain aldehyde (RCHO) and reduced flavin mononucleotide ($FMNH_2$), resulting in the production of oxidised flavin (FMN) and a long chain fatty acid (RCOOH), along with the emission of blue-green light. Since $FMNH_2$ production depends on functional electron transport, only viable cells produce light. This relationship between light emission and cellular viability forms the basis of the assay and it forms the link between toxicity and the observed response.

These bioluminescence tests are standardised (International Standardization Organization, 1998) and available as commercial devices by several suppliers.

Since the biochemical and genetic mechanisms of bacterial bioluminescence are well understood, and because of the possibilities created by recent evolutions in molecular biology, major research efforts are directed to the development of genetically modified organisms carrying the *lux* operon. In this way, toxicity tests can be developed with a wide range of novel indicator micro-organisms (Philp *et al.*, 2004).

Biosensors

Biosensors result from the direct coupling of biologically active elements (such as enzymes, DNA or immobilised micro-organisms) to a physico-chemical transducer (e.g. a conductivity sensor). The difference with bioassays is subtle. Biosensors attempt to integrate a bioassay in an instrument, whereas bioassays normally are conducted in a laboratory.

Whole cell bacterial biosensors have been used for toxicity monitoring. In these tests, a living organism is immobilised, and their response to toxic mixtures is monitored. A commercially available technique is based on an amperometric system. In this system, a chemical mediator deviates electrons from the respiratory system of the immobilised test organism to an amperometric carbon electrode.

The advantages of these sensors are their unattended operation, fast response and the (semi)continuous signal. These aspects makes them fit for inclusion in on-line monitoring systems.

3.3.7 TOXICITY MITIGATION

When confronted with recurring toxicity events, mitigation measures should be provided at the plant to reduce the impact of the toxic influent to the plant.

When influent toxicity is an issue, protection of the purification process is the main goal. Possibilities include:

- Calamity basins can be used to store limited volumes of toxic influents. These volumes can later be tankered away to specialised disposal sites, they can be treated on-site by the addition of chemicals, or they can be introduced into the system at a much lower loading rate.
- Equalisation basins can be used to mix influents from several sources. The mixture may have reduced toxicity (e.g. in the case of pH-dependent toxicants).
- Chemicals can be added to reduce toxicity. pH can be adjusted by adding acids or caustic chemicals; polymers or other coagulants can be added to aid the removal of colloidal or suspended pollutants; or powdered activated carbon can be dosed to remove toxic organic compounds. These chemicals can be added in the calamity/equalisation basin, or directly into the treatment process.
- Adaptation of the plant's operational parameters can help reduce the effect of toxic compounds. Increased aeration may result in faster breakdown of biodegradable

toxicants; or in the stripping of volatile compounds. Alternative measures include step-feeding (i.e. the introduction of the influent in multiple locations, in order to reduce the local concentration), rapid return sludge recycling (in order to increase the biomass concentration at the top of the reactor) and waste sludge storage and recycling. An evaluation of these last three, together with influent storage and reintroduction is given in Copp *et al.* (Copp *et al.*, 2002).

These measures can be taken manually, after detection of toxicity, but the strategies are more efficient if they can be automatically coupled to an action by inclusion in the supervision and control system of the plant. These strategies require on-line toxicity detection instruments.

A thorough procedure for effluent toxicity reduction is given by the US EPA (US EPA, 1999).

When confronted with effluent toxicity, the sources of this toxicity need to be traced in order to determine the correct remedial action. Toxicity can be introduced by the influent, or may originate in the treatment process itself, e.g. through the addition of certain chemicals. In the last case, replacement of these chemicals should be considered, or additional treatment steps should be taken to remove the compounds causing toxicity.

In case toxicity can be traced back to the influent, further follow-up through the wastewater origins is needed to tackle the toxicity at source. If this is not possible, existing process operation should be reviewed to check whether the plant is indeed performing optimally. By adjusting conventional process control parameters such as oxygen setpoint or mixed liquor suspended solids (MLSS) concentration, an increased treatment efficiency might be achieved that is able to remove the toxicants by conventional operation. If the process is performing at its best possible level, measures such as described above can be taken to reduce influent toxicity.

REFERENCES

Bixio, D., Geenens, D., Bogaert, H. and Thoeye, C. (2000) A stethoscope of the wastewater treatment plant. In: Proc. *73th WEFTEC*, 14–18 October 2000, Anaheim, USA (on CDROM).

Blackall, L.L. (2000) *Australasian Biotechnol.*, **10**(3), 29–32.

Blum, D.J.W. and Speece, R.E. (1991) *J. Water Pollut Control Fed.*, **63**(3), 198–207.

CEC (2001) Proposal for a regulation of the European Parliament and of the Council concerning the Registration, Evaluation, Authorisation and Restriction of Chemicals (REACH), establishing a European Chemicals Agency and amending Directive 1999/45/EC and Regulation (EC) (on Persistent Organic Pollutants). Brussels, 29.10.2003, COM(2003) 644 final.

Connell, D.W., Lam, P., Richardson, B. and Wu, R. (1999) *Introduction to Ecotoxicology*. Blackwell, Oxford, UK.

Copp, J.B., Spanjers, H. and Vanrolleghem, P.A. (Eds) (2002) Respirometry in control of the activated sludge process: benchmarking control strategies. Scientific and Technical Report No. 11. IWA, London, UK.

de Bel, M., Stokes, L., Upton, J. and Watts, J. (1996) *Water Sci. Technol.*, **33**(1), 289–296.

Devisscher, M., Bixio, D., Geenens, D. and Thoeye, C. (2001) In: *Proc. IWA 2nd World Water Congress*, 15–19 October 2001, Berlin, Germany.

European Commission (1992a) Methods for determination of ecotoxicity; Annex V, C.1. Acute toxicity for fish. Directive 92/69/EEC, O.J. L383A.

European Commission (1992b) Methods for determination of ecotoxicity; Annex V, C.2. Acute toxicity for Daphnia. Directive 92/69/EEC, O.J. L383A.

European Commission (1992c) Methods for determination of ecotoxicity; Annex V, C.3. Algal inhibition test. Directive 92/69/EEC, O.J. L383A.

Farré, M. and Barceló, D. (2003) *Trends Anal. Chem.*, **22**(5), 299–310.

Geenens, D. and Thoeye, C. (1998) *Water Sci. Technol.*, **37**(12), 213–218.

Gernaey, K., Verschuere, L., Luyten, L. and Verstraete, W. (1997) *Water Environ. Res.*, **69**(6), 1163–1169.

Gernaey, K., Bogaert, H., Vanrolleghem, P., Massone, A., Rozzi, A. and Verstraete, W. (1998) *Water Sci. Technol.*, **37**(12), 103–110.

Grau, P. and Da-Rin, B.P. (1997) *Water Sci. Technol.*, **36**(1–2), 1–8.

Gutiérrez, M., Etxebarria, J. and de las Fuentes, L. (2002) *Water Res.*, **36**(4), 919–924.

Hayes, E., Upton, J., Batts, R. and Spickin, R. (1998) *Water Sci. Technol.*, **37**(12), 193–196.

Hernando, M.D., Fernandez-Alba, A.R., Tauler, R. and Berceló, D. (2005) *Talanta.*, **65**(2), 358–366.

Hoffmann, C. and Christofi, N. (2001). Testing the toxicity of influents to activated sludge plants with the *Vibrio fischeri* bioassay utilising a sludge matrix. *Environ. Toxicol.,***16**(5), 422–427.

International Standardization Organization (1998) Water quality: determination of the inhibitory effect of water samples on the light emission of *Vibrio fischeri*. ISO 11348-1, 2 and 3. International Standardization Organization, Geneva, Switzerland.

Jönsson, K. (2001) Inhibition of nitrification in municipal wastewater – sources, effects, evaluation and remedies. PhD Dissertation, Lund University of Technology, Lund, Sweden.

Kelly, C.J., Lajoie, C. A., Layton, A .C. and Sayler, G. S. (1999). *Water Environ. Res.*, **71**, 31–35.

Kong, Z., Vanrolleghem, P.A., Willems, P. and Verstraete, W. (1996) *Water Res.*, **30**(4), 825–836.

La Point, T.W. and Waller, W.T. (2000) *Environ. Toxicol. Chem.*, **19**(1), 14–24.

Love, N.G. and Bott, C.B. (2000) A review and needs survey of upset early warning devices. Report No. 99-WWF-2. Water Environment Research Foundation, Alexandria (VA), USA.

Paxeus, N. (1996) *Water Res.*, **30**(5), 1115–1122.

Philp, J., French, C., Wiles, S., Bell, J., Whiteley, A. and Bailey, M. (2004) Wastewater toxicity assessment by whole cell biosensor. In: *The Handbook of Environmental Chemistry*, Vol. 5, Part I, pp. 165–225. Springer-Verlag, Berlin, Germany.

Ren, S. (2004) *Environ. Int.*, **30**, 1151–1164.

Ren, S. and Frymier, P.D. (2003) *Water Environ Res.*, **75**(1), 21–29.

Ren, S. and Frymier, P.D. (2004) *J. Environ. Engin.*, **130**(4), 484–488.

Spanjers, H. (1993) Respirometry in activated sludge. PhD Thesis, Wageningen Agricultural University, Wageningen, The Netherlands.

Thornton, I., Butler, D., Docx, P., Hession, M., Makropoulos, C., McMullen, M., Nieuwenhuijsen, M., Pitman, A., Rautiu, R., Sawyer, R., Smith, S., White, D., Wilderer, P., Paris, S., Marani, D., Braguglia, C. and Palerm, J. (2001) Pollutants in urban wastewater and sewage sludge. Report for Directorate-General Environment (http://europa.eu.int/comm/environment/waste/sludge/sludge_pollutants.htm).

Tinsley, D., Wharfe, J., Campbell, D., Chown, P., Taylor, D., Upton, J. and Taylor, C. (2004) *Ecotoxicol.*, **13**(5), 423–436.

US EPA (1991) Methods for aquatic toxicity identification evaluations: phase I, toxicity characterization procedures, 2nd Ed. EPA/600/6-91-003.

US EPA (1994) Whole effluent toxicity (WET) control policy: policy for the development of effluent limitations in national pollutant discharge elimination system permits to control whole effluent toxicity for the protection of aquatic life. EPA/833B-94/002.

US EPA (1999) Toxicity reduction evaluation guidance for municipal wastewater treatment plants. EPA/833B-99/002.

Vanrolleghem, P.A. (1994) On-line modelling of activated sludge processes: development of an adaptive sensor. PhD Thesis, Ghent University, Ghert, Belgium.

3.4
Nutrient Control

Victor Cerdà and **José Manuel Estela**

3.4.1 INTRODUCTION

Nutrients are chemical elements and compounds found in the environment that plants and animals need to grow and survive. For water-quality investigations the various forms of nitrogen and phosphorus are the nutrients of interest. The forms

Wastewater Quality Monitoring and Treatment Edited by P. Quevauviller, O. Thomas and A. van der Beken

include nitrate, nitrite, ammonia, organic nitrogen (in the form of plant material or other organic compounds) and phosphates (orthophosphate and others). Nitrate is the most common form of nitrogen and phosphates are the most common forms of phosphorus found in natural waters. High concentrations of nutrients in water bodies can potentially cause eutrophication and hypoxia. Eutrophication is a process whereby water bodies, such as lakes, estuaries, or slow-moving streams receive excess nutrients that stimulate excessive plant growth (algae, periphyton attached algae and nuisance weeds). This enhanced plant growth, often called an algal bloom, reduces dissolved oxygen in the water when dead plant material decomposes and can cause other organisms to die. Nutrients can come from many sources, such as fertilizers applied to agricultural fields, golf courses, and suburban lawns; deposition of nitrogen from the atmosphere; erosion of soil containing nutrients; and wastewater treatment plant discharges. Water with a low concentration of dissolved oxygen is called hypoxic hypoxia means 'low oxygen'. In many cases hypoxic waters do not have enough oxygen to support fish and other aquatic animals. Hypoxia can be caused by the presence of excess nutrients in water. Nutrient control is, therefore, essential for maintaining the quality of waters with the aim to avoid sanitary and eutrophication problems. This control also facilitates the implementation of strategies in wastewater treatment plants allowing them to comply with the legal requirements in effluent contents and to optimize processes that economize both the energetic and chemical reagent consumptions.

3.4.2 OCCURRENCE, IMPORTANCE AND TERMINOLOGY

In this section we will briefly discuss, on an individual basis, some properties and characteristics, as well as the terminology proposed (and usually accepted), used in nutrient analysis and assessment.

3.4.2.1 Nitrogen

Nitrogen is a bioessential element. The different nitrogen forms (nitrate, nitrite, ammonia and organic nitrogen) in addition to nitrogen gas (N_2) are biochemically interconvertible and are part of the so-called nitrogen cycle (Russell, 1994), which includes natural and anthropogenic components. It is of great complexity due to the diversity of compounds and transformations involved. From an analytical point of view, it is a habitual practice (Standard Methods Committee, 1988) to refer to organic nitrogen as N_{org}, nitrate nitrogen as NO_3^--N, nitrite nitrogen as NO_2^--N and ammonia nitrogen as NH_3-N.

Organic nitrogen is defined (Mopper and Zika, 1987) as the nitrogen organically linked in the oxidation state -3, and does not include all the organic compounds of nitrogen. It can be determined together with ammonia and in this case constitutes the so-called Kjeldahl nitrogen. Organic nitrogen includes products such as peptides, proteins, nucleic acids, urea and synthetic organic materials. Its concentration in wastewaters can be higher than 20 mg/l.

Total oxidized nitrogen is the sum of nitrite and nitrate nitrogens. Nitrate is only found in small amounts in domestic wastewaters; however, in the diluent of nitrifying biological treatment plants nitrate can be found at concentrations of up to 30 mg/l of NO_3^--N. Nitrite is an intermediate oxidation state of nitrogen and can be generated from oxidation of ammonia or reduction of nitrate. Oxidation is a common practice in wastewater treatment plants. An important supply of nitrites comes from industrial wastewaters since it is used as an additive to inhibit corrosion in industrial processes. Nitrite is the real cause of the well-known disease methahemoglobinemia (Burden, 1961; Johnson and Cross, 1990) and also the nitrous acid formed in acidic media can react with secondary amines, nitrosation *in vivo*, giving rise to nitrosamines many of which are carcinogenic (Forman *et al.*, 1985).

Ammonia is a naturally occurring compound found in wastewaters, produced by deamination of nitrogenated organic compounds and by hydrolysis of urea. In some wastewater treatment plants it is even used as an additive to react with chlorine and form combined residual chlorine (mono- and dichloramines). In wastewaters its concentration can surpass 30 mg/l NH_3–N.

In the European Community 91/271/CEE Council regulation on urban wastewaters treatment, a minimum reduction percentage of 70–80 % is required, establishing a maximum total nitrogen concentration (Kjedahl nitrogen) of 15 and 10 mg/l N for populations of 10 000–100 000 i.-e.(inhabitant-equivalent) and for more than 100 000 i.-e., respectively. The i.-e. is the biodegradable organic load with a biochemical oxygen demand for 5 days (BOD5) of 60 g oxygen per day.

3.4.2.2 Phosphorus

Phosphorus is one of the key elements necessary for the growth of plants and animals. Phosphorus in the elemental form is very toxic and is subject to bioaccumulation. Phosphates, derived from phosphorus, are present in three forms: orthophosphate, polyphosphate and organically bound phosphate. Ortho forms are produced by natural processes and are found in sediments, natural waters and sewage. Poly forms are used for treating boiler waters and in detergents. In water, they change into the ortho form. Organic phosphates play an important role in nature, and their occurrence may result from the breakdown of organic pesticides containing phosphates. They may exist in solution, as particles or in the bodies of aquatic organisms. Phosphorus in aquatic systems may originate from natural sources such as the mineralization of algae and the dissolution of phosphate minerals, from anthropogenic point source discharges of sewage and industrial effluents and from diffuse inputs from grazing and agricultural land. Studies carried out in the USA have demonstrated that phosphorus inputs to the environment have increased since 1950 as the use of phosphate fertilizer, manure, and phosphate laundry detergent increased; however, the manufacture of phosphate detergent for household laundry was ended voluntarily in about 1994 after many States established phosphate detergent bans. Total phosphorus concentrations in raw wastewater effluent contained about 3 mg/l of total phosphorus during the 1940s, increased to about 11 mg/l at the height of phosphate detergent use (1970), and have currently declined to about 5 mg/l. However, in some cases, tertiary wastewater treatment still is needed to effectively improve water quality of streams. Downward trends in phosphorus concentrations since 1970 have been identified in many streams, but median total phosphorus concentrations still exceed the recommended limit of 0.1 mg/l across much of the USA. In the European Community 91/271/CEE Council regulation on urban wastewaters treatment, a minimum reduction percentage of 80 % is required, establishing a maximum total phosphorus concentration of 1 and 2 mg/l for populations between 10 000 and 100 000 i.-e. and for more than 100 000 i.-e., respectively.

The analysis of water samples (natural, waste, etc.) are especially complex owing to the fact that phosphorus can be found in the form of different inorganic and organic species (McKelvie *et al.*, 1995), which in turn can be present in either the dissolved, colloidal or particulate form. However, the dominant species is always orthophosphate. Usually, in the analysis of water samples the analysis of the phosphorus content is carried out on aliquots of the whole sample and on aliquots of the sample previously filtered through membrane filters of 0.45 and 0.2 μm nominal pore size (Standards Australia and Standards New Zealand, 1998) or glass fibre filters (GF/F 0.7 and 1.2 μm) (Brober and Persson, 1988). The aim of this procedure is to obtain the data required for the calculation of the parameters that allow the evaluation of aspects such as the content of phosphorus in several organic and inorganic species, the eutrophication of aquatic systems or the amount of bioavailable phosphorus (BAP).

Parameters determined on the filtered fraction contain the word filterable, namely: filterable reactive phosphorus (FRP), total filterable phosphorus (TFP) and filterable acid-hydrolysable phosphorus (FAHP). However, in the literature it is indistinctly used together with the words dissolved or soluble (McKelvie, 2000). On the other hand, the term reactive refers to the phosphorus species that react with molybdate to form 12-phosphomolybdate (12PM) or phosphomolybdenum blue (PMB), the latter if a reducing agent is present in the reaction medium. Filterable condensed phosphates (FCP) are comprised of inorganic polyphosphates, metaphosphates and branched ring structures. The term acid-hydrolysable phosphorus refers to the required acidic hydrolysis for the conversion of condensed phosphates into orthophosphate. Therefore, FCP = FAHP and if the formation reaction of 12PM or PMB is used for the corresponding determination, thus, FRP + FAHP is obtained. The filterable organic phosphorus fraction [FOP = TFP − (FAHP + FRP)] consists of nucleic acids, phospholipids, inositol phosphates, phosphoamides, phosphoproteins, sugar phosphates, aminophosphonic acids, phosphorus-containing pesticides as well as organic condensed phosphates (Armstrong, 1972; Brober and Persson, 1988; Robards *et al.*, 1994; Stumm and Morgan, 1996).

The parameters obtained on the aliquots of the whole sample (without filtration processes) contain the word total, namely: total reactive phosphorus (TRP), total acid-hydrolysable phosphorus (TAHP), total phosphorus (TP) and total organic phosphorus (TOP) and are equivalent to those previously mentioned. However, they also consider the particulate fraction. Determination of FOP, TFP, TP or TOP requires a previous digestion of the sample for the conversion of the organic phosphates into the orthophosphate reactive specie.

The numerical treatment of the parameters determined on the filtrated and the whole fractions of the sample allow the evaluation of other parameters related to the contents of phosphorus in the particulate phase, namely: total particulate phosphorus (TPP = TP − TFP), particulate reactive phosphorus (PRP = TRF − FRP), particulate acid-hydrolysable phosphorus (PAP = TAHP − FAHP) and particulate organic phosphorus (POP = TOP − FOP).

Thus, both TP and FRP are the most measured parameters. TP provides a measurement of the maximum potential bioavailable phosphorus, whereas FRP, comprising mostly orthophosphate, provides an indication of the amount of most readily bioavailable phosphorus.

3.4.3 SAMPLE HANDLING AND PRESERVATION

Sample handling, preservation and the time involved until performing the analysis are steps that should be carefully considered. These steps may vary for the same analyte according to either the need to carry out a speciation or if the contents in the different fractions of the sample should be determined or not. Nutrients can easily evolve after the sample handling and, thus, the general recommendation is to carry out determination as soon as possible. There is a generalized tendency to reject the use of preservative additives harmful to the environment, such as mercury salts or certain organic solvents. Freezing of the sample is a widely accepted alternative, in agreement with the 'green chemistry' policy, and which usually only requires filtration and/or addition of less aggressive preservative additives. In wastewater treatment plants for daily nutrient control and for several operative reasons, monitors, sensors or kits are used to perform analysis *in situ*, thus avoiding both the sample preservation steps and storage. However, on some occasions, and due to the need to carry out sample collection, in order to use a generalized automatic analysis system and/or for the determination of certain parameters, one should resort to preservation and storage of the samples. Next, we will discuss the recommendations that have been published and successfully applied for several years (APHA-AWWA-WPCF, 2000) and that have been reviewed in other publications (Nollet, 2000).

3.4.3.1 Nitrogen

Ammonia

It is recommended that sample analysis be carried out as soon as possible, between 1–2 h after collection. If samples are to be analysed within 24 h of collection, refrigerate unacidified at 4 °C. Samples should be collected in LDPE (low density polyethylene) glass bottles or PTFE (polytetrafluoroethylene). The residual chlorine should be destroyed immediately with a dechlorinating agent (sodium sulfite, sodium thiosulfate, phenylarsine oxide or sodium arsenite) to hinder its reaction with ammonia. If a fast analysis is not possible, samples can be preserved for up to 28 days frozen at $-20\,^{\circ}$C unacidified or by acidifying with sulfuric acid (0.8 ml of conc. H_2SO_4/l sample is usually enough, pH = 1.5–2) and storing at $4\,^{\circ}$C. Acid neutralization is required prior to determination.

Nitrite

Samples can be collected in glass bottles or polyethylene. Determination should be carried out immediately after sample collection to avoid or minimize bacteria activity, and preservation of the samples with acid should never be used due to its rapid conversion into nitrate. Samples can be kept frozen for short periods of time (1 or 2 days) at $-20\,^{\circ}$C or stored at $4\,^{\circ}$C.

Nitrate

It is recommended that determination be carried out immediately after sample collection, in glass bottles or polyethylene. Samples can be stored up to 24 h at $4\,^{\circ}$C. Preservation for longer periods requires the addition of 2 ml of conc. H_2SO_4/l sample. In the case of using acid it should be borne in mind that the step from nitrite to nitrate will have taken place and, thus, both species cannot be determined individually. Sample preservation, prior to filtration through alumina, with mercury salts or chloroform is not recommended for environmental reasons and also because it interferes with the reduction of nitrate to nitrite if the granulated copper–cadmium method is employed.

Organic nitrogen

Samples can be collected in glass, poly(vinyl chloride) or polyethylene containers. As in previous cases, it is advisable to carry out the analysis immediately after sample collection. Otherwise, samples can be stored acidified with conc. H_2SO_4 (pH 1.5–2.0) and at $4\,^{\circ}$C. Mercury salts should not be used (i.e. $HgCl_2$) as preservatives since they interfere with ammonia elimination. Several authors (Dore *et al.*, 1996) recommend freezing the samples for preservation purposes. If this procedure is used, potential errors due to flocculation during freezing can be reduced by intensive mixing before analysis.

3.4.3.2 Phosphorus

Freezing is the most popular and general sample preservation procedure for P analysis. However, the manner in which samples are preserved depends on whether differentiation of the different forms is required or not.

If no differentiation is required, the TP content is determined; sample freezing and/or acidification (1 ml HCl/l sample) is a very common practice. If differentiation between the soluble forms – FRP, FAHP and TFP – is required, an immediate filtration of the samples through membrane filters of 0.45 μm and further freezing is recommended. Membrane filters should be washed with several portions of distilled water prior to use to avoid contamination of samples with low phosphate content. If the sample is difficult to filtrate a previous filtration through glass wool can be carried out. If the sample needs to be preserved for a long period of time $HgCl_2$ can be added prior to freezing, although this practice is not currently used for environmental reasons. Besides, it has been demonstrated that freezing maintains the stability of the samples for at least 4 months (Clementson and Wayte, 1992). It is not convenient to add either acid or $CHCl_3$ as preservatives if differentiation of the different phosphorus forms is required.

In all cases, samples should be collected in glass containers previously washed: first with dilute HCl and, then, several times with distilled water. The same washing procedure is recommended for all used glass material. If the samples are not frozen they should not be collected in plastic containers since phosphate losses take place by adsorption across the walls of the containers, especially if their P content is low. Also, detergents containing phosphates should not be used for the cleaning of the glass material employed in the analysis.

3.4.4 STANDARD RECOMMENDED METHODS OF ANALYSIS

In this section the characteristics of the different recommended methods proposed for the analytical control of nutrients will be presented. The reader is referred to the literature for procedure details (APHA-AWWA-WPCF, 2000; Nollet, 2000).

3.4.4.1 Nitrogen

Ammonia

Wastewater ammonia is found at low concentrations in good quality nitrified diluents and can exceed 30 mg/l in effluents. Sensitivity and interferences of a method are always factors that have an influence on the applicability of the method. Fortunately, the ammonia present in a wastewater sample can be separated from the sample by means of a previous distillation process (APHA Method 4500-NH_3 B, 2000), thus, making it possible to use those methods which in their direct application comply with the sensitivity requirements but do not comply with the selectivity requirements. For determination purposes the following methods are mainly recommended: the titration method (APHA Method 4500-NH_3 C, 2000), which requires previous distillation of the ammonia contained in the sample, the manual colorimetric methods based on the Nessler reaction (nesslerization) (APHA Method 4500-NH_3 C, 1992) or the Berthelot reaction (indophenol blue) (APHA Method 4500-NH_3F, 2000), both with or without previous distillation, and the ammonia selective electrode method, applicable both directly to the wastewater sample itself or to the distillate, on the basis of a direct potentiometry (APHA Method 4500-NH_3 D, 2000) with calibration curve or in a standard addition methodology (APHA Method 4500-NH_3 E, 2000). Nesslerization has been dropped (APHA-AWWA-WPCF, 2000) as a standard method, although it has been considered a classic water quality measurement for more than a century. The use of mercury in this test warrants its deletion because of the disposal problems. There are two automated methods (APHA Method 4500-NH_3 G, 2000; APHA Method 4500-NH_3 H, 2000) based on the classic Berthelot reaction in which the catalyser, a manganous salt, has been replaced by sodium nitroprusiate. One of these methods (APHA Method 4500-NH_3 G, 2000) is based on a segmented continuous flow analysis technique with an analysis throughput of 60 samples per hour. This method allows NH_3-N determination directly, prior sample filtration, in domestic and industrial wastewaters within a range of 0.02–2.0 mg/l. The other (APHA Method 4500-NH_3 H, 2000) is based on a flow injection analysis technique.

Water used in the preparation of reagents should be ammonia free, easily achieved by using ionic exchange resins; however, it is always advisable to obtain the blank. In the distillation process the sample is buffered, previously neutralized at pH 9.5 by addition of borate buffer solution and the distillate is collected over boric acid for the titration method or the Nessler reaction method and over 0.04 N H_2SO_4 for the remaining methods. As a rule, a previous distillation is recommended for colorimetric methods since the physical and chemical interferences of the sample such as turbidity, colour, formation of precipitates in the reaction media or those caused by species added to the samples for preservation purposes (i.e. if acid has been added and the Bertholet reaction method has been applied) are eliminated. If the Nessler reaction method is used, and for some domestic wastewater samples, it is possible that distillation may be avoided by pretreatment with zinc sulfate and an alkali. Nevertheless, this possibility should be previously studied analysing distillates of this same type of sample and assessing that comparable results are obtained. It should be stressed that distillation constitutes an important way of eliminating and/or maintaining interferences at low levels. During distillation, hydrolysis of urea and cyanates together with the presence of volatile organic compounds (hydrazine and amines) give rise to interference independent of the analysis method used. The titration method is mainly used for N-NH_3 concentrations higher than 5 mg/l. Colorimetric methods are used for concentrations lower than 5 mg/l NH_3-N, according to the following: the Berthelot method presents a sensitivity of 10 μg/l NH_3-N and is used up to 500 μg/l NH_3-N and the Nessler method possesses a lower sensitivity, 20 μg/l NH_3-N, and is used up to 5 mg/l NH_3-N. The selective electrode method can be applied within a considerably wide concentration range 0.03–1400 mg/l NH_3-N and constitutes a very interesting alternative since no sample pretreatment (distillation) is required. However, standards and samples should have a similar ionic content and be measured at the same temperature. The obtained results should be also confirmed by direct potentiometry applying a standard addition methodology.

Nitrite

This ion presents a sensitive and selective colorimetric method for its easy determination in many aqueous matrixes and, naturally, in domestic and industrial wastewaters. The method is based on the well-known Griess reaction (APHA Method 4500-NO_2^- B, 2000), formation of a pink azo-dye at pH 2–2.5 through reaction of diazotized sulfanylamide with *N*-(1-naphthyl)-ethylendiamine dichlorohydrate and it is applied to contents between 10 and 1000 μg/l NO_2^--N/l. The sample should contain no solids in suspension, requirement for any colorimetric method, and therefore, should be filtered through 0.45 μm membranes if necessary. The following ions: Sb^{3+}, Au^{3-}, Bi^{3+}, Fe^{3+}, Pb^{2+}, Hg^{2+}, Ag^+, $PtCl_6^{2-}$ and VO_3^{2-} precipitate in acid medium and, thus, should be avoided together with coloured ions and Cu^{2+}, which catalyses the decomposition of the diazonium salt and, therefore, can give rise to low results. This manual method has been automated (APHA Method 4500-NO_3^-I, 2000; APHA Method 4500-NO_3^- F, 2000). Both automated methods allow the determination of nitrite and nitrate (prior reduction to nitrite), individually, and of nitrite+nitrate, prior reduction of nitrate, in wastewaters. The automated method, based on a segmented continuous flow technique (APHA Method 4500-NO_3^- F, 2000), allows the determination in the range 0.5–10 mg N/l with an analysis throughput of 40 samples/h.

If for any reason the previous procedure is not feasible or other anions should also be analysed, in some cases, ion chromatography can be used. Chromatography is a multicomponent analysis technique that can eliminate, in certain cases, the need to use expensive or hazardous reagents. In its ionic modality determination and differentiation of halides (Br^-, Cl^- and F^-), SO_3^{2-}, SO_4^{2-}, NO_2^- and NO_3^- is feasible in an efficient way. Although there are many modalities of ion chromatography, the use of ion chromatography with chemical suppression of the eluent conductivity is recommended (APHA Method 4110 B, 2000). This method is applicable to treated wastewaters and some waters from industrial processes, such as boiler or refrigeration circuit waters, prior filtration of the former through membrane filters of 0.22 μm to avoid the obstruction of the columns. The detection limit is around 0.1 mg/l. The usual interferences are those due to the coincidence of the retention times and those caused by high concentrations, also the proximity of peaks belonging to other ions should be considered for giving rise to bad chromatographic resolutions. If the dilution of the sample is feasible, this procedure is a very useful tool for the elimination of interferences.

Nitrate

The approaches recommended for the analysis of nitrates include direct and indirect methods. Among the direct methods is the ion chromatographic method, in its modality of chemical suppression of the eluent conductivity (APHA Method 4110

B, 2000), previously mentioned when dealing with the analysis of nitrites, as well as the nitrate ion selective electrode method (APHA Method 4500-NO_3^- D, 2000). This latter method is valid for concentrations between 0.14 and 1400 mg/l NO_3^--N/l and requires keeping the ionic strength at a constant level in both standards and samples, since the electrode responds to activities rather than concentrations, pH control to avoid erratic electrode responses and adding reagents which eliminate interferences. The former requirements are achieved by the addition of a buffer solution which adjusts the ionic strength and pH to 3 to eliminate the HCO_3^- interference, contains $Al_2(SO_4)_3$ to complex organic acids, sulfaminic acid to eliminate NO_2^- interference, and Ag_2SO_4 to eliminate the interferences of Cl^-, Br^-, I^-, S^{2-} and CN^-. It should be pointed out that in the case of nonwastewater analysis the buffer composition is likely to be simpler. Indirect methods are based on a previous reduction of the nitrate ion to nitrite or to NH_3. Undoubtedly, the colorimetric method based on the Griess reaction, previously mentioned, which uses a granulated copper–cadmium column to carry out the reduction to nitrite, has been and probably will continue to be the most widely used (APHA Method 4500-NO_3^- E, 2000). The application range of this method is within 0.01 and 1.0 mg of NO_3^--N/l. Being a colorimetric method and in order to avoid obstructions in the reducing column the solids in suspension should be eliminated. In the case of a coloured sample, habitual procedures should be considered. The presence of large amounts of heavy metals, of grease and oxidizing species, such as residual chlorine, diminish the efficiency of the reducing process. Grease interference can be eliminated by prior extraction of the former with an organic solvent. Interferences due to metals and residual chlorine are eliminated by prior treatment of the sample with ethylendiamine tetraacetic acid (EDTA) and with sodium thiosulfate, respectively. This method has been automated and previously cited when dealing with nitrite analysis.

When differentiation of the concentrations of nitrate and nitrite is not required and bearing in mind that the use of columns is always tiresome, the use of the automated method based on a segmented continuous flow configuration can be an advantage. This method uses the Griess reaction for the colorimetric detection and carries out the nitrate reduction in homogenous phase by means of hydrazine (hydrazine sulfate)(APHA Method 4500-NO_3^- H, 2000). For the correct application of this method hydrazine sulfate concentration should be adjusted in such a way that the response obtained for a standard of 2.0 mg N-NO_3^-/l should coincide with that of a standard of 2.0 mg N-NO_2^-/l, and also possible absorptions due to the sample should be taken into account. The method can be applied to the determination of $NO_3^- + NO_2^-$ in domestic and industrial wastewaters between 0.01 and 10 mg N/l. The presence of sulfide ion at concentration lower than 10 mg/l produces variations of 10 % in the determined concentrations. Similarly, in the case of NH_3 and NO_2^- concentrations not being significant in relation to that of NO_3^- or the determination of the total content should be required the application of the method of reduction with titanous chloride would be of interest. The $TiCl_3$added to the sample reduces nitrate to NH_3which is detected and determined on the basis of potentiometric measurements carried out with an ammonia selective electrode (APHA Method 4500-NO_3^- G, 1992).

Organic and total nitrogen

Organic nitrogen is calculated from data obtained in the application of the Kjeldahl method (APHA Method 4500-N_{org}, 2000). This method consists of the transformation of the amine nitrogen present in many organic compounds, free ammonia and ammonium nitrogen into ammonium sulfate through sample digestion in a medium of sulfuric acid, potassium sulfate and a catalyser. The catalyser is usually mercuric sulfate. However, and due to environmental reasons, if not strongly required it can be substituted by copper sulfate. Effective digestion results from the use of a reagent having a salt/acid ratio of 1 g/ml with copper as catalyst, and specified temperature and time. Nitrogen in the form of azide, azine, azo, hydrazone, nitrate, nitrite, nitrile, nitrous, oxime and semicarbazone is not transformed. The method allows the determination of the so-called 'Kjeldahl nitrogen' (ammonia nitrogen+organic nitrogen) to which the NH_3-N content previously determined is deduced, and the result is the so-called organic nitrogen. Alternatively, boiling, prior to the digestion process, can eliminate NH_3-N and in this case the organic nitrogen content would be obtained directly. After digestion ammonia is distilled and collected over an adequate absorbent according to the selected ammonia determination method. The distillation equipment and determination methods are those above-mentioned when dealing with ammonia analysis with prior distillation. If the samples to be analysed contain high organic nitrogen concentrations either the macro (APHA Method 4500-N_{org} B, 2000) or semi-micro-Kjeldahl (APHA Method 4500-N_{org} C, 2000) method can be used, otherwise the macro method should be employed. The sample volumes to be taken is determined according to its organic nitrogen content and there are tables estimating such volumes. In any case, the macro method involves volumes between 25 ml (for levels between 50 and 100 mg N/l) and 500 ml (for levels lower than 1 mg N/l) and the semi-micro method between 5 ml (for levels between 40 and 400 mg N/l) and 50 ml (for levels between 4 and 40 mg N/l). During digestion chemical reactions may occur which can give rise to positive and negative interferences, some difficult to ponder and no adequate methods are available for their elimination. To the latter group of interferences belongs that of the nitrate ion at concentrations higher than 10 mg N/l, which can provoke the oxidation of ammonia generating from the organic matter digestion, giving rise to N_2O, leading to a negative interference, and a positive interference due to an excess of organic matter, which can reduce the nitrate ion to ammonia. Interferences produced by samples with high salt content or by sulfuric acid shortage, due to an excessive consumption in samples with high organic matter content, are more well known. In both cases a pyrolytic loss of nitrogen is produced and there is a methodology available to avoid them. Thus, in the first case it is recommended to add excess sulfuric acid (1 ml per g of salt present in the sample) which maintains the acid–salt equilibrium and avoids the temperature rising, and in the second case it is recommended to add an excess of 10 ml of sulfuric acid/3 g of TOC (total organic carbon) or of 50 ml of digestion reagent/g of TOC to the digestion flask. After digestion it may be necessary to add an excess of alkali in order to achieve a high pH prior to distillation.

For the N_{org} determination a flow injection method (APHA Method 4500-N_{org} D, 2000) has been proposed too. Domestic and industrial wastes are digested in a block digestor with sulfuric acid and copper as catalyst. The digestion recovers nitrogen components of biological origin, such as amino acids, proteins and peptides, as ammonia, but may not recover the nitrogenous compound of some industrial wastes such as amines, nitro compounds, hydrazones, oximes, semicarbazones, and some refractory tertiary amines. Nitrate is not recovered. The sample digested is injected onto a flow injection system. The ammonia produced is heated with salicylate and hypochlorite to produce a blue colour that is spectrophotometrically detected. The main source of interference is ammonia.

Total nitrogen can be determined by means of the persulfate method (APHA 4500-N C, 2000) or by in-line UV/persulfate digestion and oxidation with a flow injection analysis method (APHA Method 4500-N B, 2000). The spectrophotometric persulfate method determines total nitrogen by alkaline oxidation at 100–110 °C of all nitrogenous compounds to nitrate. Total nitrogen is determined by analysing the nitrate in the digestate by means of automated or manual cadmium reduction methods and Griess-type reaction. In the flow injection analysis method nitrogen compounds are digested and oxidized in-line to nitrate by use of heated alkaline persulfate and UV radiation. The digested sample is injected onto the manifold where nitrate is reduced to nitrite by a cadmium granule column, derivatized with a Griess-type reaction and detected by spectrophotometry. Both methods recover nearly all forms of organic and inorganic nitrogen, reduced and oxidized, including ammonia, nitrate and nitrite.

3.4.4.2 Phosphorus

As previously mentioned, phosphorus determination is complex and occasionally determination and differentiation between the different forms of phosphorus in the corresponding fractions of a water sample is required. In spite of the former precedents, which appear to complicate phosphorus analysis, it should be underlined that there are methods and procedures available which allow carrying out these determinations simply and with ease. Phosphorus analysis involves the performance of two steps if the sample handling and/or its preservation have already been carried out (these aspects have been previously discussed). The first step consists of the transformation of the phosphorus species into orthophosphate, by acidic hydrolysis or oxidizing digestion (APHA Method 4500-P B, 2000). The second step is the detection and determination of the formed orthophosphate by the colorimetric method of vanadomolybdophosphoric acid (APHA Method 4500-P C, 2000), which is recommended for routine analysis and is valid for concentrations ranging between 1 and 20 mg P/l, or the most sensitive colorimetric method of molybdenum blue, adequate for the determination in samples with low phosphorus content. Molybdenum blue formation can be carried out by reduction of molybdophosphoric acid, formed through reaction of orthophosphate with ammonium molybdate in acid medium with stannous chloride (APHA Method 4500-P D, 2000) or ascorbic acid (APHA Method

4500-P E, 2000), if formed by reaction with ammonium molybdate and antimonyl tartrate also in acid medium. The reduction method with ascorbic acid is more precise and enables the analysis of samples with levels ranging from 0.01 to 6 mg P/l. This same method has been automated using a segmented continuous flow technique (APHA Method 4500-P F, 2000). Thus, orthophosphate can be determined in domestic and industrial wastewaters within a range of 0.001–10 mg P/l in flow cells of 15 mm and with an analysis throughput of 30 samples/h. A flow injection analysis method for orthophosphate determination has been also proposed (APHA Method 4500-P G, 2000).

In practice, by these methods, the different phosphorus contents are determined in the sample, without filtration, and in the filtrate, and the contents in the particulate phase are calculated by subtraction. However, if necessary, the analysis of the fraction deposited on the filter can be carried out. Acid hydrolysis is a soft treatment performed at water boiling temperature and with the sample acidified with sulfuric acid and nitric acid. Through this treatment condensed phosphates and, probably, certain organic phosphates are transformed into orthophosphate, thus, the term phosphorus hydrolysable with acid is preferred over that of condensed phosphates. Oxidizing digestion is a far more energetic treatment, which transforms organic phosphates into orthophosphates and it is essential for the determination of total phosphorus. This digestion can be carried out by the perchloric acid method, the sulfonitric mixture method or the persulfate method. Of these three methods the perchloric acid method is the more efficient, but at the same time the most dangerous; there are explosion risks if the treating temperature is not controlled and also if working with inadequate material. Therefore, the sulfonitric mixture method is mostly recommended. The persulfate method is the easiest and there is a strong tendency towards its use, however, verification of the recoveries with one of the former methods is recommended. After sample treatment orthophosphate is determined applying the colorimetric methods to aliquots, and standards, subjected to the same treatment as that of the samples, are used for calibration purposes. In both methods, i.e. the vanadomolybdophosphoric acid and the reduction with stannous chloride methods, silica interferes positively only if the sample has been previously heated and arsenate, fluoride, thorium, bismuth sulfide, thiosulfate, thiocyannate or excess molybdate interfere negatively. Sulfide interference can be eliminated boiling the sample with bromine water. In the reduction method with stannous chloride interferences can be eliminated and sensitivity increased by extracting molybdophosphoric acid in benzene–isobutanol prior to reduction. Finally, in the ascorbic acid method arsenate strongly interferes, and more moderately, hexavalent chromium and nitrite ion, whereas sulfide and silica do not interfere within concentrations of 1.0 and 10 mg/l, respectively.

The determination of total phosphorus is a very usual practice and two methods based on the flow injection analysis technique have been proposed. Both methods are based on molybdenum blue chemistry and spectrophotometric detection of the orthophosphate generated. The differences between them remain in the necessary preliminary digestion: manual (APHA Method 4500-P H, 2000) or in-line (APHA Method 4500-P I, 2000). In the manual method, polyphosphates are converted to the orthophosphate form by sulfuric acid digestion and organic phosphorus is converted to orthophosphate by persulfate digestion. The in-line method needs UV radiation treatment for organic phosphorus conversion.

3.4.5 FLOW ANALYSIS METHODS

Analysis methods based on flow techniques are an attractive alternative since they allow the analysis of parameters of interest in different types of samples to be carried out with efficiency, speed, comfort, economy as well as with high degrees of automation. Previously, we have mentioned several standard methods for nutrient analysis in wastewaters that use a segmented flow analysis technique or flow injection analysis technique for the automation of a manual standard procedure. In this section we will provide the reader a perspective on the analysis methods for nutrient control in wastewaters based on the more important flow analysis techniques which have been or currently are the objective of research and development, namely: segmented flow analysis (SFA), continuous flow analysis (CFA), flow injection analysis (FIA), sequential injection analysis (SIA), multicommuted flow injection analysis (MCFIA), and multisyringe flow injection analysis (MSFIA). It should be stressed that the implementation of methods based on flow analysis techniques, in addition to the advantages previously mentioned, enables monitoring in real time of the physical and chemical indicators which facilitate water quality control with the aim to assess the impact of polluted inputs, enhancement of treatments, energetic savings, etc. The problem of monitoring has been dealt with in several publications (Trojanowicz *et al.*, 1991; Alexander *et al.*, 1996; Dimitrakopoulos *et al.*, 1996; Colin and Quevauviller, 1998; Hanrahan *et al.*, 2002). The scheme corresponding to a multiparametric monitor, based on a SIA configuration, useful for monitoring basic water contamination is shown in Figure 3.4.1. The monitor is equipped with

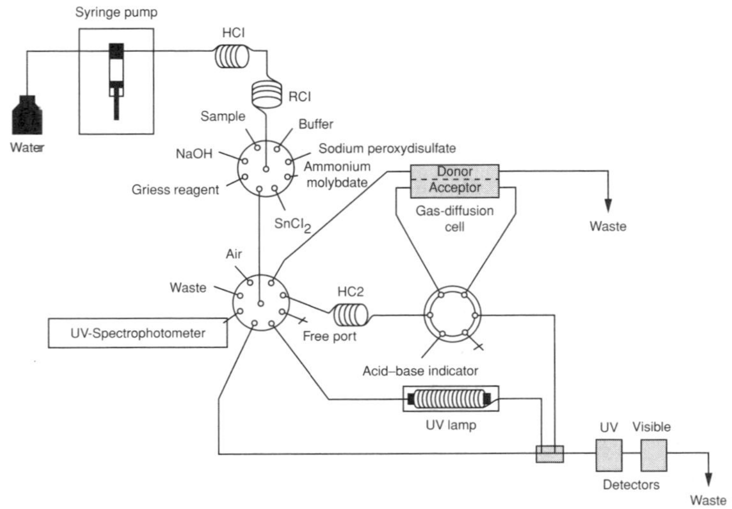

Figure 3.4.1 Wastewater quality monitoring system. HC, holding coil; RC, reaction coil

spectrophotometric detectors which allow the estimation of the dissolved organic carbonaceous pollution expressed as TOC, COD (chemical oxygen demand) or BOD (biological oxygen demand) and particulate pollution (TSS, total suspended solids) using advanced UV spectrophotometry and the determination of ammonium, nitrite, nitrate, total nitrogen, orthophosphate and total phosphate by UV–visible detection (Thomas *et al.*, 1997).

3.4.5.1 Nitrogen

Ammonium

Flow analysis techniques based on the spectrophotometric methods (Menezes-Santos *et al.*, 1992; Muraki *et al.*, 1992; Cosano *et al.*, 1993; Nobrega *et al.*, 1995) of Berthelot and Nessler have been developed and adapted to the determination of ammonium in wastewaters. Methods based on the Berthelot method, in spite of their greater complexity have proved to possess better analytical advantages. However, the use of gas diffusion units for isolation of the analyte from the wastewater matrix to a receptor solution (Andrew *et al.*, 1995; Oms *et al.*, 1996; Akse *et al.*, 1998; Shen *et al.*, 1998; Vlcek and Kuban, 1999; Mulvaney *et al.*, 2000; Wang *et al.*, 2000) constitutes an adequate strategy which allows elimination of interferences, except those caused by the presence of amines, and enables the selection of other types of detection, besides the spectrophotometric detection, such as potentiometric or conductometric detection. Direct potentiometric detection, using ammonium selective electrodes of PVC membrane containing nonactin ionophore (Moschou *et al.*, 1998) or photo-cured membranes on silver wire (Alexander *et al.*, 1997, 1998) as well as amperometric detection (Shen *et al.*, 1997; Kurzawa *et al.*, 2001) have been used in wastewater analysis. Methods with chemiluminescence detection in gas phase, with prior ammonia conversion into nitric oxide and further reaction with ozone (Aoki *et al.*, 1997), have also been successfully used. We have not found in the literature any electrochemiluminiscence, chemiluminiscence or fluorescence method in solution applied to this type of matrix. In Table 3.4.1 are summarized the analytical characteristics of several of the above-mentioned methods.

Nitrite

Several flow spectrophotometric methods have been proposed on the basis of the Griess reaction, or on modalities of the former, for the analysis of nitrites in aqueous samples of very different matrixes, among which wastewaters (Segarra-Guerrero *et al.*, 1996; Gabriel *et al.*, 1998; Hirakawa *et al.*, 1998; Van Staden and van der Merwe, 1998; Galhardo *et al.*, 2001) are found. The use of a membrane-based optical flow-through sensor (Frenzel *et al.*, 2004), using the common spectrophotometric detection scheme for nitrite based azo-dye formation, constitutes a novel option that allows, alternatively, nitrite detection to be carried out by absorption and reflectance measurements. This option does not present interferences due to colour or turbidity

Table 3.4.1 Analytical characteristics of some flow analysis methods for determination of ammonia, and nitrate and nitrite in wastewaters

Analyte	Flow system	Mode/detection technique	Reagent/sensor characteristics	Linear range (mg N/l)	RSD% (mg N/l)	Detection limit (mg N/l)	Sampling rate (/h)	Reference
NH_4^+	FIA	Spec	NaOH-CR-TB-GPM	1.6–15.6	<3.0 (7.8)	0.78	11	Wang *et al.*, 2000
	FIA	Spec	BrTB-GPM	1.0–100	3.4 (2.0)	0.6	12–13	Andrew *et al.*, 1995
	FIA	Cond	SPBB-GPM	0–49	0.46 (8)	0.008	8–9	Akse *et al.*, 1998
	FIA	Pot	Nonac-GPM-ACC	0.8–140	1.4 (2.8)	0.12	25	Shen *et al.*, 1998
	FIA	Pot	Nonac	1.4–140	3.6 (0.034)	0.014	30	Shen *et al.*, 1997
	FIA	CL	Hyp-O_3	0.0014–1.4	1,8(0.0014)	0.00014	20	Aoki *et al.*, 1997
	SIA	Cond	NaOH-Bor-GPM	Up to 62	<3.0 (7.8)	0.08	25	Oms *et al.*, 1996
NO_3^- and NO_2^-	FIA	Sim/Spec	Nitrate: Gr/Cu-Cd	0.056–11.290	2.3 (1.52)	0.056	8	Gabriel *et al.*, 1998
			Nitrite: Gr	0.015–1.520	1.1 (11.29)	0.015		
	FIA	Seq/Bia	Nitrate: NRg/UV	0.023–0.452	1.6 (NR)	0.0152	25	Gil-Torro *et al.*, 1998
			Nitrite: NRg	0.030–0.304	1.2 (NR)	0.0076		
	FIA	Seq/Spec	Nitrate: DAA/Cu-Cd	0.0000–0.0023	0.76 (0.001)	0.00002	24	Segarra-Guerrero *et al.*, 1996
			Nitrite: DAA	0.018–1.217	0.1(0.635)	0.018		
	FIA	Sim/Gp-CL	Nitrate: O_3/I^-	0.005–1.000	1.2 (0.700)	0.0007	NR	Aoki and Wakabayashi, 1995
			Nitrite: O_3/Ti^{3+}	0.003–1.000	1.4 (0.280)	0.0004		
	SIA	Seq/Spec	Nitrate: Gr/Cu-Cd	0.02–2.0	0.8 (0.10)	0.008	20	Galhardo *et al.*, 2001
			Nitrite: Gr	0.010–1.0	NR	0.006		
	FIA	Seq/Spec	Nitrate: Gr/UV	Up to 3.36	2.4 (NR)	0.112	NR	Cerdà *et al.*, 1996
			Nitrite: Gr	Up to 3.08	1.5 (NR)	0.028		
	SIA	Sim/Spec	Nitrate: Gr/Hy	Up to 5.6	3.1 (1.40)	0.045	NR	Oms *et al.*, 1995
			Nitrite: Gr	Up to 5.6	1.5 (1.40)	0.021		

RSD, relative standard deviation; NR, not reported. Mode: Sim (simultaneous) or Seq (sequential). Detection technique: Spec (spectrophotometric), Cond (conductometric), Pot (potentiometric), CL (chemiluminescence), Bia (biamperometric), Gp-CL (gas-phase chemiluminescence). Reagent/sensor characteristics: CR (cresol red), TB (thymol blue), BrTB (bromothymol blue), GPM (gas permeable membrane), SPBB (solid phase basification beds), ACC (activated carbon cartridge), Nonac (ammonium ion-selective coated wire sensor involving a nonactin-based membrane), Hyp (sodium hypochlorite), Bor (boric acid), Cu-Cd (copperized-cadmium packed-bed reactor), UV (UV-irradiation), Gr (Griess-type reaction), NRg (no reagent), DAA (3,6-diamino acridine), Hy (hydrazine).

and does not require prior treatment of the sample. Other flow methods use reactions which are not of the Griess type. Recently, spectrophotometric-FIA determination of nitrites in wastewaters based on the reaction of nitrite with thiocyannate has been described (Kuznetsov *et al.*, 2005). According to Kuznetsov *et al.* this reaction is attractive from an analytical point of view due to its high selectivity. Nitrosation of 4-iodine-*N*,*N*-dimethylaniline (Nikonorov and Moskvin, 1995), the simple formation of iodine (Miura and Kusakari, 1999), the catalytic action of nitrite in the oxidation of Victoria green stand G (Zi *et al.*, 2001) or Rodamine B (Wang and He, 1995) dyes by potassium bromate in acid medium and the reaction with fuchsine in acid medium (Zi and Chen, 2000a) in FIA and r-FIA configurations (Zi and Chen, 2000b) have been also used. Several chemiluminiscence methods have been proposed for determination of this ion in wastewaters. Thus, the use of quenching has been proposed in the oxidation of acriflavine by permanganate (Catala-Icardo *et al.*, 2001) and that based on the conversion of nitrite into nitric oxide and its reaction in gas phase with ozone (Aoki and Wakabayashi, 1995; Aoki *et al.*, 1997). The remaining detection instrumental techniques are less frequently employed. Nevertheless, methods using biamperometric detection (Gil-Torro *et al.*, 1998) and photoacoustic spectrometry (Carrer *et al.*, 1995) based on pulsed laser excitation have been proposed in FIA configurations.

Nitrate

Nitrate content in a wastewater sample is usually calculated on the basis of the difference in contents between oxidized nitrogen ($NO_2^- + NO_3^-$) and nitrite. Flow analysis methods using this approach usually carry out nitrate on-line reduction in the heterogeneous phase, using reactors filled with granulated copper–cadmium (Lapa *et al.*, 2000; Galhardo and Masini, 2001; Gabriel *et al.*, 1998; Hirakawa *et al.*, 1998; Segarra-Guerrero *et al.*, 1996) or in the homogeneous phase (Oms *et al.*, 1995) with hydrazine sulfate in the presence of copper sulfate acting as a catalyser. In both cases the use of a reaction of the Griess type allows the spectrophotometric determination of the oxidized nitrogen content. Nitrite is usually determined with the same procedure but without carrying out the reduction step. The photo-induced generation of nitrite using a mercury lamp wrapped in an aluminium foil as a light source and adding EDTA to the sample as an activator has been also proposed for the analysis of wastewaters (Cerdà *et al.*, 1995). The enzymatic reduction of nitrate as an alternative to the other systems has not been proposed for this type of sample. Also, the reduction to NO in the homogeneous phase (Aoki *et al.*, 1997) has been proposed using I^-, for nitrite, and Ti^{3+}, for nitrate and nitrite, followed by chemiluminiscence detection through the reaction of NO with ozone in the gas phase. A FIA system using a bulk acoustic wave impedance detector together with reducing columns filled with granulated Zn allows the determination and differentiation of nitrate and nitrite contents in wastewaters (Su *et al.*, 1998). The on-line reduction to NH_3 of one or the two species simultaneously, depends on the medium provided by the carrier; thus, in sulfamic acid only nitrate is reduced and in water nitrate and nitrite. Photoreduction has been also used with biamperometric detection (Gil-Torro *et al.*, 1998), detecting the triiodide formed by reaction between iodide and nitrite. As previously mentioned two measurements should be performed, with and without irradiation in order to achieve speciation. Exclusive detection of nitrate is uncommon, however, methods with electrochemical detection have been proposed for this purpose. Thus, for example, the coulometric determination of nitrate by reduction over a glassy carbon electrode, without interference of oxygen or nitrite (Nakata *et al.*, 1990) has been proposed or the potentiometric determination by means of photocured coated-wire electrodes in the flow injection potentiometric mode (Alexander *et al.*, 1998). However, it should be mentioned that only the potentiometric determination has been applied in the analysis of wastewater samples. In Table 3.4.1 the analytical characteristics of some of the above-mentioned methods can be also observed.

Organic and total nitrogen

For organic nitrogen determination sample digestion is required to transform the organic compounds containing nitrogen into nitrogen inorganic species. From the mineralized sample the total nitrogen content can be determined and by subtraction the organic nitrogen content. Sample digestion is the most tiresome and slowest step of the analysis process of these parameters and, hence, has deserved greater attention as far as researchers are concerned with the aim to achieve its automation. This digestion can be carried out in different ways, namely: Kjeldahl method, photochemical oxidation, alkaline oxidation with persulfate or combustion at high temperature. The Kjeldahl method is the recommended manual standard method and provides a parameter widely used in water characterization, the so-called Kjeldahl nitrogen. Although several flow methods may be found in the literature based on segmented flow designs, where digestion is carried out in a helicoidal reactor at controlled temperature (Davidson *et al.*, 1970), the metallic catalyser has been substituted by a sulfonitric mixture and the on-line detection is carried out by the Berthelot reaction; however, these methods have not been applied to wastewater samples. The main problems hampering the implementation of the digestion step in the case of wastewater samples are the obstruction of the flow channels in the digester together with the low recoveries attained. Due to the former reasons, the habitual analysis of this parameter in wastewaters is proposed to be performed by semiautomatic methods, whereby digestion is carried out in traditional digesters and the treatments and/or developments of the reaction for the detection in flow systems (Cerdà *et al.*, 2000). Besides, if automation of the distillation step is aimed for, the difficulty increases considerably and, thus, many researchers have looked for alternatives other than the popular Kjeldahl method. One of these alternatives is on-line UV-photooxidation in the presence of oxidizing agents such as hydrogen peroxide or potassium persulfate. Through this treatment organic nitrogen and ammonium are converted into

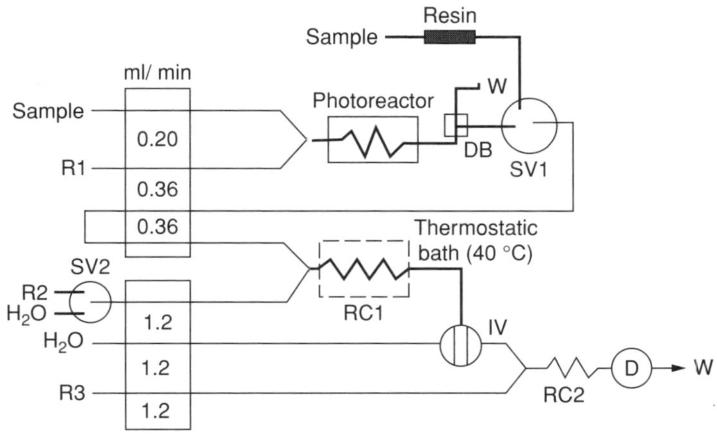

Figure 3.4.2 Flow injection arrangement for determination of nitrite, nitrate and total nitrogen. R1, peroxydisulfate alkaline solution; R2, reducing agent; R3, chromogenic reagent; Resin, amberlite XAD-7; SV, selection valve; IV, injection valve; RC, reaction coil; DB, debubbler; D, detector; W, waste. UV source: UV lamp (15 W, 254 nm)

nitrite and nitrate in a few minutes, and are spectrosphotometrically determined by a Griess-type reaction. The reactors for digestion can be made of quartz or Teflon. The latter are less fragile and easier to manipulate and have been successfully employed in the treatment of samples with very different matrixes, including wastewaters in FIA (Cerdà *et al.*, 1996), (Figure 3.4.2) or SFA (Oleksy-Frenzel and Jekel, 1996). The method of alkaline oxidation with persulfate, also known as the Korolef method (Koroleff, 1969), is another alternative for carrying out sample digestion. In this case mineralization takes place at a temperature of 120 °C and 2 bar of pressure, in an autoclave, for 30–60 min and compounds containing nitrogen are converted into nitrate. Although this method is faster and easier than that of Kjeldahl or of photo-oxidation, it also provides low recoveries with compounds containing nitrogen–nitrogen bonds or HN=C (Nidal,1978; Ebina *et al.*, 1983). The possibility of replacing the autoclave by a microwave oven has allowed a FIA method to be developed that determines the total nitrogen content (Cerdà *et al.*, 1997). In this method all steps are carried out on-line, with a total duration of less than 2 min and an analysis throughput of 45 samples/h. Digestion of the wastewater sample takes place while circulating inside the microwave oven, and at the outlet of the former the produced nitrate is reduced to nitrite with hydrazine sulfate. Nitrite is, in turn, spectrophotometrically detected using a Griess-type reaction. The joint use of alkaline oxidation with persulfate and of heated capillary reactors equipped with platinum catalysers in flow systems has also enabled total nitrogen determination to be carried out in wastewaters with efficiency and speed, achieving an analysis throughput of 15 samples/h (Aoyagi *et al.*, 1989). The last means of digestion consists of the high temperature combustion (HTC) of the sample. This combustion can be carried out in the presence or the absence of a platinum catalyser and allows all nitrogen forms to be determined using automatic equipment with sampling throughput of between 30 and 10 samples/h with high sensitivity. Nitrogen compounds are transformed into NO and this species is detected through its chemiluminescence reaction with ozone. The equipment required for the HTC implementation is more sophisticated than that used in the above-mentioned digestion methods. The procedure is more effective and it is applied to wastewater samples where the presence of refractory organic nitrogenated compounds (Cliford and McGaughey, 1982; Daughton *et al.*, 1985) can be expected.

3.4.5.2 Phosphorus

As previously stated, phosphorus analysis is complex. However, all determinations are carried out on the basis of the use of spectrophotometric methods of molybdovanadate or molybdenum blue with prior transformation into orthophosphate, if required, of the phosphorated species. Both methods have been proposed in FIA (Manzoori *et al.*, 1990; Benson *et al.*, 1996a,b; Korenaga and Sun, 1996), SIA (Muñoz *et al.*, 1997; Mas *et al.*, 1997, 2000) and MCFIA (Wang *et al.*, 1998) configurations and in different modalities for orthophosphate analysis in wastewaters. The use of Nafion or Accurel membranes in FIA configurations in combination with laser diodes and special flow cells (Korenaga and Sun, 1996) has allowed determination of orthophosphate traces. Two SIA methods using spectrophotometric detection, the first based on the formation of an ionic association between molybdovanadophosphoric acid and the green malachite dye (Muñoz *et al.*, 1997) and the second, in the electrogeneration in the tubular flow electrodes of molybdenum blue (Mas *et al.*, 2004) (Figure 3.4.3), have been proposed for orthophosphate determination in these

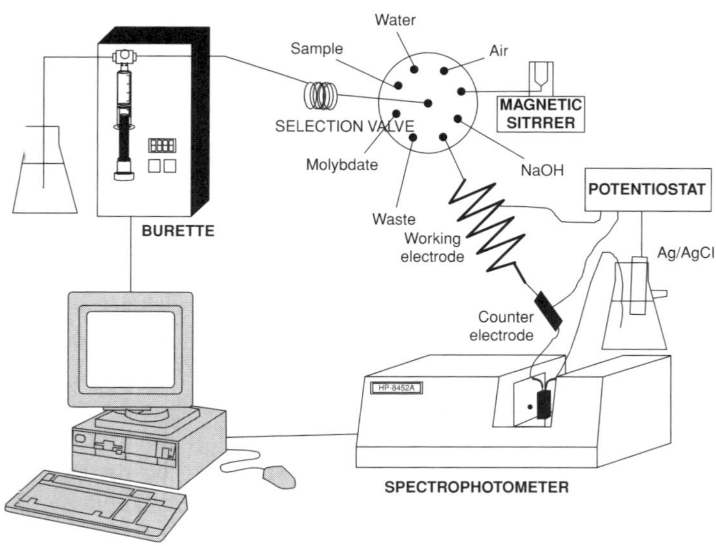

Figure 3.4.3 Schematic illustration of the sequential injection set-up devised for the spectrophotometric determination of orthophosphate based on the electrochemical generation of molybdenum blue

matrixes. Although the implementation of these new flow analysis methods has represented an important step forward in the application and automation of orthophosphate analysis methods, undoubtedly, the most interesting aspect is the possibility of also carrying out the required on-line pretreatments, following methodologies with high degrees of automation, which facilitate the determination of parameters such as dissolved organic phosphorus (DOP) or dissolved total phosphorus (DTP). Thus, FIA methods have been proposed with spectrophotometric detection, which use the molybdenum blue formation reaction allowing the determination of DOP (Higuchi *et al.*, 1998) and DTP (Williams *et al.*, 1993; Halliwell *et al.*, 1996) in wastewaters. In the former case the photo-oxidative and the acid hydrolysis methods are carried out on-line. In this context it is worthwhile mentioning the FIA method (Benson *et al.*, 1996) which enables determination of total phosphorus (TP) and implies the use of a combined photo-oxidation and thermal digestion system with which conversion of condensed and organic phosphates into orthophosphates is carried out in the soluble and particulate phase. Also, flow injection gel filtration techniques have been used for speciation of phosphorus compounds in wastewaters (McKelvie *et al.*, 1993). FIA methods (Miyazaki and Bansho, 1989; Manzoori *et al.*, 1990) which use combined spectrophotometry and inductively coupled plasma spectroscopy with optical detection techniques (FIA-ICP-AES) have been proposed to carry out rapid differential determination of orthophosphate and total phosphate in wastewaters. As regards to electric techniques, the following should be outlined: a FIA-potentiometric method (De Marco *et al.*, 1998), which uses a second-species cobalt wire ISE relied upon cobalt phosphate determination for orthophosphate precipitation in wastewaters and a FIA-amperometric method for the determination of total phosphorus in domestic wastewaters, which uses continuous microwave oven decomposition with subsequent detection of orthophosphate (Hinkamp and Schwedt, 1990). In Table 3.4.2 are summarised the analytical characteristics of several of the above-mentioned methods.

3.4.6 CHROMATOGRAPHIC METHODS

Analysis of nutrients in their inorganic form can be carried out in a simultaneous, efficient and rapid way by application of a chromatographic method. Undoubtedly, methods based on ion chromatography (IC) in its modality of ionic exchange with eluent conductivity suppression, suppressed ion chromatography (SIC), have been and currently are the most widely used since their introduction (Small *et al.*, 1975). On the other hand, it should be mentioned that this method became a standard method for determination of chloride, bromide, nitrite, nitrate, phosphate and sulfate in water and wastewaters. In wastewater analysis the only pretreatment of the sample consists in its filtration through 0.45 μm membranes and NO_3^-, NO_2^- and PO_4^{3-} contents are determined by SIC, and NH_4^+ content by automated wet chemistry, e.g. FIA, SIA, etc. In this context Matsui *et al.* (Matsui *et al.*, 1997) have proposed a method for the determination of ammonium, nitrite, nitrate, chloride and sulfate in wastewaters. Ammonium is spectrophotometrically detected in a FIA system by

Table 3.4.2 Analytical characteristics of some flow analysis methods for orthophosphate determination in wastewaters

Flow system	Detection technique	Reagents	Linear range (mg P/l)	RSD% (mg P/l)	Detection limit (mg P/l)	Sampling rate (/h)	Reference
FIA	Spec	Mo-V	Up to 200	2 (10)	0.8	8	Manzoori *et al.*, 1990
FIA	Spec	Mo/Sn-Hy	0–25	0.4 (8,75)	0.05	20	Benson *et al.*, 1996a,b
FIA	Spec	Mo-Sb/Asc	0.001–0.05	1.0 (0.020)	0.0006	12	Korenaga and Sun, 1996
FIA	Spec	DR: Perox + H2SO4, Mo-Sb/Asc-NaDS	0.10–1.0	2.25–0.13(0.024–3.03)	0.001	20	Higuchi *et al.*, 1998
FIA	Spec	MWD in HNO_3 medium, Mo/Asc	Up to 6.53	<5.0 (0.033–6.53)	0.033	30	Williams *et al.*, 1993
FIA	Spec	DR: Perox + HClO4, Mo/Sn-Hy	0–18	≤2.0 (10.2)	0.15	32	Benson *et al.*, 1996a,b
IC-FIA	Spec	TD in H_2SO_4 medium, Mo/Sn-Hy	Or: 0.010–1.00 Pyr and Tri: 0.020–2.00	≤3.0 (1.00)	Or: <0.01 Pyr and Tri: 0.020	5	Halliwell *et al.*, 1996
SIA	Spec	Mo-V	Up to 18.00	2.1(5.00)	0.15	30	Muñoz *et al.*, 1997
SIA	Spec	Mo-V-MG	0.05–0.40	18(0.10)	0.01	30	Muñoz *et al.*, 1997
SIA	Spec	Mo/Sn	0.05–4.00	1.7(2.50)	0.01	30	Muñoz *et al.*, 1997
SIA	Spec	Mo/SSTFTE	0.3–20	1.8(10)	0.1	18	Mas *et al.*, 2004
SIA	Spec	Mo-V	Up to 12	1.4 (9)	0.2	23	Mas *et al.*, 1997
SIA	Spec	Mo-V	0.8–15	2.1(5.0)	0.23	30	Mas *et al.*, 2000
MCFIA	Spec	Mo-Sb/Asc	Up to 3	1.4(2.47)	NR	180	Wang *et al.*, 1998
FIA	Spec + ICP-AES	Spec: Mo-V	Up to 200 Or and TP	ICP: 2.01 (10)	Spec: 0.8 ICP:0.5	80	Manzoori *et al.*, 1990
FIA	Pot	Pht-CoW	3.1–310	4.0 (31)	0.093	NR	De Marco *et al.*, 1998
FIA	Amp	MWD/DR: Perox or $HClO_4$	Up to 30	3 (5.0)	0.10	21	Hinkamp and Schwedt, 1990

RSD, relative standard deviation; NR, not reported Detection technique: Spec (spectrophotometric), ICP (inductively coupled plasma), AES (atomic emission spectrophotometry), Pot (potentiometric), Amp (amperometric). Reagents: Mo (potassium ammonium molybdate), V (ammonium vanadate), Sn [tin (II)], Hy (hydrazine), Sb (antimony tartrate), Asc (ascorbic acid), DR (digestion reagent), Perox (sodium peroxydisulfate), NaDS (sodium dodecylsulfate), MWD (microwave digestion), TD (thermal digestion), MG (malachuite green), SSTFTE (stainless steel tubular flow-through electrode), Pht (phthalate buffer), CoW (cobalt wire electrode), Or (orthophosphate), Pyr (pyrophosphate), Tri (triphosphate), TP (total phosphorus).

a postcolumn derivatization reaction using the indophenol reaction, and the other ions by conductimetric detection. In other studies (Karmarkar, 1998, 1999) the use of this strategy is also proposed for nutrient analysis in wastewaters. In Karmarkar's first study (Karmarkar, 1999) a sequential IC-FIA method is used which allows determination in only one injection of NO_3^-, PO_4^{3-} and NH_4^+. Ammonium is determined at the outlet of the column in the void volume by a FIA system and the remaining analytes with a conductimetric detector in the usual SIC way. In Karmarkar's second study (Karmarkar, 1998) F^-, Cl^-, NO_3^-, Br^-, HPO_4^{2-} and SO_4^{2-} are analysed in wastewaters by enhanced IC with sequential FIA. The use of on-line dialysis has been proposed for automation of sampling and pretreatment of wastewater samples in order to carry out the analysis of ions and small molecules by FIA and chromatography, in a fast economical way and without analyte loss (Frenzel, 1997). Laubli *et al.* (Laubli *et al.*, 1999) have determined F^-, Cl^-, NO_2^-, NO_3^-, Br^-, PO_4^{3-} and SO_4^{2-} in wastewaters by SIC using a Metrosep Anion Dual 2 column, and a mixture of $NaHCO_3$ and Na_2CO_3 as eluent, in combination with a sample pretreatment in an on-line dialysis unit and using a stop-flow technique.

3.4.7 CAPILLARY ELECTROPHORESIS METHODS

This technique presents sensitivity, low sample consumption, high resolution and it is fast in relation to chromatographic methods. However, there are few literature data with regard to the application of this technique to nutrient analysis in wastewaters. One of the few applications, which can illustrate the potential of this technique, is that described by Pantsar-Kallio *et al.* (Pantsar-Kalio *et al.*, 1997). These authors propose a method which allows separating and determining a total of nine organic acids and seven inorganic anions (Cl^-, SO_4^{2-}, NO_2^-, NO_3^-, F^-, PO_4^{3-} and CO_3^{2-}) in wastewaters. The method uses pyridine-2,6-dicarboxylic acid as electrolyte, tetradecyltrimethylammonium bromide as electro-osmotic flow modifier and the analytes were detected by measuring indirect UV absorption.

REFERENCES

Alexander, P.W., Dimitrakopoulos, T. and Hibbert D.B. (1997) Electroanalysis, **9**(17), 1331–1336.

Alexander, P.W., Dimitrakopoulos, T. and Hibbert, D.B. (1998) Electroanalysis, **10**(10), 707–712.

Alexander, P.W., Di Benedetto, L.T., Dimitrakopoulos, T., Hibbert, D.B., Ngila, J.C., Sequeira, M. and Shiels, D. (1996) Talanta, **43**(6), 915–925.

Andrew, K.,. Worsfold, P.J. and Comber, M. (1995) Anal. Chim. Acta, **314**(1–2), 33–43.

Aoki, T. and Wakabayashi, M. (1995) Anal. Chim. Acta, **308** (1–3), 308–312.

Aoki, T., Fukuda, S., Hosoi, Y. and Mukai, H. (1997) Anal. Chim. Acta, **349**(1–3), 11–16.

Aoyagi, M., Yasumasa, Y. and Nishida, A. (1989) Anal. Sci., **5** (2), 235–236.

Akse, J.R., Thompson, J.O., Sauer, R.L. and Atwater, J.E. (1998) Microchem. J., **59**(3), 372–382.

American Public Health Association-American Water Works Association-Water Pollution Control Federation (APHA-AWWA-WPCF). (2000) Standard Methods for the Examination of Water and Wastewater, 20th ed. APHA, Washington, DC.

APHA Method 4110 B. (2000) Chromatographic Method.

APHA Method 4500-N B. (2000) In-line UV/Persulfate Digestion and Oxidation with Flow Injection Analysis.

APHA Method 4500-N C. (2000) Persulfate Method.

APHA Method 4500-NH_3 B (2000) Preliminary Distillation Step.

APHA Method 4500-NH_3 C. (1992) Nesslerization method. Standard Methods for the Examination of Water and Wastewater, 17th ed. American Public Health Association-American Water Works Association-Water Pollution Control Federation (APHA-AWWA-WPCF).APHA, Washington, DC.

APHA Method 4500-NH_3 C. (2000) Titrimetic Method.

APHA Method 4500-NH_3 D. (2000) Ammonia Selective Electrode Method.

APHA Method 4500-NH_3 E. (2000) Ammonia Selective Electrode Method Using Known Addition.

APHA Method 4500-NH_3F. (2000) Phenate Method.

APHA Method 4500-NH_3 G. (2000) Automated Phenate Method.

APHA Method 4500-NH_3 H. (2000) Flow Injection Analysis.

APHA Method 4500-NO_2^- B. (2000) Colorimetric Method.

APHA Method 4500-NO_3^- D. (2000) Nitrate Electrode Method.

APHA Method 4500-NO_3^- E. (2000) Cadmium Reduction Method.

APHA Method 4500-NO_3^- F. (2000) Automated Cadmium Reduction Method.

APHA Method 4500-NO_3^- G. (1992) Method of reduction with titanuos chloride. Standard Methods for the Examination of Water and Wastewater, 17th ed. American Public Health Association-American Water Works Association-Water Pollution Control Federation (APHA-AWWA-WPCF). APHA, Washington, DC.

APHA Method 4500-NO_3^- H. (2000) Automated Hydrazine Reduction Method.

APHA Method 4500-NO_3^-I. (2000) Cadmium Reduction Flow Injection Method.

APHA Method 4500-N_{org}. (2000) Nitrogen Organic Method.

APHA Method 4500-N_{org}B. (2000) Macro-Kjeldhal Method.

APHA Method 4500-N_{org}C. (2000) Semi-Micro Kjeldhal Method.

APHA Method 4500-N_{org}D. (2000) Block Digestion and Flow Injection Analysis.

APHA Method 4500-P B. (2000) Sample Preparation.

APHA Method 4500-P C. (2000) Vanadomolybdophophoric Acid Colorimetric Method.

APHA Method 4500-P D. (2000) Stannous Chloride Method.

APHA Method 4500-P E. (2000) Ascorbic Acid Method.

APHA Method 4500-P F. (2000) Automated Ascorbic Acid Reduction Method.

APHA Method 4500-P G. (2000) Flow Injection Analysis for Orthophosphate.

APHA Method 4500-P H. (2000) Manual Digestion and Flow Injection Analysis for Total Phosphorus.

APHA Method 4500-P I. (2000) In-line UV/Persulfate Digestion and Flow Injection Analyis for Phosphorus.

Armstrong, D.E. (1972) In: M. Halmann (Ed.), Analytical Chemistry of Phosphorus Compounds. Wiley/Interscience, New York, pp. 744–769.

Benson, R.L., McKelvie, I.D., Hart, B.T., Truong, Y.B. and Hamilton, I.C. (1996a) Anal. Chim. Acta, **326**(1–3), 29–39.

Benson, R.L., Truong, Y.B., McKelvie, I.D. and Hart, B.T., (1996b) Water Res., **30**(9), 1959–1964.

Brober, O. and Persson, G. (1988) Hydrobiologia, **170,** 61.

Burden, E.H.W.J. (1961) Analyst **86**, 429–433.

Carrer, I., Cusmai, P., Zanzottera, E., Martinotti, W. and Realini, F. (1995) Anal. Chim. Acta, **308** (1–3), 20–27.

Catala-Icardo, M., Martinez-Calatayud, J. and Garcia-Mateo, J.V. (2001) Analyst, **126**(8), 1423–1427.

Cerdà, A, Oms, M.T. and Cerdà, V. (2000) Determination of Organic Nitrogen: Handbook of Water Analysis. Marcel Dekker Inc., New York.

Cerdà, A., Oms, M.T., Cerdà, V. and Forteza, R. (1995) Anal. Methods Instrum., **2**(6), 330–336.

Cerdà, A., Oms, M.T., Forteza, R. and Cerdà, V. (1996) Analyst, **121**(1), 13–17.

Cerdà, A., Oms, M.T., Forteza, R. and Cerdà, V. (1997) Anal. Chim. Acta, **351**(1–3), 273–279.

Clementson, L.A. and Wayte, S.E. (1992) Water Res., **26**(9), 1171–1176.

Cliford, D.A. and McGaughey, L.M. (1982) Anal. Chem., **54**(8), 1345–1350.

Colin, F. and Quevauviller, P. (Eds) (1998) Monitoring of Water Qualitity, Proceedings of the European Workshop on Standards, Measurements and Testing. Elsevier, Amsterdam.

Cosano, J.S., Luque de Castro, M.D. and Valcárcel, M. (1993) J. Autom. Chem., **15**(4), 147–150.

Daughton, C.G., Jones, B.M. and Sakaji, R.H. (1985) Anal. Chem., **57**(12), 2326–2333.

Davidson, J., Mathieson, J. and Boyne, A.W. (1970) Analyst, **95**(127), 181–193.

De Marco, R., Pejcic, B. and Chen, Z.L. (1998) Analyst, **123**(7), 1635–1640.

Dimitrakopoulos, T., Alexander, P.W. and Hibbert. D.B. (1996) Electroanalysis, **8**(5), 438–442.

Dore, J.E., Houlihan, T.,. Hebel, D.V, Tien, G., Tupas, L. and Karl, D. M. (1996) Mar. Chem., **53**(3/4), 173–185.

Ebina, J., Tsutsui, T. and Shirai, T. (1983) Water Res., **17**(12), 1721–1726.

Forman, D., Al-Dabbaghand, S. and Doll, R. (1985) Nature, **313** (6004), 620–625.

Frenzel, W. (1997) GIT Labor-Fachzeitschrift, **41**(7), 734, 736–738, 740–741.

Frenzel, W., Schulz-Brussel, J. and Zinvirt, B. (2004) Talanta, **64**(2), 278–282.

Gabriel, D., Baeza, J., Valero, F. and Lafuente, J. (1998) Anal. Chim. Acta, **359**(1–2), 173–183.

Galhardo, C.X. and Masini, J.C. (2001) Anal. Chim. Acta, **438**(1–2), 39–48.

Gil-Torro, I., García-Mateo, J.V. and Martínez-Calatayud, J. (1998) Anal. Chim. Acta, **366**(1–3), 241–249.

Halliwell, D.J., McKelvie, I.D., Hart, B.T. and Dunhill, R.H. (1996) Analyst, **121**(8), 1089–1093.

Hanrahan, G., Ussher, S., Gledhill, M., Achterberg, E.P. and Worsfold, P.J. (2002) Trends Anal Chem., **21**(4), 233–239.

Higuchi, K., Tamanouchi, H. and Motomizu, S. (1998) Anal. Sci., **14**(5), 941–946.

Hinkamp, S. and Schwedt, G. (1990) Anal. Chim. Acta, **236**(2), 345–350.

Hirakawa, K. Yoshida, I. and Ishii, D. (1998) Bunseki Kagaku, **47**(6), 341–348.

Johnson, C.J. and Cross, B.C. (1990) Am. J. Ind. Med., **18**(4), 449–456.

Karmarkar, S.V. (1998) Am. Environ. Lab., **10**(2), 6–7.

Karmarkar, S.V. (1999) J. Chromatogr. A, **850**(1 and 2), 303–309.

Korenaga, T. and Sun, F.S. (1996) Talanta, **43**(9), 1471–1479.

Koroleff, F. (1969) Determination of total nitrogen in natural waters by means of persulfate oxidation. Congress for Exploration of the Sea (ICES), Poster C8.

Kurzawa, C., Schuhmann, W., Wang, L., Orth, H., Schwendtke, I., Gerberding, H., Stadler, H. and Grundig, B. (2001) GIT Labor-Fachzeitschrift, **45**(2), 156–160.

Kuznetsov, V.V., Zemyatova, S.V. and Ermolenko, Y.V. (2005) J. Anal. Chem., **60**(3), 289–296.

Lapa, R.A.S., Lima, J.L.F.C. and Pinto, I.V.O.S. (2000) Analusis, **28**(4), 295–301.

Laubli, M., Sugai, T., Ono, J. and Schafer, H. (1999) Kogyo Yosui, **493**, 32–34.

Manzoori, J.L., Miyazaki, A. and Tao, H. (1990) Analyst, **115**(8), 1055–1058.

Mas, F., Estela, J.M., Miro, M., Cladera, A. and Cerda, V. (2004) Anal. Chim. Acta, **510**(1), 61–68.

Mas, F., Muñoz, A., Estela, J.M. and Cerdà, V. (1997) Analyst, **122**(10), 1033–1038.

Mas, F., Muñoz, A., Estela, J.M. and Cerdà, V. (2000) Int. J. Environ. Anal. Chem., **77**(3), 185–202.
Matsui, M., Gotoh, T., Ishibashi, T. and Nishikawa, M. (1997) Kankyo Kagaku, **7**(1), 23–30.
McKelvie, I.D. (2000) Phosphates: Handbook of Water Analysis. Marcel Dekker, Inc., New York.
McKelvie, I.D., Peat, D. and Worsfold, P.J. (1995) Anal. Proc., **32** (10), 437–445.
McKelvie, I.D., Hart, B.T., Cardwell, T.J. and Cattrall, R.W. (1993) Talanta, **40**(12), 1981–1993.
Menezes-Santos, M., Reis, B.F., Bergamin, H. and Baccan, N. (1992) Anal. Chim. Acta, **261**(1–2), 339–343.
Miura, Y. and Kusakari, K. (1999) Anal. Sci., **15**(9), 923–926.
Miyazaki, A. and Bansho, K. (1989) Kogai, **24**(2), 87–93.
Mopper, K. and Zika, R.G. (1987) Nature, **325**(6110), 246–249.
Moschou, E., Chaniotakis, N. A., Papandroulakis, N. and Divanach, P. (1998) Am. Environ. Lab., **10**(7), 10–12.
Mulvaney, R.L., Strle, K. and Horgan, B.P. (2000) J. Environ. Qual., **29**(6), 1890–1895.
Muñoz, A., Mas, F., Estela, J.M. and Cerdà, V. (1997) Anal. Chim. Acta, **350**(1–2), 21–29.
Muraki, H., Higuchi, K., Sasaki, M., Korenaga, T. and Toei, K. (1992) Anal. Chim. Acta, **261**(1–2), 345–349.
Nakata, R., Terashita, M., Nitta, A. and Ishikawa, K. (1990) Analyst, **115**(4), 425–430.
Nidal, F. (1978) Water Res., **12**(12), 1123–1130.
Nikonorov, V.V. and Moskvin, L.N. (1995) Anal.Chim Acta, **306**(2–3), 357–360.
Nobrega, J.A., Mozeto, A.A., Alberici, R.M. and Guimaraes, J.L. (1995) J. Braz. Chem. Soc., **6**(4), 327–330.
Nollet, L.M.L. (ED.) (2000) Handbook of Water Analysis. Marcel Dekker, Inc., New York.
Oleksy-Frenzel, J. and Jekel, M. (1996) Anal.Chim. Acta, **319**(1–2), 165–175.
Oms, M.T., Cerdà, A. and Cerdà, V. (1995) Anal. Chim. Acta, **315**(3), 321–330.
Oms, M.T., Cerdà, A. and Cerdà, V. (1996) Electroanalysis, **8**(4), 387–390.
Pantsar-Kallio, M., Kuitunen, M. and Pentti, P.K.G. (1997) Chemosphere, **35**(7), 1509–1518.
Robards, K., McKelvie, I.D., Benson, R.L., Worsfold, P.J., Blundell, N. and Casey H. (1994) Anal. Chim. Acta, **287**(3), 147–190.
Russell. S. (1994) Ammonia. WRc Instrument Handbooks, Swindon, UK.
Segarra-Guerrero, R., Gomez-Benito, C. and Martinez-Calatayud, J. (1996) Talanta, **43**(2), 239–246.
Shen, H., Cardwell, T.J. and Cattrall, R.W. (1997) Analyst, **122**(1), 89–93.
Shen, H., Cardwell, T.J. and Cattrall, R.W. (1998) Anal. Chim. Acta, **367**(1–3), 193–199.
Small, H., Stevens, T.S. and Bauman, W.C. (1975) Anal. Chem., **47**(11), 1801–1809.
Standard Methods Committee. (1988) American Public Health Association-American Water Works Association-Water Pollution Control Federation (APHA-AWWA-WPCF).
Standards Australia and Standards New Zealand. (1998) Australian/New Zealand, Water Quality-Sampling. Part 1: Guidance on the Design of Sampling Programs, Sampling Techniques and the Preservation and Handling of Samples.
Stumm, W. and Morgan, J.J. (1996) Aquatic Chemistry, 3rd ed. Wiley, New York.
Su, X.L., Chen, P., Qu, X.G., Wei, W.Z. and Yao, S.Z. (1998) Microchem. J., **59**(3), 341–350.
Thomas, O., Theraulaz, F., Cerdà, V., Constant, D. and Quevauviller, Ph. (1997) Trends Anal. Chem., **16**(7), 419–424.
Trojanowicz, M., Benson, R.L. and Worsfold, P.J. (1991) Trends Anal. Chem., **10**(1), 11–17.
Van Staden, J.F. and van der Merwe, T.A. (1998) Mikrochim. Acta, **129**(1–2), 33–39.
Vlcek, J. and Kuban, V. (1999) Coll. Czech. Chem. Commun., **64**(12), 1966–1974.
Wang, J. and He, R. (1995) Huanjing Kexue, **16**(1), 71–73.
Wang, L., Cardwell, T.J., Cattrall, R.W., Luque de Castro, M.D. and Kolev, S.D. (2000) Anal. Chim. Acta, **416**(2), 177–184.

Wang, X.D., Cardwell, T.J., Cattrall, R.W., Dyson, R.P. and Jenkins, G.E. (1998) Anal. Chim. Acta, **368**(1–2), 105–111.

Williams, K.E., Haswell, S.J., Barclay, D.A. and Preston, G. (1993) Analyst, **118**(3), 245–248.

Zi, Y. and Chen, L. (2000a) Fenxi Ceshi Xuebao, **19**(2), 70–72.

Zi, Y. and Chen, L. (2000b) Fenxi Huaxue, **28**(3), 395.

Zi, Y., Duan, L. and Lu, Ch. (2001) Fenxi Huaxue, **29**(2), 186–188.

4.1
State Estimation for Wastewater Treatment Processes

Olivier Bernard, Benoît Chachuat, and **Jean-Philippe Steyer**

Wastewater Quality Monitoring and Treatment Edited by P. Quevauviller, O. Thomas and A. van der Beken

4.1.1 INTRODUCTION

A major bottleneck in the application of advanced monitoring and optimization strategies for wastewater treatment plants (WWTPs) lies in the difficulty of measuring chemical and biological variables. Even though considerable progress in on-line sensor technology has been made over recent years, it is still often difficult to measure many of the key state variables such as biomass concentration, specific bacterial activity or intermediate product concentration. Hence, the idea arose of using *observers*, also called *software sensors*, that combine a number of readily available on-line measurements (gaseous flow rates, pH, dissolved gases, etc.) with a process model for estimating the values of (unmeasured) state variables.

Quite a few methods have been proposed to design such observers. In this chapter, we shall only focus on those approaches that are relevant to the field of WWTPs. It is worthwhile noting that the principles underlying the design of observers can be quite different. Therefore, the choice of an observer inherently depends on the specificities of the problem at hand. In practice, this choice is strongly guided by the reliability of the process model as well as the amount and accuracy of the data. If a reliable process model is available and if this model has been thoroughly identified and validated, either an (extended) *Kalman filter* or a high gain observer can be developed. When the process model is not accurate enough, an *asymptotic observer* relying on mass-balance principles, but not on the uncertain kinetics, shall be used instead. Finally, if bounds are known for the uncertain inputs and/or parameters, an *interval observer* can be used for predicting intervals in which the unmeasured variable are guaranteed to belong (instead of point-wise estimates).

The type of observer to be constructed should not only be based on the model quality, but it must also account for the objectives to be achieved. Indeed, an observer can have other purposes than simply monitoring a WWTP. It can be developed with the objective of applying a control action that needs an estimate of some internal state; it can also be used for diagnosing whether a failure occurred during process operation or not.

The remainder of this chapter is organized as follows. A number of useful definitions and results are given in Section 4.1.2. Observers that require a full-model description of the process are presented in Section 4.1.3, with emphasis placed on linear systems. The design of observers relying on the mass-balance principles is discussed in Section 4.1.4. Interval observers that exploit knowledge of bounds on the model uncertainty are presented in Section 4.1.5. Finally, Section 4.1.6 concludes the chapter.

4.1.2 PRELIMINARIES

This section gives an overview of the main theoretical concepts in system observability. These concepts are useful in the analysis conducted later on. The interested reader is referred to Luenberger (Luenberger, 1979) and Gauthier and Kupka (Gauthier and Kupka, 2001) for additional information.

It is first fundamental to study the *observability* property of a system prior to designing an observer. Intuitively, observability consists of determining whether the measured signals contain sufficiently rich information to estimate the unmeasured state variables; a system is then said to be *observable* if it satisfies this property from a theoretical point of view. The problem to address next is to derive an observer for the problem at hand, i.e. an auxiliary dynamic system that provides the state estimates. At this point, it should be noted that the problems of observability and observer design are very different in nature. In particular, the observability property does not give any clue on how to build an observer.

The theory of observation has been extensively developed in the linear case. Several methods also exist in the nonlinear case, but are tailored to specific classes of models.

4.1.2.1 Notion of Observability

We consider the following general model driving the process dynamics:

$$\begin{cases} \dfrac{\mathrm{d}x(t)}{\mathrm{d}t} = f\,[x(t), u(t)]\,; \quad x(0) = x_0 \\ y(t) = h[x(t)] \end{cases} \tag{S}$$

where $u \in \Re^m$ is the input vector, $y \in \Re^p$ is the output vector and $x \in \Re^n$ is the state vector made up of the concentrations of the various species inside the liquid phase; x_0 is the vector of initial conditions. The applications f and h provide the dynamics of the state variables and the links between the state variables and the measurements, respectively.

The objective is to estimate $x(t)$ from the measurements $y(t)$. Observability is a structural property of a system that states whether this is possible or not.

Property 1. The system (S) is said to be observable, if $x(t)$ can be uniquely determined from

$$y(t), \frac{\mathrm{d}y(t)}{\mathrm{d}t}, \frac{\mathrm{d}^2y(t)}{\mathrm{d}t^2}, \ldots, \frac{\mathrm{d}^{n_y}y(t)}{\mathrm{d}t^{n_y}}$$

and

$$u(t), \frac{\mathrm{d}u(t)}{\mathrm{d}t}, \frac{\mathrm{d}^2u(t)}{\mathrm{d}t^2}, \ldots, \frac{\mathrm{d}^{n_u}u(t)}{\mathrm{d}t^{n_u}}$$

for some (possibly infinite) $n_y \geq 0$ and $n_u \geq 0$.

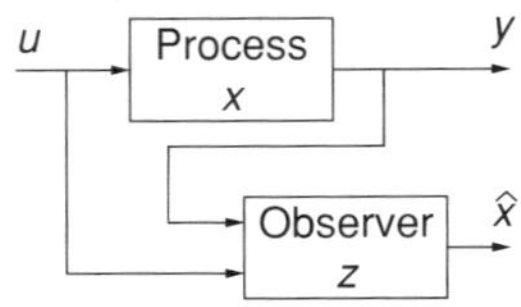

Figure 4.1.1 Observer principle

The reader is referred to Luenberger (Luenberger, 1979) and Gauthier and Kupka (Gauthier and Kupka, 2001) for more details.

4.1.2.2 General Definition of an Observer

Once a system has been shown to be observable, the next step is to design an observer that estimates the state variable x based on a model and a set of input/output measurements. The principle of an observer is presented in Figure 4.1.1. Roughly speaking, an observer is an auxiliary dynamic system coupled to the original system via the measured inputs and outputs. This is formalized in the following definition.

Definition 1. An observer is an auxiliary system (O) coupled to the original system (S) as:

$$\begin{cases} \dfrac{\mathrm{d}z(t)}{\mathrm{d}t} = \hat{f}\,[z(t), u(t), y(t)]\,; \quad z(0) = z_0 \\ \hat{x}(t) = \hat{h}\,[z(t), u(t), y(t)] \end{cases} \qquad (O)$$

where $z \in \Re^q$ denotes the state of the observer, $\hat{f}$ is the observer dynamics and $\hat{h}$ relates z to the estimate $\hat{x}$ of the real system. An observer has the property that the observation error converges to zero asymptotically:

$$\lim_{t\to\infty} \|\hat{x}(t) - x(t)\| = 0$$

A desirable property for an observer is the ability to tune the convergence rate in order for the estimates to converge more rapidly than the original dynamics of the system. Another desirable property is that the estimate $\hat{x}(t)$ should remain equal to $x(t)$ under proper initialization, i.e. when it is initialized with the true value $x(0)$. This easily justifies that the following structure is often used to design observers in practice:

$$\begin{cases} \dfrac{\mathrm{d}\hat{x}(t)}{\mathrm{d}t} = f\,[\hat{x}(t), u(t)] + k\,\{z(t), [h\,[\hat{x}(t) - y(t)]]\} \\ \dfrac{\mathrm{d}z(t)}{\mathrm{d}t} = \hat{f}\,[z(t), u(t), y(t)] \quad \text{with} \quad k\,[z(t), 0] = 0 \end{cases}$$

This observer consists of a replica of the original dynamics corrected by a term that depends on the discrepancy between both the measured and predicted outputs. Note also that the correction amplitude is tuned via the function k that is often referred to as the observer *gain* (internal tuning of the observer).

4.1.3 OBSERVERS FOR LINEAR SYSTEMS

For time-invariant, linear systems, the general system (S) simplifies to:

$$\begin{cases} \dfrac{\mathrm{d}x(t)}{\mathrm{d}t} = Ax(t) + Bu(t) \\ y(t) = Cx(t) \end{cases} \tag{S_L}$$

with $A \in \Re^{n\times n}$ $(n \geq 2)$ and $C \in \Re^{p\times n}$. A well-known observability criterion for (S_L) is given by the rank condition:

$$\text{rank}\begin{pmatrix} C \\ CA \\ \vdots \\ CA^{n-1} \end{pmatrix} = n$$

4.1.3.1 Luenberger Observer

Theorem 1. If the pair *(A,C)* is observable, a Luenberger observer for (S_L) is obtained as (Luenberger, 1966):

$$\frac{\mathrm{d}\hat{x}(t)}{\mathrm{d}t} = A\hat{x}(t) + Bu(t) + \boldsymbol{K}\,[C\hat{x}(t) - y(t)] \tag{O_L}$$

where $\boldsymbol{K}$ is a $n \times n$ gain matrix that can be used for tuning the convergence rate of the observer, and can be chosen in order for the observation error to converge to zero arbitrarily fast.

Proof. The dynamics of the observation error $e(t) = \hat{x}(t) - x(t)$ is given by:

$$\frac{\mathrm{d}e}{\mathrm{d}t} = (A + \boldsymbol{K}C)e$$

and is independent of the input $u(t)$. The result follows from the pole placement theorem which guarantees that the error dynamics can be chosen arbitrarily.

In principle, the gain matrix $\boldsymbol{K}$ can be chosen in such a way that the observation error converges to zero as quickly as desired. However, the larger the gain of the observer, the more sensitive it becomes to external perturbations (measurement noise for example). A good compromise must thus be sought that ensures both stability and accuracy at the same time. The Kalman filter, discussed in the subsection below, proposes a way of achieving such a compromise.

4.1.3.2 The Linear Case up to an Output Injection

A particular situation wherein a linear observer can be designed for a nonlinear system arises in the simple case where the nonlinearities depend on the output y only:

$$\begin{cases} \dfrac{\mathrm{d}x(t)}{\mathrm{d}t} = Ax(t) + \phi\,[t, y(t)] + Bu(t) \\ y(t) = Cx(t) \end{cases}$$

with ϕ being a (known) nonlinear function in $\Re^n$. The following 'Luenberger-like' observer has linear dynamics with respect to the observation error:

$$\frac{\mathrm{d}\hat{x}(t)}{\mathrm{d}t} = A\hat{x}(t) + \phi\,[t, y\,(t)] + Bu(t) + \boldsymbol{K}\,[C\hat{x}(t) - y(t)]$$

In particular, the error dynamics can be chosen arbitrarily, provided that the pair (A,C) is observable. Here again, however, an adequate choice for the gain vector $\boldsymbol{K}$ is one that guarantees a fast enough convergence of the observer, while keeping it stable.

4.1.3.3 Kalman Filter

The Kalman filter is notorious in the field of linear systems (Lewis, 1986). Loosely speaking, a Kalman filter can be seen as a Luenberger observer with a time varying gain. More specifically, the gain is chosen in such a way that the variance of the observation error is minimized (or, equivalently, the integral between t_0 and t of the squared errors is minimized); for this reason the Kalman filter is often referred to as the *optimal estimator*.

Consider an observable continuous-time system in the following stochastic representation:

$$\frac{\mathrm{d}x(t)}{\mathrm{d}t} = Ax(t) + Bu(t) + G\omega(t); \quad x(t_0) = x_0 \tag{1}$$

where $\omega \sim [0, Q(t)]$ is a white noise process with zero mean and covariance $Q(t)$. Suppose that initial state x_0 is unknown, but there is available a priori knowledge that $x_0 \sim (\bar{x}_0, P_0)$. Suppose also that measurements are given at discrete times t_k according to:

$$y_k = Cx(t_k) + v_k \tag{2}$$

where $v_k \sim (0, R_k)$ is uncorrelated with $\omega(t)$ and x_0.

Besides *initialization*, a continuous/discrete Kalman filter for system (1,2) consists of two steps: a *propagation* step (between two successive measurements), followed by a *correction* step (at measurement times):

Initialization ($t = t_0^+$)

$$P(t_0) = P_0, \quad \hat{x}(t_0) = \bar{x}_0$$

Propagation ($t_{k-1}^+ \leq t \leq t_k^-, k \geq 1$)

$$\begin{cases} \dfrac{\mathrm{d}P(t)}{\mathrm{d}t} = AP(t) + P(t)A^T + GQ(t)G \\ \dfrac{\mathrm{d}\hat{x}(t)}{\mathrm{d}t} = A\hat{x}(t) + Bu(t) \end{cases}$$

Correction ($t = t_k^+, k \geq 1$)

$$\begin{cases} K_k = P(t_k^-)C^T\left[CP(t_k^-)C^T + R_k\right]^{-1} \\ P(t_k^+) = [I - \boldsymbol{K}_k]\,P(t_k^-) \\ \hat{x}(t_k^+) = \hat{x}(t_k^-) + K_k\left[z_k - C\hat{x}(t_k^-)\right] \end{cases}$$

At this point, we shall emphasize several points. Note first that the foregoing Kalman filter can be applied to time-varying linear system, i.e. with matrices $\boldsymbol{A}$, $\boldsymbol{B}$, $\boldsymbol{C}$ and $\boldsymbol{G}$ depending on time. One should however keep in mind that observability must be proven for such systems prior to constructing the observer. Note also that Kalman filters can be extended by adding a term $-\theta P(t)$, $\theta > 0$, in the propagation equation of P. This exponential forgetting factor allows us to consider the case where $Q = 0$. Finally, estimating the positive definite matrices $\boldsymbol{R}$, $\boldsymbol{Q}$ and $\boldsymbol{P_0}$ often proves to be tricky in practice, especially when the noise properties are not known precisely.

4.1.3.4 The Extended Kalman Filter

Consider a continuous-time nonlinear system of the form:

$$\frac{\mathrm{d}x(t)}{\mathrm{d}t} = f\left[x(t), u(t)\right] + G\omega(t); \quad x(t_0) = x_0 \tag{3}$$

with measurements at discrete time t_k given by:

$$y_k = h[x(t_k)] + v_k \tag{4}$$

The idea behind the Extended Kalman Filter (EKF) is to linearize the nonlinear system (3,4) around its current state estimate $\hat{x}(t)$ (Lewis, 1986). By doing so, the problem becomes equivalent to building a Kalman filter for a nonstationary linear system (1,2) with A and C taken as:

$$A(t) = \left(\frac{\partial f}{\partial x}\right)_{x(t)} \qquad C(t_k) = \left(\frac{\partial h}{\partial x}\right)_{x(t_k)}$$

The EKF is used routinely and successfully in many practical applications, including WWTPs, even though few theoretical guarantees can be given as regards its convergence (Lewis, 1986; Bastin and Dochain, 1990). Note also that multirate versions of the EKF have been developed to handle those (rather frequent) situations where measurements are available at different samplings rates (Gudi *et al.*, 1995). An application of EKF in alternating activated sludge WWTPs is detailed hereafter. Other applications to the activated sludge process can be found (Zhao and Kümmel, 1995; Lukasse *et al.*, 1999).

4.1.3.5 Application to an Alternating-activated-sludge Plant

We consider an alternating-activated-sludge (AAS) WWTP similar to the one shown in Figure 4.1.2. AAS plants degrade both organic and nitrogenous compounds by alternating aerobic and anoxic phases in the bioreactor. Besides dissolved oxygen (DO) concentration that is routinely measured in activated sludge WWTPs, both nitrate and ammonia concentrations can also be measured on-line (at a lower frequency than DO, though). The objective here is to estimate the concentration of COD in the bioreactor based on these measurements.

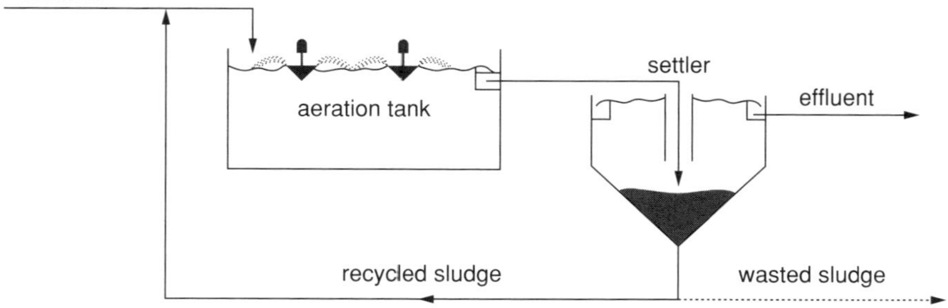

Figure 4.1.2 Typical small-size alternating activated sludge treatment plant

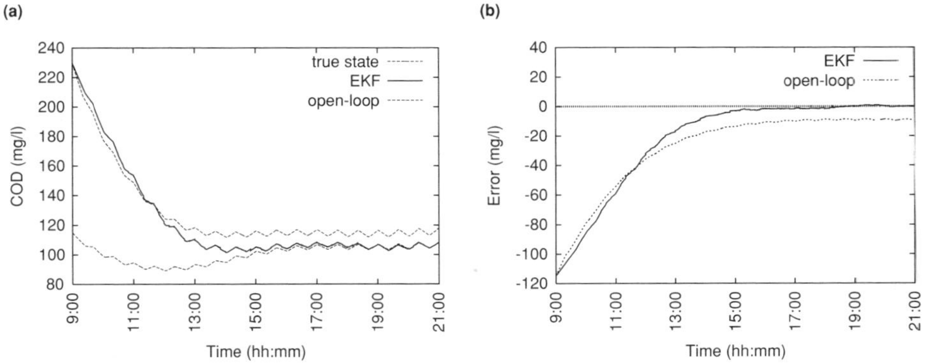

Figure 4.1.3 Estimated COD concentration (a) and observation error (b)

A multirate EKF is developed based on the reduced nonlinear model given in Chachuat *et al.* (Chachuat *et al.*, 2003). This five-state model describes the dynamics of COD, nitrate, ammonia, organic nitrogen and DO, and was shown to be observable under both aerobic and anoxic conditions (with the aforementioned measurements).

Numerical simulations have been performed by using a set of synthetic data produced from the full ASM1 model (Henze *et al.*, 1987) corrupted with white noise. The DO, nitrate and ammonia measurements are assumed to be available every 10 s, 10 min and 10 min, respectively. The results are shown in Figure 4.1.3; for the sake of comparison, the EKF estimates are compared with the open-loop estimates (i.e. without correction).

These results show satisfactory performance of the EKF for COD estimation. However, it should be noted that the COD estimates are very sensitive to model–parameter mismatch, which is hardly compatible with the fact that some parameters are time-varying and/or badly known in real applications. This motivates the development of mass-balance-based observer that are independent of the uncertain kinetic terms.

4.1.4 OBSERVERS FOR MASS-BALANCE-BASED SYSTEMS

The underlying structure of many WWTP models consists of two parts (Bastin and Dochain, 1990): (1) a linear part based on mass-balance considerations; and (2) a number of nonlinear term that describes the biological reaction rates (kinetics). These latter kinetic terms are often poorly known in practice, and there is little hope to construct a reliable observer by accounting for such uncertain terms. In contrast to the previous section wherein a full-model structure was used in the observer design, we shall show, in this section, how to take advantage of the foregoing two-fold structure to design observers that are independent of the uncertain reaction rates, via a linear change of variables. As we shall see, the main condition to design a so-called *asymptotic observer* is that enough variables are measured. Note also that asymptotic observers fall into the broad class of observers with unknown inputs (Kudva *et al.*, 1980; Hou and Müller, 1991; Darouach *et al.*, 1994), whose principle relies on cancellation of the unknown part via a change of variables.

4.1.4.1 Preliminaries

Throughout this section, we consider mass-balance models for WWTPs of the following form (Bastin and Dochain, 1990):

$$\frac{\mathrm{d}x(t)}{\mathrm{d}t} = Kr(x) - D(t)x + D(t)x^{in}(t) - Q(x) \tag{5}$$

where x, $x^{in} \in \Re^n$ represent the concentrations in the reactor and the influent, respectively, $D \in \Re^{n\times n}$, the dilution rate matrix, $Q(x) \in \Re^n$, the gaseous exchange between the reaction medium and the environment, $r(x) \in \Re^p$, the reaction rates and $K \in \Re^{n\times p}$, a constant pseudo-stoichiometric coefficient matrix. In this representation, $Kr(x)$ stands for the biological and biochemical conversions in the reactor (per unit of time) according to the underlying macroscopic reaction network.

Suppose that the set of available measurements y corresponds to is partitioned into y_1 and y_2 such that:

- y_1 is a set of $q \leq n$ measured state variables; without loss of generality, we assume that y_1 corresponds to the first q components of x, $y_1 = [x_1, \ldots, x_q]^T$.
- y_2 consists of the measured gaseous flow rates, $y_2 = [q_1(x), \ldots, q_n(x)]^T$.

The measurements y_1 induce a partition of the state variables $x = [x_1, x_2]$ with $x_1 = y_1$. Accordingly, Equation (5) can be rephrased as:

$$\begin{aligned}\frac{\mathrm{d}x_1}{\mathrm{d}t} &= \boldsymbol{K_1}r(x) - Dx_1 + D\boldsymbol{x_1}^{in} - Q_1(x)\\ \frac{\mathrm{d}x_2}{\mathrm{d}t} &= \boldsymbol{K_2}r(x) - Dx_2 + D\boldsymbol{x_2}^{in} - Q_2(x)\end{aligned} \tag{6}$$

where matrices $\boldsymbol{K_1}$ and $\boldsymbol{K_2}$, vectors $\boldsymbol{x_1}^{in}$, $\boldsymbol{x_2}^{in}$, $\boldsymbol{q_1}$ and $\boldsymbol{q_2}$ are such that

$$\boldsymbol{K} = \begin{pmatrix}\boldsymbol{K_1}\\ \boldsymbol{K_2}\end{pmatrix}, \boldsymbol{x}^{in} = \begin{pmatrix}\boldsymbol{x_1}^{in}\\ \boldsymbol{x_2}^{in}\end{pmatrix}, \boldsymbol{q} = \begin{pmatrix}\boldsymbol{q_1}\\ \boldsymbol{q_2}\end{pmatrix}$$

4.1.4.2 Asymptotic Observers

Constructing an asymptotic observers for system (5) requires the following two technical assumptions to hold:

(1) There are more measured quantities than reactions, i.e. $q \geq p$.

(2) Matrix $\boldsymbol{K}_1$ has full rank.

These assumptions guarantee that a nonzero r cannot cancel the term $K_1 r$ and the $q \times p$ matrix $\boldsymbol{K}_1$ has a left inverse. Accordingly, there exists a $p \times q$ matrix $\boldsymbol{G}$ such that:

$$\boldsymbol{G}\boldsymbol{K}_1 = I_p$$

Let us denote $A = -\boldsymbol{K}_2\boldsymbol{G}$ and $M = (A\ I_{n-p})$. An observer for subsystem (6) can be obtained as indicated in the following.

Property 2. (Bastin and Dochain, 1990). If D is positive definite, the solution $\hat{x}_2$ of the auxiliary system

$$\begin{cases} \dfrac{\mathrm{d}\hat{\zeta}_2}{\mathrm{d}t} = -D(\hat{\zeta}_2 - Mx^{in}) - My_2 \\ \hat{x}_2 = \hat{\xi}_2 - Ay_1 \end{cases}$$

converges to the solution x_2 of subsystem (6), asymptotically.

4.1.4.3 Application to an Anaerobic Digester

We consider a very simple model of the anaerobic digestion process that accounts for a single 'global' degradation step of the soluble COD (S_t) by the biomass X_t (Andrews, 1968):

$$k_t S_t \xrightarrow{\mu_t(.)X_t} X_t + k_m \mathrm{CH_4}$$

where k_t and k_m are the yield coefficients associated with COD degradation and methane production, respectively; $\mu_t(\cdot)$ stands for bacterial growth rate. The corresponding mass-balance model reads:

$$\frac{\mathrm{d}S_t}{\mathrm{d}t} = -k_t\mu_t(\cdot)X_t - D(S_t - S^{in}) \tag{7}$$

$$\frac{\mathrm{d}X_t}{\mathrm{d}t} = \mu_t(\cdot)X_t - \alpha DX_t$$

$$q_{\mathrm{CH_4}} = k_m\mu_t(\cdot)X_t \tag{8}$$

where q_{CH_4} stands for the methane outflow rate, D is the dilution rate and α is the fraction of bacteria not attached onto a support (i.e. being affected by the dilution rate D in the reactor).

The objective is to design an asymptotic observer for the biomass X_t, based on COD measurements, without knowing the reaction rates $\mu_t(\cdot)$. In this case, we have $x_1 = y_1 = S_t$, $x_2 = X_t$ and $Q(x) = 0$. Moreover, $K_1 = -k_t$ and $K_2 = 1$, hence a possible choice for matrices $\boldsymbol{A}$ and $\boldsymbol{G}$ is: $\boldsymbol{G} = -1/k_t$, $\boldsymbol{A} = 1/k_t$. Finally, property 1 provides the following auxiliary differential system:

$$\frac{\mathrm{d}\hat{\zeta}_2}{\mathrm{d}t} = -\alpha D\left(\hat{\zeta}_2 - \frac{S^{in}}{\alpha k_t}\right) - (1-\alpha)D\frac{y_1}{k_t} \tag{9}$$

$$\hat{X}_t = \hat{\zeta}_2 - \frac{y_1}{k_t} \tag{10}$$

Equations (9) and (10) are an observer for the biomass concentration in the digester. However, this observer may provide poor estimates in practice, because of the large errors made on S^{in} and k_t. A more appropriate solution is then to use an interval observer for coping with uncertainty, as discussed in the following section.

4.1.5 INTERVAL OBSERVERS

Usual observers rely on the implicit assumption that the process model is a good approximation of the real plant. Nevertheless, we have seen that WWTP models are often corrupted. In such situations where large modelling and measurement errors prevail, one can no longer construct an exact observer (i.e. with the guarantee that the observation error converges to zero asymptotically and that the convergence rate can be tuned). Instead, the observation principle must be weakened. In this section, we explain how to derive rigorous bounds enclosing the estimated states by accounting for the uncertainty in the process model (Rapaport and Gouzé, 1999; Gouzé *et al.*, 2000).

4.1.5.1 Principle

Interval observers require knowledge of time-varying bounds enclosing the uncertainty. These bounds are used to calculate time-varying bounds enclosing the state variable to be estimated.

Consider the following general system:

$$\begin{cases} \dfrac{\mathrm{d}x(t)}{\mathrm{d}t} = f\,[x(t), u(t), \omega(t)]; \quad x(t_0) = x_0 \\ y(t) = h[x(t), \upsilon(t)] \end{cases} \tag{S_I}$$

where known lower and upper bounds are available for the uncertain quantities $\omega \in \Re^r$ and $\upsilon \in \Re^s$:

$$\omega^-(t) \leq \omega(t) \leq \omega^+(t) \quad \forall t \geq t_0$$
$$\upsilon^-(t) \leq \upsilon(t) \leq \upsilon^+(t) \quad \forall t \geq t_0$$

Based on the fixed model structure (S_I) and on the set of measured data, an auxiliary dynamic system (O_I) can be designed such that it provides a lower bound and an upper bound for the state variables:

$$\begin{cases} \dfrac{\mathrm{d}z^-}{\mathrm{d}t} = f^-(z^-, z^+, u, y, \omega^-, \omega^+, \upsilon^-, \upsilon^+); \quad z^-(t_0) = g^-\left(x_0^-, x_0^+\right) \\ \dfrac{\mathrm{d}z^+}{\mathrm{d}t} = f^+(z^-, z^+, u, y, \omega^-, \omega^+, \upsilon^-, \upsilon^+); \quad z^+(t_0) = g^+\left(x_0^-, x_0^+\right) \\ x^- = h^-(z^-, z^+, u, y, \omega^-, \omega^+, \upsilon^-, \upsilon^+) \\ x^+ = h^+(z^-, z^+, u, y, \omega^-, \omega^+, \upsilon^-, \upsilon^+). \end{cases} \qquad (O_I)$$

Definition 2 (*interval estimator*). System (O_I) is said to be an interval estimator of system (S_I) if for any pair of initial conditions $x^-(t_0) \leq x(t_0) \leq x^+(t_0)$, there exists bounds $z^-(t_0)$, $z^+(t_0)$ such that the solutions of the coupled system (S_I, O_I) verify:

$$x^-(t) \leq x(t) \leq x^+(t) \quad \forall t \geq t_0$$

Interval estimators result from the coupling of two estimators which provide both an underestimate $x^-(t)$ and an overestimate $x^+(t)$ of $x(t)$. Of course, such bounds can be very large, thus making interval observers practically useless in some cases. However, for particular classes of systems (e.g. linear systems up to an output injection), theoretical guarantees can be given that the time-varying intervals $[x^-(t), x^+(t)]$ converge to a 'limit' interval of finite magnitude (Gouzé *et al.*, 2000). Moreover, the convergence rate towards this limit interval can be tuned if certain properties hold (Rapaport and Gouzé, 2003). Note that these ideas find their origin in the theory of positive systems (Smith, 1995). More recently, *probabilistic observers* have been formulated for a class of uncertain biological processes (Chachuat and Bernard, 2006); these observers take advantage of the knowledge of probability density functions (PDFs) for the uncertain parameters to calculate the PDFs of the unmeasured state variables.

An application of interval observers to an anaerobic WWTP is presented next; another application to an activated sludge process can be found in Hadj-Sadok and Gouzé (Hadj-Sadok and Gouzé, 2001).

4.1.5.2 Application to an Anaerobic Digester

We consider the same reduced model of an anaerobic digestion plant as in Section 4.1.4.3. The objective is to design an interval observer that estimates the COD concentration S_t from on-line methane measurements, with $\mu_t(\cdot)$ and α being unknown.

Using Equation (8) of the methane flow rate, Equation (7) can be rewritten as:

$$\frac{\mathrm{d}S_t}{\mathrm{d}t} = -\gamma q_{\mathrm{CH_4}} - D\left(S_t - S_t^{in}\right)$$

where the uncertain parameter $\gamma = k_t/k_m$ is such that $\gamma^- \leq \gamma \leq \gamma^+$; the inlet COD concentration $S_t^{in}(t)$ fluctuates between known bounds as $S_t^{in-}(t) \leq S_t^{in}(t) \leq S_t^{in+}(t)$, $\forall t \geq t_0$; and the initial COD concentration is bounded as $S_t^-(t_0) \leq S_t(t_0) \leq S_t^+(t_0)$.

Property 3. The auxiliary dynamic system

$$\begin{cases} \dfrac{\mathrm{d}S_t^-}{\mathrm{d}t} = -\gamma^+ q_{\mathrm{CH_4}} - D\left(S_t^- - S_t^{in-}\right) \\ \dfrac{\mathrm{d}S_t^+}{\mathrm{d}t} = -\gamma^- q_{\mathrm{CH_4}} - D\left(S_t^+ - S_t^{in+}\right) \end{cases}$$

is an interval observer for S_t, i.e. guarantees that $S_t^-(t) \leq S_t(t) \leq S_t^+(t)$ at each $t \geq t_0$.

Proof. It is easily verified that both lower bound $e^- = S_t - S_t^-$ and upper bound $e^+ = S_t^+ - S_t$ on the observation error remain positive [see Bernard and Gouzé (Bernard and Gouzé, 2004) for details].

In practice, it was found that the upper bound $S_t^+(t)$ is weak when large uncertainties are considered for γ and $S_t^{in}(t)$. These considerations motivate the following improvements.

Improvements. The idea for reducing prediction intervals consists of using structured kinetic models for $\mu_t(\cdot)$, despite uncertainty. We suppose here that the process does not operate in a region where inhibition phenomena occur, and use a Monod kinetic model for obtaining an estimate of $q_{\mathrm{CH_4}}$ as:

$$\hat{q}_{\mathrm{CH_4}}(S_t) = k_m \mu_t \frac{S_t}{S_t + K_{S_t}} \hat{X}_t$$

where $\hat{X}_t = (S_t^{in} - S_t)/\alpha k_t$. Then, these estimates are used in the following robust observer (Bernard and Gouzé, 2004):

$$\frac{\mathrm{d}S_t}{\mathrm{d}t} = -\gamma q_{\mathrm{CH_4}} - D(S_t - S_t^{in}) + \lambda \left[q_{\mathrm{CH_4}} - \hat{q}_{\mathrm{CH_4}}(S_t)\right] \tag{11}$$

Finally, an interval observer is derived from Equation (11) as:

$$\begin{cases} \dfrac{\mathrm{d}S_t^-}{\mathrm{d}t} = \phi_\varepsilon(S_t^-)\left\{-\gamma^+ q_{\mathrm{CH_4}} - D(S_t^- - S_t^{in-}) + \lambda(S_t^{in-} - S_t^-)\left[q_{\mathrm{CH_4}} - \hat{q}_{\mathrm{CH_4}}(S_t^-)\right]\right\} \\ \dfrac{\mathrm{d}S_t^+}{\mathrm{d}t} = \phi_\varepsilon(S_t^+)\left\{-\gamma^- q_{\mathrm{CH_4}} - D(S_t^+ - S_t^{in+}) + \lambda(S_t^{in+} - S_t^+)\left[q_{\mathrm{CH_4}} - \hat{q}_{\mathrm{CH_4}}(S_t^+)\right]\right\} \end{cases}$$

where the function $\phi_\varepsilon(S_t^-) = S_t^-/(S_t^- + \varepsilon)$ {with $0 < \varepsilon \ll \inf[S_t(t)]$} is used to enforce $S_t \geq 0$; and the bounds $\hat{q}_{\mathrm{CH_4}}^-$ and $\hat{q}_{\mathrm{CH_4}}^+$ on $\hat{q}_{\mathrm{CH_4}}$ are calculated by considering the bounds μ_t^- and μ_t^+ on μ_t. This observer is applied to a real process in the next paragraph.

Application to real measurements. The interval observer given in Equation (11) was implemented on a pilot-scale fixed bed up-flow anaerobic digester used for wine wastewater processing. Both the dilution rate D and methane outflow rate $q_{\mathrm{CH_4}}$ were measured on-line at a high frequency. Besides on-line measurements, a COD sensor was also used to validate the observer predictions. Details on the plant configuration and the experiments can be found in Bernard *et al.* (Bernard *et al.*, 2001) and Steyer *et al.* (Steyer *et al.*, 2002). The predictions of the interval observer are presented in Figure 4.1.4, together with the off-line COD measurements.

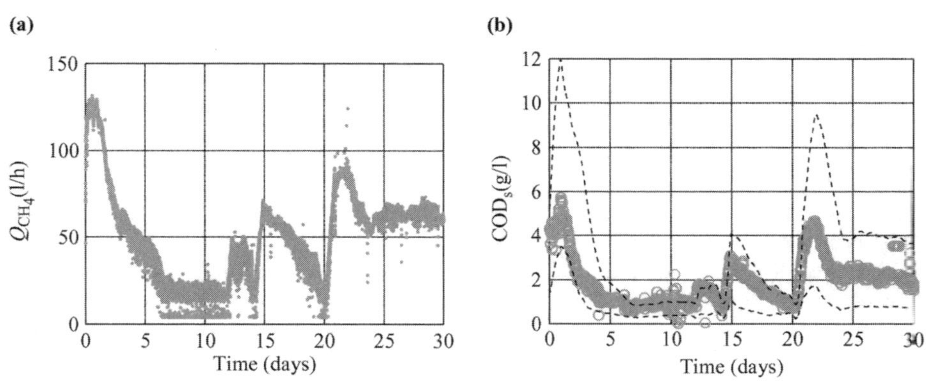

Figure 4.1.4 Methane flow rate (a) and interval observer (- -) for soluble COD (b)

4.1.6 CONCLUSIONS

In this chapter, we have provided the key ideas on how to build observers for WWTPs. Depending on the reliability of the process model at hand, the available measurements and the level of uncertainty associated with the influent concentrations, different classes of observers have been considered. In particular, a distinction has been made between those observers relying on a full model description (e.g. the extended Kalman filter), and those based on a mass-balance model wherein the biological kinetics are assimilated to unknown inputs (e.g. the asymptotic observer); moreover, if bounds are known for the uncertainties, then interval observers can be designed.

Clearly, other techniques exist, and we did not pretend to be exhaustive. The observers presented herein assumed constant parameter values in the models. In some cases, however, the parameters can evolve during process operation, and specific algorithms must be used to estimate these parameters at run-time, hence leading to adaptive observers.

The implementation of an observer requires a discretization to be performed as regards the continuous-time equations, e.g. a Euler-type algorithm. Although not difficult, discretization must be performed carefully. In the case of low measurement frequency, for example, the use of a continuous/discrete observer shall be preferred to full discretization of a continuous-time observer.

Finally, it is worth insisting on the fact that an observer should always be validated prior to using it in a real treatment plant. In particular, systematic and extensive comparisons should be made between the observer predictions and direct on-site measurements (other than those used to calibrate the observer).

REFERENCES

Andrews, J. (1968). A mathematical model for the continuous culture of microorganisms utilizing inhibitory substrate. *Biotechnol. Bioengng.*, 10:707–723.

Bastin, G. and Dochain, D. (1990). *On-line Estimation and Adaptive Control of Bioreactors*. Elsevier, Amsterdam, The Netherlands.

Bernard, O. and Gouzé, J. L. (2004). Closed loop observers bundle for uncertain biotechnological models. *J. Process Contr.*, 14(7):762–774.

Bernard, O., Hadj-Sadok, M. Z., Dochain, D., Genovesi, A. and Steyer, J.-P. (2001). Dynamical model development and parameter identification for an anaerobic wastewater treatment process. *Biotechnol. Bioengng.*, 75(4):424–438.

Chachuat, B. and Bernard, O. (2006). Probabilistic observers for a class of uncertain biological processes. *Int. J. Robust Nonlinear Control*, 16(3):157–171.

Chachuat, B., Roche, N. and Latifi, M. A. (2003). Reduction of the ASM1 model for optimal control of small-size activated sludge treatment plants. *Rev. Sci. Eau.*, 16(1):5–26.

Darouach, M., Zasadzinski, M. and Xu, S. J. (1994). Full-order observers for linear systems with unknown inputs. *IEEE Trans. Automat. Contr.*, 39(3):1068–1072.

Gauthier, J. P. and Kupka, I. (2001). *Deterministic Observation Theory and Applications*. Cambridge University Press, New York.

Gouzé, J.-L., Rapaport, A. and Hadj-Sadok, M. Z. (2000). Interval observers for uncertain biological systems. *Ecol. Modell.*, 133:45–56.

Gudi, R. D., Shah, S. L. and Gray, M. R. (1995). Adaptive multirate state and parameter estimation strategies with application to a bioreactor. *AIChE J.*, 41:2451–2464.

Hadj-Sadok, M. Z. and Gouzé, J.-L. (2001). Estimation of uncertain models of activated sludge process with interval observers. *J. Process Contr.*, 11(3):299–310.

Henze, M., Grady, C. P. L., Gujer, W., Marais, G. v. R. and Matsuo, T. (1987). Activated Sludge Model No. 1. Technical report, IAWQ, London.

Hou, M. and Müller, P. (1991). Design of observers for linear systems with unknown inputs. *IEEE Trans. Automat. Contr.*, 37(6):871–875.

Kudva, P., Viswanadham, N. and Ramakrishna, A. (1980). Observers for linear systems with unknown inputs. *IEEE Trans. Automat. Contr.*, 25(1):113–115.

Lewis, F. L. (1986). *Optimal Estimation (with an Introduction to Stochastic Control Theory)*. John Wiley & Sons, Ltd, New York.

Luenberger, D. G. (1966). Observers for multivariable systems. *IEEE Trans. Automat. Contr.*, 11:190–197.

Luenberger, D. G. (1979). *Introduction to Dynamic Systems: Theory, Models and Applications*. John Wiley & Sons, Ltd, New York.

Lukasse, L., Keesman, K. and van Straten, G. (1999). A recursively identified model for short-term predictions of NH_4/NO_3 concentrations in alternating activated sludge processes. *J. Process Contr.*, 9(1):87–100.

Rapaport, A. and Gouzé, J.-L. (1999). Practical observers for uncertain affine outputs injection systems. In *Proc. ECC'99*, Karlsruhe, Germany. European Union Control Association (CD-ROM).

Rapaport, A. and Gouzé, J.-L. (2003). Parallelotopic and practical observers for non-linear uncertain systems. *Int. J. Control*, 76(3):237–251.

Smith, H. L. (1995). *Monotone Dynamical Systems: an Introduction to the Theory of Competitive and Cooperative Systems*. American Mathematical Society, Providence, Rhode Island.

Steyer, J.-P., Bouvier, J.-C., Conte, T., Gras, P. and Sousbie, P. (2002). Evaluation of a four year experience with a fully instrumented anaerobic digestion process. *Wat. Sci. Technol.*, 45 (4–5):495–502.

Zhao, H. and Kümmel, M. (1995). State and parameter estimation for phosphorus removal in an alternating activated sludge process. *J. Process Contr.*, 5(5):341–351.

4.2 Industrial Wastewater Quality Monitoring

Olivier Thomas and **Marie-Florence Pouet**

4.2.1 REGULATORY CONTEXT

Among the regulation texts related to the management of industrial wastewater, the European directive of 21 May 1991 (see Chapter 1.1) concerning urban wastewater treatment (91/271/EEC) (Annex I, C) states that industrial wastewater entering collecting systems and urban wastewater treatment plants shall be subject to such pre-treatment as is required in order to:

- protect the health of staff working in collecting systems and treatment plants;
- ensure that collecting systems, wastewater treatment plants and associated equipment are not damaged;
- ensure that the operation of the wastewater treatment plant and the treatment of sludge are not impeded;

Wastewater Quality Monitoring and Treatment Edited by P. Quevauviller, O. Thomas and A. van der Beken

- ensure that discharges from the treatment plants do not adversely affect the environment, or prevent receiving water from complying with other Community Directives;
- ensure that sludge can be disposed of safely in an environmentally acceptable manner.

When an industrial wastewater discharges directly, after treatment, in the receiving medium, it shall respect the regulation compliance depending on the industry nature, the effluent characteristics, the minimum efficiency of treatment and the support capacity of the receiving medium. For example, these considerations are precised in the USA, in the National Pollution Discharge Elimination System (NPDES) program and in the total maximum daily load (TMDL) of the Clean Water Act (see Chapter 1.1).

4.2.2 CHARACTERISTICS OF INDUSTRIAL WASTEWATER

Contrary to the treatment of industrial wastewater, there exist very few studies and books about industrial wastewater characteristics. The composition of wastewater varies with the industrial activity (Table 4.2.1), but also with the size and location of enterprises. Small and medium enterprises (SMEs) of urban area are generally connected to the sewer network after or without wastewater treatment, but larger industries are often outside the urban area and equipped with their own treatment plant with a specific discharge in the receiving medium. Moreover, there is a trend to recycle water in processes and thus to reduce and minimise the effluent load (Trebuchon *et al.*, 2000; Gomes *et al.*, 2006). For all industries, the regulation gives threshold limits in concentration and daily load, depending on the existence of a treatment plant, parameters and industrial activity. A way of characterised industrial wastewater is also to consider the ratio of pollution [for example, biological oxygen demand (BOD)] on the production unit (for example ton of paper). This ratio is the specific load.

One key characteristic of industrial wastewater is its daily, weekly or seasonal variation in composition and load which can be expressed by a statistical distribution, plotting the value of a given parameter (concentration, load or specific load) against the percent of time for which the value of the parameter is equal or less than a given one (Eckenfelder, 2001). The composition variation can also be estimated by the variability explained in Section 4.2.4.

If the characteristics of raw wastewater of a given enterprise exceed the limits, the industry must install a pretreatment step before the discharge into the sewer network. Nevertheless, considering the composition of some industrial wastewater, its impact on sewer and treatment plant can be effective, particularly if the industry is close to the treatment plant. The main impact is related to shock loads and toxicity effects on

Table 4.2.1 Concentration of some parameters for industrial wastewater (in mg/l). (Adapted from Eckenfelder, 2001; Baurès, 2002; Metcalf and Eddy, 2003; Pons *et al.*, 2004; Degrémont, 2005)

	Domestic	Textile	Petrochemical	Fine chemicals	Pulp and paper	Agro-food	Metal industry
BOD	100–400	100–3000	100–300	500–20 000	250–15 000	100–7000	
COD	300–800	250–1500	150–3500		500–100 000	10–10 000	300–1200
TSS	100–500	100–800	30–1000	1000–170 000	100–2500	30–7000	200–1000
TN	20–100	10–50	25–60 (NH_4^+)			1–5	25–6000 (NH_4^+)
TP	5–20					1–10	
Oil and grease	50–100	30	200–3000	0–2000			100–600
Phenols			0–300	0–5000	0–800		0–4000
Others		S^{2-} : 0–50 Cr(VI) : 1–4	S^{2-} : 0–15	As, Ba, Cd	Se, Zn		CN^- : 100–400 Fe^{2+} : 50–300

BOD, biological oxygen demand; COD, chemical oxygen demand; TSS, total suspended solids; TN, total nitrogen; TP, total phosphorus.

biological treatments. The potential risk must be evaluated from sampling campaign and measurement, analysis and tests.

Before considering the monitoring of industrial wastewater, it is important to point out that other nondomestic pollution sources must be considered in a sewer network. Hospitals and other services (health centers, restaurants, schools, commercial centers, etc.) must be considered as point sources of specific pollution with the discharge of emerging pollutants, such as pharmaceutical products, the impact of which is more and more studied (Barcelo, 2005).

4.2.3 MONITORING OF INDUSTRIAL WASTEWATER

The monitoring of industrial wastewater quality does not differ significantly from that of urban wastewater. The objectives, the main parameters and the methodologies are similar. However, there are some differences due to the nature and production mode of industrial wastewater:

- The parameter values (concentration or other physico-chemical parameters) are somewhat higher (Table 4.2.1), increasing the difficulty for the detection of micropollutants (matrix effect).
- The variability of parameters (including the pollution loads) can be much higher than for urban wastewater.
- The toxicity of some effluents can be rather high depending on the toxic substances.
- The monitoring is either difficult to be carried out for SMEs connected to the sewer network or is integrated to the process control for large industries equipped with a treatment plant.

The monitoring of industrial discharges is often made directly on the industrial site, for regulation monitoring and/or for the calculation of the discharge tax (once a year for example). In the case of an existing wastewater treatment plant of a large industry, the procedure is the same as that for an urban wastewater treatment plant. Sampling campaigns are planned with the use of automatic samplers, at the inlet and outlet of the treatment plant and connected to flow rate measurement for loads calculation (see Chapter 1.2). Samples are analyzed in the laboratory for the main and specific parameters (see hereafter). For large industries like refineries, petrochemical and chemical plants, on-line sensors/analyzers are frequently installed at the inlet of the treatment plant. Parameters like total organic carbon (TOC), total oxygen demand (TOD), total petroleum hydrocarbons (TPH), pH, conductivity, turbidity or UV characteristics (UV_{254}, or UV spectrum) can be considered for on-line measurement.

If the industry is connected to an urban sewer, monitoring can also be carried out directly from the sewer in order to have a better knowledge of the pollution to be treated. In this case, critical control points must be selected both for diagnosis and for the control of the efficiency of further actions (industrial fluxes reduction). The use of hazard analysis and critical control point (HACCP) methodology (see Chapter 1.2) allows some relevant critical control points to be defined. After identifying the location of industries at risk (regarding wastewater quality), connections and branches of interest are chosen from the sewer map. Then, the monitoring program is carried out from sampling campaigns based on the use of automatic samplers, connected to flow rate measurement for loads calculation, and with on-site measurement of relevant parameters for sampling assistance (see Chapter 1.2). The results of the survey should establish possibilities for water conservation and reuse and the variation in flow and load to undergo wastewater treatment (Eckenfelder, 2001).

Generally, the same parameters as for urban wastewater are acquired [temperature, pH, conductivity, turbidity, TOC, chemical oxygen demand (COD), BOD, total suspended solids (TSS), N and P forms], plus some specific parameters depending on the nature of the expected industrial wastewater discharges, such as oil and grease, surfactants and phenols (Table 4.2.1). Among these main parameters, the physico-chemical ones (temperature, pH and particularly conductivity), are very easy to measure and the main changes along the sewer may be related to an industrial discharge. This list can be completed by complementary specific and other parameters:

- Mineral substances, including heavy metals, depending on the industry, such as sulfides, cyanides, chromium (hexavalent and total), iron, nickel, copper, zinc, arsenic, selenium, cadmium, barium, mercury and lead.
- Organic pollutants including priority (benzene and derivatives, polyaromatic hydrocarbons, halogenated organics, etc.) and emerging pollutants (for example pharmaceuticals and phthalates).
- Toxicity tests and measurement for the estimation of global toxicity, for example with the Microtox or the SOS chromotest (Jolibois and Guerbet, 2005), up to the characterisation and identification of toxic organic constituents by fractionation and chromatographic analysis (Reemtsma *et al.*, 1999).
- Nonparametric measurements such as the use of a UV spectrum for global characterisation of wastewater (regarding absorbing substances, see Chapter 1.5), or for the treatability estimation (Castillo *et al.*, 1999a,b; Muret *et al.*, 2000).

4.2.4 VARIABILITY

One key point of industrial wastewater monitoring is that composition and load variability can be important, depending on the size and activity of the industry. For industries with a single line production (pulp and paper, metal industry and some agro-food plants) the variability can be rather low, contrary to more complex industries (refinery, fine chemicals and textile) or SMEs with a very variable production.

The impact of the variability of industrial wastewater quality depends on the sewer system. For an industrial activity connected to an urban network, mainly for

SMEs or industry located in an industrial area with a common treatment plant, the variability is smoothed, on one hand, by the regulation compliance (see Section 4.2.1) involving a pretreatment step before the discharge, and, on the other hand, by a buffer effect due to the dilution between the discharge and the final treatment plant. For a more important industry with several production lines and equipped with a wastewater treatment plant, the variability is related to the different mixtures of workshop discharges, from upstream to the treatment plant and to the occurrence of internal accidental pollution.

The variability is generally expressed from measurements of some relevant parameters (for example TOC, COD, conductivity) as the ratio between the standard deviation and the average value for a given flux and for a given period of time. This ratio provides a good indicator of variability only if the flow rate is constant during the period. Thus, the best way is to consider loads instead of concentrations, which exclude physical parameters like conductivity.

Another method for variability estimation is to use a nonparametric measurement (see Chapter 1.4) like a UV spectrum. The examination of the UV spectra set corresponding to samples of a given flux generally shows an isosbestic point (Gallot and Thomas, 1993) where several spectra cross at the same wavelength. That means that there is a qualitative and quantitative conservation of the composition between two states. In the case of flow rate variation, a normalisation step (Vaillant *et al.*, 2002) can reveal hidden isosbestic points meaning that there is only a qualitative conservation of the composition (Pouet *et al.*, 2004).

Without the knowledge of the wastewater composition, the variability can be simply expressed as the ratio between the number of spectra crossing together at the isosbestic point and the number of total spectra (Thomas *et al.*, 2005).

An example is given in the study of variability of a wastewater network in a refinery (Baurès, 2002; Thomas *et al.*, 2005). The UV spectra of a lot of samples from different points along the wastewater network up to the treatment plant have been acquired. The sets of spectra are rather different from one point to another and characterised by peaks or shoulders related to the presence of specific substances like sulfides (around 240 nm) or phenols (around 270–280 nm). These figures tend to decrease and even vanish along the sewer system. The evolution of the variability from upstream to the discharge of treated wastewater shows a strong decrease under the effect of storage tanks and treatment steps, from 70 % close to the production workshop to 30 % for the storage tank and finally, 10 % at the outlet of the treatment plant (Thomas *et al.*, 2005).

4.2.5 ACCIDENT DETECTION AND SOURCE IDENTIFICATION

The analysis of industrial pollutants from urban wastewater samples is not easy (Castillo *et al.*, 1999a,b; Barcelo, 2005). The detection of accidental pollution in an industrial wastewater network is preferably undertaken through the use of on-line analyzers, which are frequently installed at the inlet of industrial treatment plants, particularly for the protection of biological processes. Accidental pollution is due to human factor (intentional or accidental) or to technical problems (leaks for example). In the case of exceeding a threshold value, complementary treatment steps start or wastewater can be bypassed towards a storage tank. The parameters to be measured on-line are usually the organic pollution with TOC or TOD (sometimes used in the chemical industry), the mineral matrix with conductivity and pH and the estimation of suspended solids from the turbidity.

These parameters can be completed by the use of UV spectrophotometry, particularly for chemical, petrochemical and the pulp and paper industry where absorbing substances are numerous in wastewater. A first study (El Khorassani *et al.*, 1999) based on the use of UV spectrophotometry has developed a general procedure for the rapid identification of industrial pollutants in petrochemical and refinery wastewater. The same authors have carried out a study involving the measurement of TOD at the inlet of the wastewater treatment plant of a petrochemical site and the acquisition of UV spectra in order to explain the high values of TOD (El Khorassani *et al.*, 2000). This procedure has been validated from samples analysis in the laboratory using chromatographic techniques, and has led to the identification of the main source of incidents and to the reduction of the pollution load (and consequently to the protection of the biological treatment plant).

Thus, as for the variability estimation, UV spectrophotometry is a good technique for industrial wastewater characterisation. A study carried out at a chemical site (Thomas *et al.*, 1999) confirmed the interest of this technique, complementary to the measurement of TOC. Another use of UV spectrophotometry is the examination of outliers in a set of spectra (Gallot and Thomas, 1993). Outliers are spectra which do not cross with others at the isosbestic point, even after normalisation (Pouet *et al.*, 2004). This property means that the composition of wastewater is very different from that of other samples and that there is probably a supplementary absorbing substance, not normally present, perhaps the sign of an accident.

For urban wastewater networks where industrial discharges may lead to shock load or accidental pollution which may damage the treatment plant, it is more difficult to detect abnormal situations, because sewers do not generally display on-line analyzers. The literature gives very few examples of procedures for the purpose.

A study based on the use of conductivity measurement and UV spectrophotometry (Naffrechoux *et al.*, 1991) proposes a rapid method for the identification of industrial wastewater in an urban wastewater. Another one has chosen a more classical approach with the analysis of the main and specific parameters (oil and grease, for example) for the characterisation of industrial inputs (Gorgun *et al.*, 1999). Bourgeois (Bourgeois and Stuetz, 2002) has developed a chemical sensor array for detecting organic pollutants (such as volatile hydrocarbons) in domestic wastewater. The sensor array (consisting of eight conducting polymers) has been tested to continuously monitor the presence or absence of industrial pollutants in the headspace of wastewater generated from an on-line flow-cell. Response patterns between the sensors were used to detect the presence of organic compounds in wastewater. Correlations between the sensor response patterns or fingerprints were also analyzed using principal component analysis.

Complementary to the chemical procedure, some biological tools can be used. For example, a study carried out in the urban sewer of Turin, Italy (Romano and Lorenzi, 2000) has shown the possibility of a protection device for the wastewater treatment plant composed of an alarm system based on toxicity and TOC monitoring. The screening of industrial wastewater has also been included in the control activity, in order to prevent nitrification inhibition in active sludge oxidation basins by heavy metals and organic compounds, using the MINNTOX test (Juliastuti *et al.*, 2003).

A last interesting approach for the monitoring of industrial impact on urban wastewater is the toxicity identification evaluation (TIE), applied in a sewer study of a Chinese town (Hongxia *et al.*, 2004). Proposed in the 1990s, The TIE general procedure is carried out in three steps: (i) the characterization of the classes of toxicants; (ii) the identification of the specific toxic compounds within these classes; and (iii) the validation that these substances are the key toxicants. Unfortunately, the numerous and complex manipulations and analysis of this approach are limitations to its application.

Finally, further efforts are needed in research and development for the characterisation and monitoring of industrial wastewater.

REFERENCES

Barcelo, D. (2005) *Emerging Organic Pollutants in Wastewater and Sludge. The Handbook of Environmental Chemistry*, vol. 5, parts I and O. Springer-Verlag, Berlin.

Baurès, E. (2002) La mesure non paramétrique, un nouvel outil pour l'étude des effluents industriels: application aux eaux résiduaires d'une raffinerie. PhD Thesis, University of Aix Marseille III.

Bourgeois, W. and Stuetz R.M. (2002) *Water Res.*, **36**(12), 4505–4512.

Castillo, L., El Khorassani, H., Trebuchon, P. and Thomas, O. (1999a) *Water Sci. Technol.*, **39** (10–11), 17–23.

Castillo, M., Barcelo, D. and Aquino Neto, F.R. (1999b) *Trends Anal. Chem*, **18**(1), 26–36.

Degrémont (2005) *Mémento technique de l'eau*, 10th Edn. Paris.

Eckenfelder, W.W. (2001) *Industrial Water Pollution Control. McGraw Hill Series in Water Resources and Environmental Engineering*, 3rd Edn. McGraw Hill, Boston.

El Khorassani, H., Trebuchon, P., Bitar, H. and Thomas, O. (1999) *Water Sci. Technol.*, **39**(10–11), 77–82.

El Khorassani, H., Trebuchon, P., Bitar, H. and Thomas, O. (2000) *Water Sci. Technol*, **42**(5–6), 15–22.

Gallot, S. and Thomas, O. (1993) *Fresenius J. Anal. Chem.*, **346**, 976–983.

Gomes, J.F.S., Queiroz, E.M. and Pessoa, F.L. (2006) *J. Clean. Prod.*, available at www.sciencedirect.com

Gorgun, E., Babuna, F., Orhon, D., Ozbasaran, M. and Seckin, N. (1999) *Water Sci. Technol*, **40**(1), 13–21.

Hongxia, Y., Jing, C., Yuxia, C., Huihua, S., Zhonghai, D. and Hongjun J. (2004) *Ecotoxicol. Environ. Saf.*, **57**, 426–430.

Jolibois, B. and Guerbet, M. (2005) *Mutat. Res.*, **565**, 151–162.

Juliastuti, S.R., Baeyens, J. and Creemers, C. (2003) *Environ. Engin. Sci.*, **20**(2), 79–90.

Metcalf and Eddy (2003) *Wastewater Engineering, Treatment and Reuse*, 4th Edn. McGraw Hill, Boston.

Muret, C., Pouet, M.F., Touraud, E. and Thomas, O. (2000) *Water Sci. Technol*, **42**(5–6), 47–52.

Naffrechoux, E., Mazas, N. and Thomas, O. (1991) *Environ. Technol.*, **12**, 325–332.

Pons, M.N., Spanjers, H., Baetens, D., Nowak, O., Gillot, S., Nouwen, J. and Schuttinga, N. (2004) *Wastewater Characteristics in Europe – A Survey*, available at http://www.ewaonline.de/journal/2004_04.pdf

Pouet, M.-F., Baurès, E., Vaillant, S. and Thomas, O. (2004) *Appl. Spectrosc.*, **58**(4), 486–490.

Reemtsma, T., Putschew, A. and Jekel, M. (1999) *Waste Manag.*, **19**, 181–188.

Romano P. and Lorenzi E. 2000. Control of industrial discharges into the sewerage system. *Water Supply*, **18**, 685–689.

Thomas, O., El Khorassani, H., Touraud, E. and Bitar, H. (1999) *Talanta* **50**, 743–749.

Thomas, O., Baurès, E. and Pouet, M.-F. (2005) *Water Qual. Res. J. Can.*, **40**(1), 51–58.

Trebuchon, P., Bitar, H. and Thomas, O. (2000) *Water Sci. Technol*, **42**(5–6), 15–22.

Vaillant, S., Pouet, M.-F. and Thomas, O. (2002) *Urban Water*, **4**, 273–281.

5.1
Quality Survey of Wastewater Discharges

Marie-Florence Pouet, Geneviève Marcoux and **Olivier Thomas**

5.1.1 CHARACTERISTICS AND IMPACT OF WASTEWATER DISCHARGES

A wastewater discharge concerns, most of the time, the treated effluents of a wastewater treatment plant, in a receiving medium (river, lake, sea). Sometimes, in the case of heavy rainfall, raw wastewater can also be discharged, without treatment, from combined sewer overflows. The aim of this section is to present the main characteristics and impacts of wastewater discharges and the monitoring tools available, particularly the alternative ones for an on-site/on-line use.

Wastewater discharges are mixtures of organic and mineral compounds, mainly dissolved and colloidal. However, depending on the characteristics of raw wastewater and on the type and efficiency of the applied treatment (physico-chemical or biological), particulate matter (coarse colloids and suspended matter) may represent

Wastewater Quality Monitoring and Treatment Edited by P. Quevauviller, O. Thomas and A. van der Beken

a significant fraction of organic matter of the discharge (loss of biomass, colloidal matter not removed by coagulation-flocculation) (Vaillant, 2000):

- Water treated by physico-chemical treatment contains mainly soluble and fine colloids, initially present in raw water. The composition of the effluent is not stable and may lead to post flocculation with suspended solid increase in the receiving medium (Pouet *et al.*, 1999).
- One consequence of the biological treatment is the modification of the grain size distribution of the organic matter. Thus, the initial colloidal fraction is removed for a great part as well as 80 % of the organic soluble fraction, depending on the mass load applied (Boller, 1993). After a biological treatment, the effluent is essentially composed of matter below 1 μm (Levine *et al.*, 1991), grouping colloidal and soluble matter. The soluble fraction is composed of about 40–50 % of humic like substances, hardly biodegradable (Levine *et al.*, 1985), and soluble products produced by the biomass and composed of heteropolysaccharides (Hejzlar and Chudoba, 1986). Of these microbial by-products, 50 % have a molecular weight $>$10 000 Da, and 40 % $<$1000 Da (Pitter and Chudoba, 1990). Thus, the effluent of a biological treatment plant is essentially soluble, composed of products slowly biodegradable or nonbiodegradable, giving a rather stable and low concentrated matrix (Vaillant, 2000). Typical residual constituents of treated wastewater effluents and their potential impacts on receiving medium are summarized in Table 5.1.1.

Table 5.1.1 Main residual constituents found in treated wastewater effluents and their impacts. (Adapted from Metcalf and Eddy, 2003)

Residual constituents	Impacts on the receiving medium
Suspended and colloidal solids (inorganic and organic)	May cause sludge deposits or trouble receiving water clarity May deplete oxygen resources May affect aquatic life (fishes)
Dissolved organic matter	
Total organic carbon	May deplete oxygen resources (if biodegradable)
Refractory organics	Toxic to humans, carcinogenic
Volatile organic compounds	Toxic to humans, carcinogenic, form photochemical oxidants
Pharmaceutical compounds	Impact aquatic species (sexual change)
Surfactants	Cause foaming
Dissolved inorganic matter	
Ammonia	May deplete oxygen resources, impact aquatic species (toxicity)
Nitrate	Stimulates, with phosphorus, growth of algae and macrophytes
Phosphorus	leading to eutrophication
Biological	
Bacteria	
Protozoan cysts and oocysts	Can cause disease, and limit the use of aquatic resources
Viruses	

The impact of a wastewater discharge in a receiving medium depends generally on several factors (adapted from US/EPA):

(1) quantities, composition, and potential bioaccumulation or persistence of the pollutants to be discharged;

(2) potential transformation (including degradation) and transport of the pollutants and their by products by biological, physical or chemical processes;

(3) composition and vulnerability of potentially exposed biological communities;

(4) importance of the receiving water area to the surrounding biological community, e.g. spawning sites, migratory pathways;

(5) potential direct or indirect impacts on human health;

(6) existing or potential recreational and commercial fishing.

Among the potential impacts on receiving medium, some pollutants like ammonia, nitrate, phosphate and emerging pollutants have to be highlighted.

The ecological impact of ammonia in aquatic ecosystems is, on the one hand, acute toxicity depending on concentration and pH (see Chapter 2.1), and on the other hand, chronic toxicity regarding fishes and benthic invertebrate populations (reduced reproductive capacity and growth of young) (Environment-Canada, 2001). The zone of impact varies greatly with discharge conditions, river flow rate, temperature and pH. Under estimated average conditions, some municipal wastewater discharges could be harmful for 10–20 km (Environment-Canada, 2001). Severe disruption of the benthic flora and fauna has been noted below municipal wastewater discharges. Recovery may not occur for many (20–100) kilometres. It is not clear whether these impacts are solely from ammonia or from a combination of factors, but ammonia is a major, potentially harmful constituent of municipal wastewater effluents.

The consequence of discharges of nitrates and phosphorus is eutrophication. The impact depends on the support capacity of the receiving medium (Zabel *et al.*, 2001). In Europe, the Water Framework Directive (European Commission, 2000) indicates that:

- Estuarine and coastal waters with a high dispersion capacity may receive primary treatment.
- Wastewater discharges into river must receive at least a biological treatment.
- For wastewater discharges in sensitive areas [defined by the risk of eutrophication and exceeding the drinking water standard for nitrate (50 mg $N\text{-}NO_3/l$)], the removal of nutrients is required.

These considerations mean that the control of the discharge quality and impact on the receiving medium is evaluated from the same parameters used for the evaluation of the performance of a wastewater treatment plant.

Other important pollutants have also to be considered for their impacts on the receiving medium. Substances such as antibiotics, antitumor drugs, anesthetics or disinfectants from hospital effluents are not totally removed by treatment plants and are not detected by a classical survey of discharges (Kümmerer, 2001; Kümmerer *et al.*, 2004). These pharmaceutical compounds and personal care products, but also surfactants, and gasoline additives are grouped as emerging organic pollutants and must be taken into account for the discharges survey because of their ecotoxicological potential (Barcelo, 2005). For example, the occurrence and fate of pharmaceutical products in the aquatic environment is recognized as one of the emerging issues in environmental chemistry, in particular in urban areas (Heberer, 2002; Heberer *et al.*, 2002).

Finally, wastewater discharges monitoring needs additional qualitative or quantitative information (e.g. pollutants size distribution, wastewater fractionation, detection of incidents) in order to achieve an optimized treatment and to protect the receiving medium.

Moreover, the survey of wastewater discharges quality and the control of impact on a receiving medium implies the coupling between physico-chemical and biological approaches.

5.1.2 CHEMICAL MONITORING

Urban discharges characterization is usually achieved using aggregate parameters analysis or measurement from samples [biological oxygen demand (BOD), chemical oxygen demand (COD), total organic carbon (TOC), total suspended solids (TSS), N forms and P forms]. The minimum requirements for discharges quality, according to European directive of 21 May 1991 concerning urban wastewater treatment (European Commission, 1991) (see Chapter 1.1) or other regulation texts, concern mainly these parameters.

Chemical monitoring is related to the monitoring of chemical and physico-chemical parameters. It can be achieved with biosensors, chemical or physico-chemical systems. Passive sampling, with biomonitoring and on-line continuous monitoring attempt also to overcome the problems associated with spot analysis (Allan *et al.*, 2006). Passive samplers are being considered as emergent tools for monitoring a range of priority pollutants and coupling with bio-markers or bio-indicators could provide, in the future, information relative to the toxicological potential of effluents (Allan *et al.*, 2006).

In this section, methods allowing the determination of aggregate and specific parameters are presented. The first systems developed for the measurement of regulated parameters were chemical or physico-chemical systems (Table 5.1.2):

- For nutrients determination, *in-situ* methods like ready-to-use test kits (see Section 1.4) have been used for more than 20 years. Some of them provide semi-quantitative results (strip tests) and others, based on colorimetric methods, can lead to a good estimation of nutrients. On-line devices are also available.

Table 5.1.2 Physico-chemical measurement of parameters for wastewater monitoring discharges. (Adapted from Greenwood *et al.*, 2004; Thomas and Pouet, 2005)

Parameters	Main principles	Other principles
Ammonium	Ion-selective electrode Colorimetry Titrimetry	UV spectrophotometry (after photooxidation) (Ionic chromatography) (Chemiluminescence)
BOD	Respirometry	(UV spectrophotometry)
COD	Titrimetry (after oxidation) Colorimetry (after microwave oxidation)	UV spectrophotometry Photometry IR (after catalytical oxidation)
Conductivity	Electrical	
Dissolved oxygen	Electrochemistry	Luminescence
Heavy metals	Electrochemistry	UV photometry (cold steam method) for mercury
Nitrate	UV spectrophotometry	(Ion selective electrode)
Organic matter	UV spectrophotometry	
PAH	Fluorimetry	NDIR photometry Optical: light intensity reflection (UV spectrophotometry)
pH	Electrochemistry	Electronic (ISFET)
Phosphate	Colorimetry	UV spectrophotometry (Ionic chromatography) Titrimetry
TOC	NDIR photometry (after oxidation)	UV spectrophotometry
Total nitrogen	Colorimetry (after digestion)	UV spectrophotometry (after photooxidation) (Chemiluminescence)
Total phosphorous	Colorimetry	UV spectrophotometry
Turbidity	Nephelometry	UV spectrometry

NDIR, nondispersive infra-red; PAH, polycyclic cromatic hydrocarbons.

- For aggregate parameters, on-line methods are commercially available for wastewater discharges monitoring, based on optical or electrochemical principles. Limitations of their use are related to fouling problems and maintenance costs. The majority of existing systems are based on electrochemical and optical methods. Electrochemical systems are often proposed with specific electrodes but interferences can cause poor quality of results if they are not taken in account. Spectrophotometric devices are increasingly used for the determination of regulated aggregate and specific parameters because there are easy to use, robust and give rapid results (Thomas and Constant, 2004; Thomas and Pouet, 2005). UV spectrophotometry is particularly interesting because the interferences due to the presence of colloids and particles are reduced by deconvolution methods (Thomas and Constant, 2004). Moreover, UV spectrophotometry gives further interesting qualitative information from the exploitation of the whole UV spectrum (see Chapter 1.4).

Biosensors can be also valuable tools for on-site/on-line wastewater monitoring. Considered as emerging tools as they are not yet really used (most monitoring devices being based on physico-chemical principles), they can detect specific compounds or measure aggregate parameters (Table 5.1.3), but overall, they can obviously give biological information (see Section 5.1.3). The need to develop biosensors is to complete the variety of substances that physico-chemical systems can detect. Because of their biological nature, they can give relevant measurements of parameters like BOD. A lot of chemical substances can be detected with biosensors such as pesticides, phenol, aromatic amines, naphthalene and pharmaceuticals. The detection of these compounds is mostly related to industrial wastewaters.

In Table 5.1.3, biosensors able to detect nitrate (NO_3^-), ammonia (NH_4^+) and aggregate parameters (BOD, COD) are described. The table contains the name of the compound, the category of sensor, the principle of the biosensor and its application. The classification used is the one proposed in Section 1.5: biocatalytic, bioaffinity and microbe-based systems. These systems are linked to electrochemical, optical or acoustic transducers.

5.1.3 BIOLOGICAL MONITORING

Biosensors are actually mostly developed because of the need in sanitary requirements to monitor pathogen micro-organisms and fecal pollution. However, one of the main applications of biological monitoring is the measurement of wastewater toxicity. Even if no regulation concerning toxicity of wastewater exists, it is of great interest. Since a complete characterization of wastewater is impossible, the toxicity measurement is a way of having an idea of the degree of wastewater pollution. Toxicity can hence detect the effect on living organisms or parts of organisms of the major pollutants found in wastewater, but can also detect the effects of emerging organic pollutants such as personal care and pharmaceutical products, endocrine disruptors and antibiotics, that cannot all be detected yet.

The inhibition of respiration is a form of toxicity. Instead of giving a measurement of toxicity, it gives a measurement of a difference between what is supposed to be and what is in reality. For example, if nitrification is inhibited in a given wastewater, the inhibitor is not known but its effect is visible. It is then possible to conclude that at least one inhibitor is present.

There exist several commercial devices for toxicity measurement based on respirometry in the presence of a microbial biocatalyst or on an optical recognition method (bioluminescence, fluorescence) with genetically engineered micro-organisms (GEMs) (Allan *et al.*, 2006). Some commercialized biological tools are ready-to-use test kits and others are measuring instruments. The tests kits are used to determine the presence of specific compounds such as pesticides, PAH, BTEX or PCB. Other biosensors are measuring instruments that can be installed on-line (Allan *et al.*, 2006).

In Table 5.1.4 potential alternative biological tools able to detect the global toxicity or specific toxicity of wastewaters are described. The description concerns the type

Table 5.1.3 Main parameters measured by biosensors in wastewater

	Category/Method[a]	Principle	Reference
COD	Microbial biocatalyst/Respiration	Gas analysis of CO_2 concentration in wastewater	Vaiopoulou *et al.*, 2005
BOD	GEMs/Electronic recognition with conductimetric biosensors	Use of salt-tolerant yeast *Arxula adeninivorans* LS3	Lehmann *et al.*, 1999
	GEMs/Photocatalytic biosensor	Use of *Pseudomonas putida* SG10 with semiconductor TiO_2	Chee *et al.*, 2005
	GEMs/Optical fibre optic biosensors	Use of activated sludge and *Bacillus subtilis* to monitor dissolved oxygen with luminescence intensity variation	Kwok *et al.*, 2005
	GEMs/Amperometric biological recognition	Use of mediator-less microbial fuel cell as sensor	Chang et al., 2004
	GEMs/Electronic recognition with conductimetric biosensors	Based on a pre-tested, synergistic formulated microbial consortium	Rastogi *et al.*, 2003
	Microbial biocatalyst/Respiration	Use of thermally killed cells of complex macrobial culture	Tan and Lim, 2005
	GEMs/Electronic recognition with conductimetric biosensors	Based on an immobilized mixed culture of micro-organisms in combination with a dissolved oxygen electrode	Liu *et al.*, 2000
NH_4^+	GEMs/Amperometric biological recognition	Use of glutamate dehydrogenase (GlDH) which consumes ammonium and glutamate oxidase (GXD) which consumes dissolved oxygen	Kwan *et al.*, 2005
	Microbial biocatalyst/Respiration	Bacterial oxidation of ammonia with oxygen	Bollmann and Revsbech, 2005
NO_3^-	GEMs/Photocatalytic biosensor	Fluorescence measurement of intracellular nicotinamide adenine dinucleotide (NADH)	Farabegoli *et al.*, 2003
	GEMs/Electronic recognition with conductimetric biosensors	Diffusion of nitrate/nitrite through a tip membrane into a dense mass of bacteria	Larsen *et al.*, 2000
	GEMs/Amperometric biological recognition	Microbial nitrate reductase from *Pseudomonas stutzeri* (NaR, EC 1.7.99.4)	Kirstein *et al.*, 1999

Table 5.1.3 Main parameters measured by biosensors in wastewater (*Continued*)

	Category/Method[a]	Principle	Reference
Pesticides	GEMs/Amperometric biological recognition	Use of the screen-printed four-electrode system with immobilized tyrosinase, peroxidase, acetylcholinesterase and butyrylcholinesterase	Solna *et al.*, 2005
	Enzymes/Catalytic transformation of pollutant	Catechol detection with the immobilization of Cl-catechol 1,2-dioxygenase (CCD) in nanostructured films	Zucolotto *et al.*, 2006
	GEMs/Optical fibre immunosensor	Based on solid-phase fluoroimmunoassay	Rodriguez-Mozaz *et al.*, 2004
Phenols	GEMs/Amperometric biological recognition	Use of the screen-printed four-electrode system with immobilized tyrosinase, peroxidase, acetylcholinesterase and butyrylcholinesterase	Solna *et al.*, 2005
	GEMs/Amperometric biological recognition	Use of laccase from *Rigidoporus lignosus*	Vianello *et al.*, 2004
Naphthalene	GEMs/Optical recognition with bioluminescence	Use of *Pseudomonas fluorescens* HK44	Valdman *et al.*, 2004

[a] See Chapter 1.5.

of toxicity detected, the category of biosensor and the method of transduction, the principle of the biosensor and its application. Global toxicity is due to a mix of compounds. The effect of one compound is not known but the effect of the whole is measured. Specific toxicity is the opposite of global toxicity. Specific toxicity is due to the presence of a known compound. In Table 5.1.4, the biological tools are able to measure either a global toxicity, or a specific toxicity, or an inhibition of respiration.

The principles and applications of the biological tools described in Table 5.1.4 do not aim to replace bioassays such as the Daphnia or Microtox® tests, largely discussed in Section 1.5, but are given to help on-line monitoring for the detection of toxicity or respiration inhibition, which could lead to further investigation for the characterization of the pollutants.

Toxicity can also be evaluated using a more classical approach based on on-line respirometry. A recent study carried out on wastewater discharges and comparing respirometry and bioluminescence inhibition with *Vibrio fisheri* (Microtox®) has shown that respirometry inhibition is more adapted when using activated sludge micro-organisms (Kungolos, 2005).

Table 5.1.4 Alternative methods for toxicity measurement in wastewater

	Category/Method[a]	Principle	Application	Reference
Direct toxicity assessment	Microbial biocatalyst/ Respiration	Based on manometric bacterial respirometry	Direct toxicity assessment measurement of wastewater	Tzoris *et al.*, 2005
Inhibitors of nitrification	Microbial biocatalyst/ Respiration	Consists of a Clark oxygen probe as a transducer and an immobilized mixed nitrifying culture as the microbial component	Measurements of inhibitors of nitrification in environmental samples from wastewater	König *et al.*, 1998
NH_4^+ oxidation inhibition	Microbial biocatalyst/ Respiration	Use of immobilized cells of pure culture of *Nitrosomonas europaea*	Measurement of nitrogen ammonia oxidation inhibition in wastewater	Cui *et al.*, 2005
Toxicity due to organic toxicants	Microbial biocatalyst/ Respiration inhibition	Detection of modifications of the feed based on the response of the acetoclastic methanogens	Detection of toxicity due to organic toxicants	Pollice *et al.*, 2001
Toxicity due to aromatic amines	Microbial biocatalyst/ DNA electrochemical sensor	Based on intercalative or electrostatic collection of aromatic amines onto an immobilized dsDNA or ssDNA layer	Determination of toxicity due to aromatic amines	Chiti *et al.*, 2001
Toxicity due to phenol	GEMs/Optical recognition with bioluminescence	Use of whole cell genetically modified bioluminescent biosensors and their immobilization	Discrimination of toxicity by phenols in industrial wastewater	Philip *et al.*, 2003
Global toxicity	GEMs/Optical recognition with bioluminescence	Based on chlorophyll fluorescent signals from photosynthetic enzymatic complexes	Determination of the global toxicity of wastewater	Boucher *et al.*, 2005
Global toxicity	GEMs/Amperometric biological recognition	Use of *Escherichia coli*	Detection of global toxicity in wastewater	Farre *et al.*, 2001
Global toxicity	GEMs/Optical recognition with bioluminescence	Use of *E. coli* HB101 pUCD607	Diagnostic of effluent type and composition by toxicity fingerprinting of pollutants	Hernando *et al.*, 2005

[a] See Chapter 1.5.

Sampling and identification of benthic macroinvertebrates is a biological monitoring method that has been used since the 1970s, in order to evaluate the degradation of a receiving medium under the influence of wastewater discharge. Methods have been proposed for the rapid assessment of wastewater discharge impact on river water quality (Uvanik *et al.*, 2005).

5.1.4 IN PRACTICE

Wastewater discharge monitoring generally requires at least a survey of the quality of treated wastewater and of the receiving medium, upstream and downstream of the discharge (Figure 5.1.1). For the monitoring of treated wastewater discharge, online measurement of physico-chemical parameters (TOC, TSS, nitrate) is completed with permanent analysis of specific parameters in the case of industrial discharge (for example daily analysis of phenols for a refinery) or of toxicity, from composite samples, needing flow rate measurement. For river water quality monitoring, permanent analysis is planned (for example weekly), but complementary integrative procedures can be chosen.

A first method can be the use of natural passive samplers like aquatic moss. A recent study (Figueira and Ribeiro, 2005) has shown that *Fontinalis antipyretica* can be considered as a good concentrator for mineral compounds (Ca, K, Mg, Cu, Fe, Ni, Zn, Pb). An up and coming methodology is the use of passive samplers, the development of which is important (Vrana *et al.*, 2005). Limited by matrix effects and the need for a complex calibration for raw wastewater, their use is more adapted for dilute medium such as treated wastewater and surface water. Several systems exist for the preconcentration of organic compounds and/or trace metals (Petty *et al.*, 2004; Alvarez *et al.*, 2005).

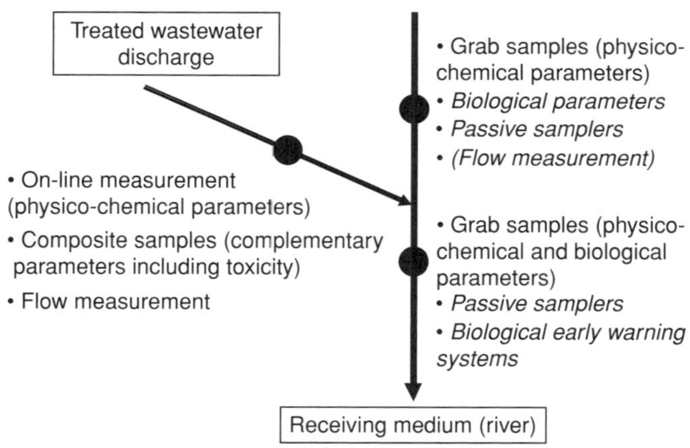

Figure 5.1.1 Wastewater discharge monitoring

Another way of integrative monitoring is based on biological early warning systems or bioindicators. For example, a recent study has shown that Zebra mussel (*Dreissena polymorpha*) and common carp (*Cyprinus carpio*) can be considered for the study of wastewater discharges impact (Smolders *et al.*, 2004). Depending on the experimental conditions (*in situ* and laboratory), the toxicological impact of effluents, in terms of growth and condition related endpoints (i.e. condition, growth, lipid budget) can vary because of food availability. In this study, Zebra mussel has shown to be a better toxicity indicator than the common carp.

There exist very few applications of wastewater discharge monitoring. One, based on the use of an on-line respirometric biosensor using activated sludge microorganisms for toxicity measurement from respirometric inhibition (Kungolos, 2005) has shown that the toxicity is generally higher during the evening and at weekends, probably due to he discharge of partially treated wastewater from some units or to washing streams. Another study, using benthic macroinvertebrate-based parameters, has shown that the results of biological index (Biological Monitoring Working Party, Trent Biotic Index, Chandler Score) and classical parameters (COD, BOD, dissolved oxygen) were in good agreement and coherent with the existence of a wastewater discharge (Uvanik *et al.*, 2005).

A quite different application has been carried out on a small river receiving two wastewater discharges, one urban and one industrial (El Khorassani *et al.*, 1998). The use of UV spectrophotometry has been proposed for TOC and nitrate estimation and for the calculation of the dilution factors of the discharges. The results have been confirmed by laboratory analysis.

REFERENCES

Allan, I.J., Vrana, B., Greenwood, R., Mills, G.A., Roig, B. and Gonzalez, C. (2006) *Talanta*, **69**, 302–322.

Alvarez, D.A., Stackelberg, P.E., Petty, J.D., Huckins, J.N., Furlong, E.T., Zaugg, S.D. and Meyer, M.T. (2005) *Chemosphere*, **61**, 610–622.

Barcelo, D. (2005) *Emerging Organic Pollutants in Wastewater and Sludge. The Handbook of Environmental Chemistry*, vol. 5, parts I and O. Springer-Verlag, Berlin.

Boller, M. (1993) *Water Sci. Technol.*, **27**(11), 167–183.

Bollmann, A. and Revsbech, N.P. (2005) *Sensors Act. B: Chemical*, **105**(2), 412–418.

Boucher, N., Lorrain, L., Rouette, M.-E., Perron, E., Deziel, N., Tessier, L. and Bellemare, F. (2005) *Am. Lab.*, **37**(6), 34–38.

Chang, I.S., Jang, J.K., Gil, G.C., Kim, M., Kim, H.J., Cho, B.W. and Kim, B.H. (2004) *Biosensors Bioelectr.*, **19**(6), 607–613.

Chee, G.J., Nomura, Y., Ikebukuro, K. and Karube, I. (2005) *Biosensors Bioelectr.*, **21**(1), 67–73.

Chiti, G., Marrazza, G. and Mascini, M. (2001) *Anal. Chim. Acta*, **427**(2), 155–164.

Cui, R., Chung, W.J. and Jahng, D. (2005) *Biosensors Bioelectr.*, **20**(9), 1788–1795.

El Khorassani, H., Theraulaz, F. and Thomas, O. (1998) *Acta Hydrochim. Hydrobiol.*, **26**, 296–299.

Environment-Canada (2001) Canada Gazette Part I, 135, 25, available at http://www.ec.gc.ca/Ceparegistry/documents/notices/g1-13525_n1.pdf.

European Commission (1991) Council directive of 21 May 1991 concerning urban waste water treatment (91/271/EEC).

European Commission (2000) Council directive of 23 October 2000 establishing a framework for Community action in the field of water policy (2000/60/EC).

Farabegoli, G., Hellinga, C., Heijnen, J.J. and Van Loosdrecht, M.C.M. (2003) *Water Res.*, **37**, 2732–2738.

Farré, M., Pasini, O., Alonso, M.C., Castillo, M. and Barcelo, D. (2001) *Anal. Chim. Acta*, **426**(2), 155–165.

Figueira, R. and Ribeiro, T. (2005) *Environ. Pollut.*, **136**, 293–301.

Greenwood, R., Roig, B. and Allan, I.J. (2004) Draft report: Operational manual, overview of existing screening methods, available at http://www.swift-wfd.com

Heberer, Th. (2002) *Toxicol. Lett.*, **131**, 5–17.

Heberer, Th., Reddersen, K. and Mechlinski, A. (2002) *Water Sci. Technol.*, **46**(3), 81–88.

Hejzlar, J. and Chudoba, J. (1986) *Water Res.*, **20**(10), 1209–1216.

Hernando, M. D., Fernandez-Alba, A. R., Taulera, R. and Barcelo, D. (2005) *Talanta*, **65**(2), 358–366.

Kirstein, D., Kirstein, L., Scheller, F., Borcherding, H., Ronnenberg, J., Diekmann, S. and Steinruecke, P. (1999) *J. Electroanal. Chem.*, **474**(1), 43–51.

König, A., Riedel, K. and Metzger, J. W. (1998) *Biosensors Bioelectr.*, **13** (7–8), 869–874.

Kümmerer, K. (2001) *Chemosphere*, **45**, 6–7, 957–969.

Kümmerer, K., Alexy, R., Hüttig, J. and Schöll, A. (2004) *Water Res.*, **38**, 2111–2116.

Kungolos, A. (2005) *J. Environ. Sci. Health*, **40**(4), 869–880.

Kwan, R.C.H., Hon, P.Y.T. and Renneberg, R. (2005) *Sensors Act. B: Chemical*, **107**(2), 616–622.

Kwok, N.Y., Dong, S., Lo, W. and Wong, K.Y. (2005) *Sensors Act. B: Chemical*, **110**(2), 289–298.

Larsen, L.H., Damgaard, L.R., Kjaer, T., Stenstrom, T., Lyngaard-Jensen, A. and Revsbech, N.P. (2000) *Water Res.*, **34**(9), 2463–2468.

Lehmann, M., Chan, C., Lo, A., Lung, M., Tag, K., Kunze, G., Reidel, K., Gruendig, B. and Renneberg, R. (1999) *Biosensors Bioelectr.*, **14**(3), 295–302.

Levine, A.D., Tchobanoglous, G. and Asano, T. (1985) *J. Water Pollut. Contr. Fed.*, **57**(7), 805–816.

Levine, A.D., Tchobanoglous, G. and Asano, T. (1991) *Water Res.*, **25**(8), 911–922.

Liu, J., Bjornsson, L. and Mattiasson, B. (2000) *Biosensors Bioelectr.*, **14**(12), 883–893.

Metcalf and Eddy (2003) *Wastewater Engineering, Treatment and Reuse*, 4th Edn. McGraw Hill, Boston.

Petty, J.D., Huckins, J.N., Alvarez, D.A., Brumbaugh, W.G., Cranor, W.L., Gale, R.W., Rastall, A.C., Jones-Lepp, T.L., Leiker, T.J, Rostad, C.E. and Furlong, E.T. (2004) *Chemosphere*, **54**, 695–705.

Philip, J.C., Balmand, S., Hajto, E., Bailey, M.J., Wiles, S., Whiteley, A.S., Lilley, A.K., Hajto, J. and Dunbar, S.A. (2003) *Anal. Chim. Acta*, **487**(1), 61–74.

Pitter, P. and Chudoba, J. (1990) *Biodegradability of Organic Substances in the Aquatic Environment.* CRC Press, Boca Raton, FL.

Pollice, A., Rozzi, A., Tomei, M.C., Di Pinto, A.C. and Laera, G. (2001) *Water Res.*, **35**(5), 1179–1190.

Pouet, M.F., Muret, C., Touraud, E., Vaillant, S. and Thomas, O. (1999) *Proc. Interkama-Isa Conference on CD rom*, 18–20 October 1999, Düsseldorf, Germany.

Rastogi, S., Rathe, P., Sexena, T.K., Mehra, N.K. and Kumar, R. (2003) *Current Appl. Phys.*, **3**(2–3), 191–194.

Rodrigez-Mozaz, S., Reder, S., Lopez de Alba, M., Gauglitz, G. and Barcelo, D. (2004) *Biosensors Bioelectr.*, **19**(7), 633–640.

Smolders, R., Bervoets, L. and Blust, R. (2004) *Environ. Pollut.*, **132**, 231–243.

Solna, R., Sapelnikova, S., Skladal, P., Winther-Nielsen, M., Carlsson, C., Emneus, J. and Ruzgas, T. (2005) *Talanta*, **65**(2), 349–357.

Tan, C. and Lim, E.W.C. (2005) *Sensors Act. B: Chemical*, **107**(2), 546–551.

Thomas, O. and Constant, D. (2004) *Water Sci.Technol.*, **49**(1), 1–8.

Thomas, O. and Pouet, M-F. (2005) In: *The Handbook of Environmental Chemistry*, vol.5, part O, Barcelo, D. (Ed.). Springer-Verlag, Berlin, pp. 245–272.

Tzoris, A., Fernandez-Perez, V. and Hall, E.A.H. (2005) *Sensors Act. B: Chemical*, **105**(1), 39–49.

US EPA. ocean discharge guidelines: www.epa.gov/owow/oceans/regulatory/403.html

Uvanik, S., Yilmak, G., Yesilnacar, M.I., Aslan, M. and Demir, O. (2005) *Fresenius Environ. Bull.*, **14**(4), 268–272.

Valdman, E., Battaglini Leite, S.G.F. and Valdman, B. (2004) *Sensors Act. B: Chemical*, **103**(1–2), 7–12.

Vaillant, S. (2000) La matière organique des eaux résiduaires: caractérisation et évolution. PhD Thesis, University of Pau, France.

Vaiopoulou, E., Melidis, P., Kampragou, E. and AIivasidis, A. (2005) *Biosensors Bioelectr.*, **21**(2), 365–371.

Vianello, F., Cambrai, A., Ragusa, S., Cambria, M. T., Zennaro, L. and Rigo, A. (2004) *Biosensors Bioelectr.*, **20**(2), 315–321.

Vrana, B., Mills, G.A., Allan, I.J., Dominiak, E., Svensson, K., Knutsson, J., Morrison, G. and Greenwood, R. (2005) *Trends Anal. Chem.*, **24**(10), 845–868.

Zabel, T., Milne, I. and Mckay, G. (2001) *Urban Water*, **3**, 25–32.

Zucolotto, V., Pinto, A.P.A., Tumolo, T., Moraes, M.L., Baptista, M.S., Riul Jr, A., Araújo, A.P.U. and Oliveira Jr, O.N. (2006) *Biosensors Bioelectr.*, **21**(7), 1320–1326.

5.2
Monitoring for Water Quality Modelling

Véronique Vandenberghe, Ann van Griensven and **Peter Vanrolleghem**

Wastewater Quality Monitoring and Treatment Edited by P. Quevauviller, O. Thomas and A. van der Beken

5.2.1 INTRODUCTION

Modelling and monitoring are instruments that can help in describing and understanding a system by gathering information and knowledge on its status and functioning. Good water management should thus make maximal use of both modelling and monitoring. An important issue in water management is the protection of the aquatic ecosystems. A healthy ecology community needs a healthy environment that is described by the status of the chemical composition of the water, often referred to as the water quality. While professionals with different backgrounds and at different institutes often perform water quality monitoring and modelling, these actions are often done with a limited dialogue between them. A joint use of monitoring and modelling should thus be stimulated.

In Europe, the EU Water Framework Directive (EU WFD) (European Commission, 2000) formulates water policy. A clear goal is to get all water bodies to good status, with the 'pristine' condition as target. A good status primarily needs to be presented by a good ecological function, which is evaluated by the chemical status and hydro-physical aspects. A good chemical status is only a target for the 'heavily modified water bodies'.

The EU WFD creates new challenges for monitoring and modelling, but it also promotes the interaction between these two (European Commission, 2004). The water bodies are highly complex systems as they hold many unknowns and uncertainties due to the incomplete understanding of the processes, to scaling aspects and to the high variability of the variables in time and space. Therefore, model outcomes become also uncertain and these uncertainties can come from different sources:

- input uncertainty;
- model uncertainty;
- uncertainty in the estimated model parameter values;
- mathematical uncertainty.

Consequently, it is not possible to develop one perfect model or to design an optimal monitoring network with the present information and knowledge. A dynamic approach is therefore needed: models may indicate errors and inadequacies in the monitoring network. Conversely, the model is revised and updated as new data become available.

5.2.2 WATER QUALITY MODELLING

To assist water quality managers in evaluating the impact of conventional pollutant loads on the receiving water, e.g. to evaluate the effectiveness of wastewater treatment, basic water quality models are used. A pressure-impact analysis is important as it facilitates the evaluation of different scenarios. As such, point pollution reduction by improving wastewater treatment versus diffuse pollution reduction by changing land management practices can be simulated and the predictions together with uncertainty bounds around them give an estimation of the possible obtained improvement of the receiving water quality.

5.2.2.1 Water Quality Model Concepts

Two main trends in river water quality modelling can be identified. The traditional QUAL2E model that is based on Streeter–Phelps equations and the recently developed RWQM1 (River Water Quality Model No. 1) is based on the ASM (Activated Sludge Model) for wastewater treatment plants (WWTPs).

Qual2E

During the 1980s and 1990s the standard model in water quality was QUAL2E (Brown and Barnwell, 1987; Shanahan *et al.*, 1998). QUAL2E is an example of a multi-constituent river ecosystem model. This model is able to predict a variety of water quality constituents including conservative substances, algal biomass and chlorophyll-a, ammonia, nitrite, nitrate, phosphorus, carbonaceous biological oxygen demand (cBOD), sediment oxygen demand (SOD), dissolved oxygen (DO), coliforms and radio nuclides.

The main state variables are BOD and DO. By adding new variables and constituents, the model became a three-layer model (Masliev *et al.*, 1995):

- The phenomenological level: the traditional Streeter–Phelps state variables (BOD and DO).
- The biochemical level: the extended Streeter–Phelps model variables (ammonia, nitrate, nitrite and SOD).
- The ecological level: the algae model variables (organic nitrogen, organic phosphorus, dissolved phosphorus and chlorophyll-a).

The QUAL2E model is currently the most widely known model for river water quality. A large number of eutrophication models are based on its process descriptions and it is implemented in a large number of simulation programs, e.g. SWAT (Arnold *et al.*, 1996), QUAL2K, WASP5 (Ambrose and Martin, 1993) ISIS, DUFLOW-EUTRO and MIKE11. QUAL2E was designed for steady-state conditions, but it is also coupled to complex hydrodynamic models to be applicable for unsteady-state conditions, e.g. ESWAT (van Griensven and Bauwens, 2001), CE-QUAL-RIV1, CE-QUAL-W2, CE-QUAL-ICM and WASP5.

Masliev *et al.* (Masliev *et al.*, 1995) and Shanahan *et al.* (Shanahan *et al.*, 1998) point toward problems with the use of QUAL2E, like nonclosed mass balances, e.g. the decay of algal biomass is not included in the BOD and processes in the sediments are not linked to the river column processes. Further, the variable BOD as a measure for organic carbon, only has a biological meaning and not a quantitative mass value, and is thus hard to estimate.

RWQM1

In order to overcome some of the problems with QUAL2E, a new model has recently been developed, the RWQM1 (Reichert and Vanrolleghem, 2001). The main goal was to formulate a set of standardised, consistent river water quality models and guidelines for their use. Moreover, RWQM1 was aimed to be compatible with the existing ASM models since they are both COD-based models. RWQM1 introduced the bacterial biomasses as model components. In this way, bacterial concentration can vary in time, allowing a better description of the observed water quality changes without modifications of the parameters. It also introduces some new processes that were not included in QUAL2E like pH equilibrium reactions, precipitation and predation processes. RWQM1 is designed to have closed mass and elemental balances. For every organic component a fixed composition is given, described by the mass fractions αC (carbon), αH (hydrogen), αO (oxygen), αN (nitrogen), αP (phosphorus) and αX (all other elements), which sum to one. With the aid of the chemical oxidation reaction and a choice of a reference compound for every element considered, the COD of each form of organic matter can be determined. For all reactions the ionic charge balance is closed as well.

The RWQM1 model is implemented in the following software platforms: West (Hemmis NV), Aquasim (Reichert, 1994) and ESWAT (van Griensven and Bauwens, 2001).

5.2.2.2 Integrated Water Quality Modelling

Integrated water quality modelling can be interpreted in various ways. Most strictly, 'integrated' refers to considering the complete causal chain in the model. This means that for all processes, the inputs are also taken into account. It contains rainfall runoff models, urban drainage models, land use and management practices models and pesticide fate models. However, it is also used in another context, an integration of models that simulate two different aspects of water quality, e.g. a model that simulates conventional pollutants and organic contaminants at the same time (Deksissa and Vanrolleghem, 2004) or an integration of the natural and social system, e.g. a river water quality model that is linked to a socio-economic model (Cai *et al.*, 2003). Integrated modelling is complicated for different reasons. The integrated model assembles information from different disciplines in one model, knowledge about all the physical processes need to be present and links from one model to the other need to be made because water quality variables from one model do not have necessarily the same physical meaning in the other model, or translations of physical variables to costs, social indicators and biological meaning need to be established.

Such integration can be realised by model integration where models for the assessment of the pressures (such as wastewater treatment models and land erosion models) are linked to a water quality model. Alternatively, a model for the receiving water model can load inputs (point pollution, runoff, diffuse pollution) that are calculated by other models or are measured in the field.

5.2.3 DATA NEEDS IN WATER QUALITY MODELLING

To build and calibrate integrated water quality models, an enormous amount of data needs to be gathered. The overall goal is to describe the chemical status and its interaction with the ecology. Hence, all variables that affect ecological life are to be described and for all important scales in space and time. For a continuously varying environment, this is a utopia. Therefore, the EU WFD requires 'sufficient' monitoring of the 'key variables'.

When models are used for management purposes, they typically have the role of:

- filling gaps in information (e.g. spatially distributed results versus the point estimates);
- providing predictions capacity with regard to programmes of measures (scenarios).

Where the WFD monitoring programmes will primarily serve the legal requirements and focus on the 'key variables', models can benefit from a much broader range of data types, including the so-called soft data (e.g. data gathered through remote sampling), with the only goal to reduce the uncertainty in the model output to be used for decision making.

Different processes influence variables used to indicate the state of the water. According to the purpose of the model that depends on the kind of evaluation to be made, more or less data will be needed. For example, if the aim of the data gathering is to set up a detailed physical model for evaluating oxygen depletion due to eutrophication in a river, monthly, weekly or even daily measurements will not be enough. The algae processes cause sub-daily variations in the oxygen concentration and hence hourly measurements are needed. If, however, evaluations are wanted of pollution effects of BOD coming from the outflow of a WWTP, the average quality state of a river is sufficient and data taken on a monthly basis indicate that well enough.

The data needs also depend on the model concept. RWQM1 is best served with organic C or COD observations, while QUAL2E will take more advantage of BOD estimations. In RWQM1, all variables are quantifiable and can be expressed in masses and all parameters have a direct meaning in the expressions for the process rates, while in QUAL2E, several variables have a phenomenological meaning (such as BOD) and several parameters express a conceptual meaning (e.g. sediment oxygen depletion). As the sensitivities of the parameters to observations are different for RWQM1 and QUAL2E (Vandenberghe *et al.*, 2006), additional observations will give a different reduction on the parameter uncertainties for these concepts.

5.2.4 CASE STUDY

To show further the methods used in monitoring for water quality modelling we use the river Dender in Flanders as a case study. The model covers an area of around 700 km^2. The water quality is modelled for the main channel with a length of 50 km.

An extended version of the Soil Water and Assessment Tool (SWAT) (Arnold *et al.*, 1996) was used to build a water quality model on a catchment scale including diffuse and point source pollution. The SWAT model codes were extended to simulate hourly water quality variables based on the QUAL2E concept (van Griensven and Bauwens, 2001). The water quality calibration objectives included oxygen, BOD, ammonia, nitrate and phosphate at a the outlet of the river (Denderbelle). More information on the Dender river and the model is given in the literature (van Griensven and Bauwens, 2001; Vandenberghe *et al.*, 2002, 2005a, 2006).

5.2.5 MONITORING FOR THE IDENTIFICATION OF WATER QUALITY MODEL PARAMETERS

5.2.5.1 Optimal Experimental Design

Parameters of a river water quality model are not always identifiable due to a lack of data or due to the fact that data are taken in periods or in places that are not suitable for calibration of the model. Optimal experimental design (OED) techniques are a useful tool to construct experiments that may deliver the information needed for the calibration of a model of the system under consideration. There exist applications of OED in many disciplines like modelling of WWTPs (Vanrolleghem *et al.*, 1995), modelling of pyruvate production (Zelic *et al.*, 2004), groundwater modelling experiments (Catania *et al.*, 2004), systems biology (Faller *et al.*, 2003), food technology (Nahor *et al.*, 2001), pharmacology (Fedorov and Leonov, 2001) and electrical engineering (Ko *et al.*, 2004). A common element in all those studies is that the experimental conditions such as temperature, time, pH, measurement frequency, initial concentration, etc., are controllable. When the experimental design gives a set of conditions that is found to be optimal to do experiments for extracting a lot of information about the process, these conditions can be artificially introduced. For a natural river system, things become more complicated as a combination of different factors like temperature, flow and concentration cannot be manipulated and is not occurring at the desired moments and as such, a method has to be found which maximises the information content of the experiments, without knowing the exact timing under which those optimal circumstances will occur. External conditions like weather, discharges in the river or diffuse pollution into the river can change year after year, so measurements that are likely to be optimal for a particular year can appear to be sub-optimal the next year. All those reasons mean that a normal straightforward OED cannot be used for river water quality modelling and extensions of those designs are needed to find a good measurement set-up.

Here we show how to design a good set-up for measurements for the calibration of a river water quality model, based on a set of previous measurements and a model calibrated with those measurements. It is assumed that the calibrated model gives good results but that the uncertainty bounds related to the model outcomes are too wide to draw reliable conclusions for management decisions. The results of the OED can further be used to find a cost-effective solution, by relating model improvements that are obtained by using more or better measurements to the costs and practical considerations.

5.2.5.2 Methodology

For the 'traditional' use of OED in river water quality modelling, three problems arise:

(1) The experimental circumstances cannot be directed; data for all possible circumstances are not available.

(2) The behaviour of the system varies randomly according to the natural variability of external circumstances.

(3) The model used for the OED contains itself a lot of uncertainty: the parameters can be under- or overestimated.

Therefore, it is not possible to find one exclusive best experimental design but the experimental design can help in giving information about how, were and when the measurements are the best taken, because there exist periods and circumstances under which the model is more sensitive to parameter changes than others.

Wald (Wald, 1974) has demonstrated that when experiments are carried out in sequence a smaller number of them are required, on average, than when they are performed simultaneously. This is true even where no use is made of information gained in one experiment for planning the next one. The gain in this case accrues entirely from the ability to terminate the experimentation precisely at the most appropriate point at which most information can be obtained. If, in addition, one is able to design each experiment in the light of the results of the previous ones, the gain in efficiency can be even more impressive. This can be generalised when considering an experiment as a collection of data.

As such, the OED method in river water quality modelling gives the best results when it is used in an iterative way: year after year the model parameters can be updated by calibrating it with new available data and the parameters will finally converge to one best parameter set.

The purpose in this study here is to maximise the practical identifiability of the model parameters by defining an optimal experiment that increases the information content of the data. Different experiments (sampling schemes) will reveal more or less information and more or less parameter reliability, e.g. schemes that lack dynamics will provide less information than schemes with more. Optimal sampling design techniques aim at the identification of sampling schemes to improve different aspects of the mathematical modelling process, according to explicitly stated objectives (Dochain and Vanrolleghem, 2001; De Pauw and Vanrolleghem, 2004). The objective considered here is to increase the precision of the parameters for the water quality module of ESWAT.

The D-optimal experimental design (Goodwin and Payne, 1977; Walter and Pronzato, 1999) was used, because it is the most general method for minimising the error on all estimated parameters.

In a D-optimal experimental design, considering the determinant of the inverse of the covariance matrix of the parameter estimates ($\boldsymbol{C}$) or Fisher information matrix (**FIM**) (Godfrey and Distefano, 1985) assesses the precision of the parameters:

$$\boldsymbol{C}(\boldsymbol{b}) = \sigma^2(\boldsymbol{S}^T\boldsymbol{Q}\boldsymbol{S})^{-1} \quad \mathbf{FIM}(\mathbf{b}) = \boldsymbol{C}^{-1}(\boldsymbol{b})$$

with $\boldsymbol{b}$ representing the model parameter vector, σ the standard deviation of the parameters, $\boldsymbol{Q}$ a diagonal matrix, the elements being the squares of the observation weights and $\boldsymbol{S}$ the sensitivity matrix of the outputs to the parameters in comparison to the observations. Calculation of the covariance matrix based on the Jacobian matrix instead of the Hessian is acceptable when assuming linearity and assuming observations with constant standard deviations (Bard, 1974). The determinant of the **FIM**, det(**FIM**), is proportional to the volume of the confidence region. Thus, by maximising det(**FIM**), the volume of the confidence ellipsoids, and, correspondingly, the geometric average of the parameter errors is minimised. D-optimal experiments also have the advantage of being invariant with respect to any scaling of the parameters (Petersen, 2000). An extra aspect to be considered here is that for nonlinear models the **FIM** is parameter dependent. The OED technique thus requires an initial data set to calibrate the model. Nonaccurate parameter estimates may therefore lead to an inefficient experimental layout. This means that for the processes related to the nonaccurate parameters better measurements could be identified. The design can only be approached by an iterative process of data collection and design refinement, known as a 'sequential design' (Casman *et al.*, 1988). Figure 5.2.1 shows the iterative scheme that is used to find the optimal measurements starting with a model that is calibrated with the currently available data. Next the different steps are explained in more detail.

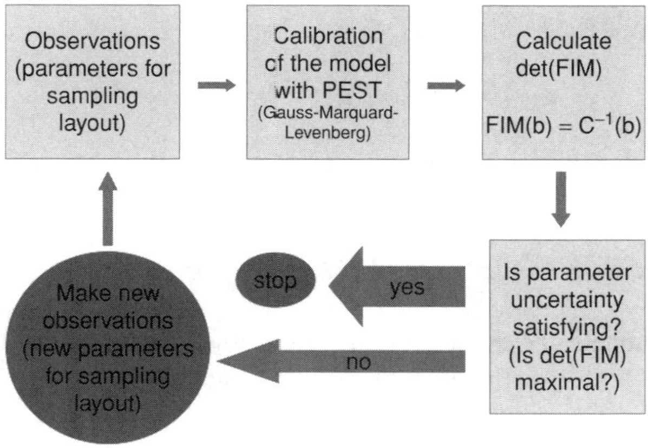

Figure 5.2.1 Optimal experimental design for river water quality modelling. [PEST, Parameter ESTimation model (Doherty, 2000)]

Generating synthetic data series

The evaluation of different sampling schemes requires the availability of a long time series of high frequency water quality data at different places along the river. Because such historical series were not available, a simulation generated synthetic 'observation' series with the Dender model using ESWAT. For realism, the output series were subsequently altered by addition of pseudo-random noise. Noise was generated from a normal distribution with variations that are consistent with the accuracy of the measuring devices used to measure the variables (Vandenberghe *et al.*, 2005a): 3 % for DO, 10 % for BOD and 5 % for NO_3 and NH_4. Then the parameters for the sampling layout are defined. Examples of such parameters are the sampling frequency, e.g. every 2 h, location of the measurements, e.g. downstream and 6 km more upstream, and the kind of measured variables, e.g. DO + NH_3.

Calibration of the model

With the data selected from the synthetic time series on the basis of a certain sample layout, the model can be calibrated again. To be sure that the calibration process does not end in a local optimum, the initial parameter values are taken in the neighbourhood of the final parameter values obtained during the calibration with the available data. The purpose of this step is not to find the parameter values but rather to obtain the Jacobian matrix during the calibration with a derivative-based method because the inverse of the Jacobian matrix is the **FIM**. Here the PEST (Parameter ESTimation) program (Doherty, 2000) is used. The parameter estimation in the PEST program is done by a minimisation of the objective function (J) by finding the optimal choice of the parameters θ:

$$J(\theta) = [y - y(\theta)]^T \times \boldsymbol{Q} \times [y - y(\theta)]$$

with y being the output of the model and $\boldsymbol{Q}$ a weighing factor for the model outputs.

Calculating the determinant of the ***FIM***

For nonlinear models, the expected value of the objective functional for a parameter set slightly different from the optimal one can be approximated by:

$$E\left[J\left(\theta+\delta\theta\right)\right] \cong \delta\theta \underbrace{\left[\frac{\partial y^T}{\partial\theta}\boldsymbol{Q}\frac{\partial y}{\partial\theta}\right]}_{\text{Fisher information matrix}} \partial\theta + \mathrm{c}$$

with E the expected value of the objective functional, y the model outputs, $\boldsymbol{Q}$ a weighing matrix for the model outcomes and c a small constant. The PEST program calculates the covariance matrix of the parameters at the best estimate which means that also the **FIM** can be determined. det(**FIM**) is then inversely proportional to the volume of the confidence region around the parameters.

*Maximisation of the det(**FIM**) by changing the sampling layout*

In a loop, different observations characterised by a different sampling layout of the program can be selected. The shuffled complex methods (SCE-UA) (Duan *et al.*, 1992) is used here to maximise the det(**FIM**) in order to optimise the parameters of the sample layout. The parameters of the sample layout are the parameters that are changed to obtain a maximisation of the objective function, the det(**FIM**). After several evaluations of the det(**FIM**), the shuffled complex method finds the optimum very fast because the method searches the whole parameter space in an efficient and effective manner.

5.2.5.3 Results and Discussion

The methodology has been applied for an OED at the Dender river. As an illustration of the applicability of the method, a simple case, whereby only DO is considered at one specific location is presented first. The synthetic 'observation' series consists of 1 year of hourly data. The optimisation is here limited to the measuring frequency, the number of samples and the period of the year for sampling. The sampling time step was allowed to vary between 1 h and 2 days; the minimum number of samples is 1 and the maximum number is 8760 (365×24). Samples could be taken during

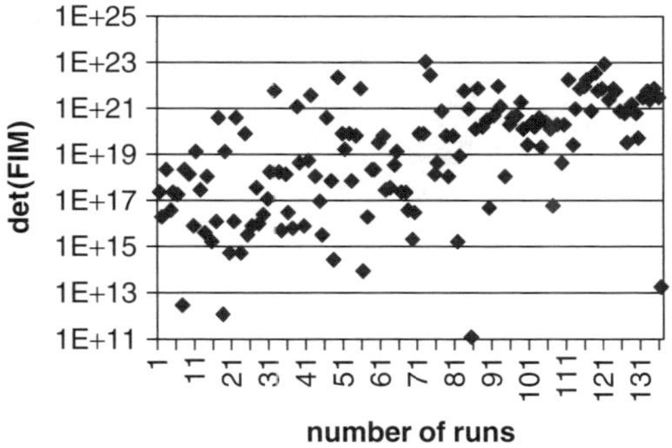

Figure 5.2.2 Optimisation of the det(**FIM**) (three parameters of sampling layout)

winter, summer or a mixed summer–winter period, depending on the start of the period and the total number of samples that are taken.

In Figure 5.2.2, the optimisation process is shown. SCE-UA used 136 runs to find the optimum for which the det(**FIM**) is the largest. As could be expected, the results show that the uncertainty in the parameters became minimal for the smallest sampling interval (Figure 5.2.3a), a very large number of samples (Figure 5.2.3b) and a large period, mainly spring and summer months (data not shown). A sample every hour, starting in February and ending on 30 August, representing a total of 5804 samples appears to provide the best results.

A second example supports a more complex planning, whereby in addition also the data type (only DO or combined DO-NO_3, DO-NO_3-BOD or DO-NO_3-BOD-NH_4) and sample locations (four possible combinations of three possible locations: upstream, halfway, downstream) are considered as parameters for the sampling layout.

A substantial increase in the number of iterations for the optimisation is observed (Figure 5.2.4). The best way to take samples is on an hourly basis (Figure 5.2.5a), over nearly the whole year (8730 samples) (Figure 5.2.5b), on two locations (data not shown) and with measurement of four variables (data not shown). This is again a very logical result. However, looking at Figure 5.2.5 it can be depicted that other sampling schemes could be defined that provide a quasi-similar accuracy, with fewer samples

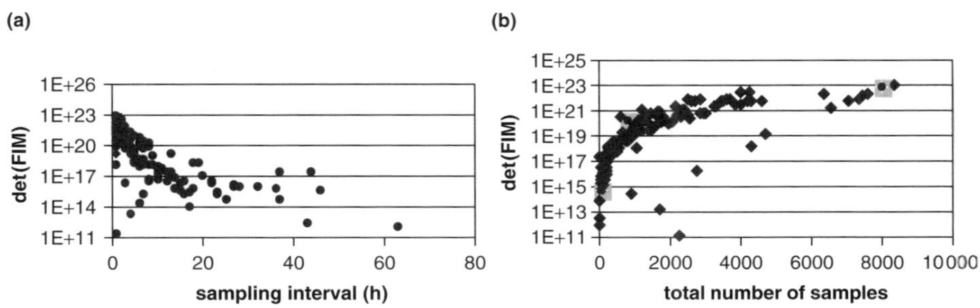

Figure 5.2.3 det(**FIM**) as a function of the sampling interval (a) and total number of samples (b)

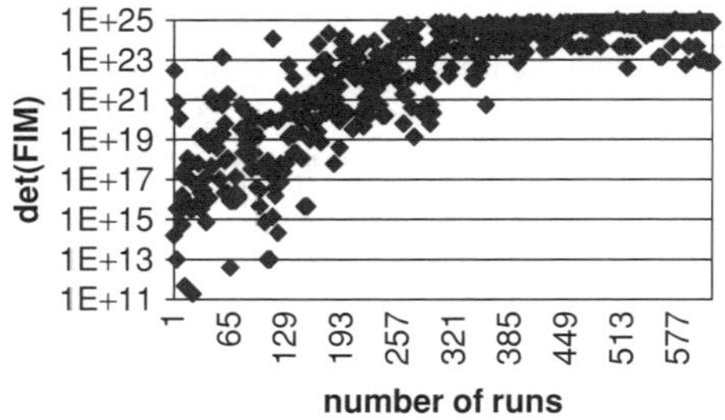

Figure 5.2.4 The optimisation of det(**FIM**) with variation of five parameters

or a lower frequency. det(**FIM**) is not changing between 5000 and 8000 samples which means that the confidence regions around the parameters do not differ very much in that range. This is explained by other factors that influence the accuracy, such as the period of the year during which the sampling takes place.

Alternatively, some sampling schemes clearly appear as nonoptimal (such schemes are indicated by squares in Figure 5.2.5): these schemes require a lot of samples, but due to the wrong choice of other factors, the information content of these schemes is poor. More details on these schemes are given in Table 5.2.1. The reason for the bad performance of these schemes is related to the sampling place (upstream) and to the fact that the sampling period does not include the spring period, which seems here to be important for the calibration process.

The search for the optimal experimental design including practical considerations

The value of the det(**FIM**) has no physical meaning. A further analysis is needed to check the improvement of the calibration with the optimal set of measurements in contrast with a – for calibration purposes – less good measurement set, which is however characterised by lower cost and effort or that is more practically feasible.

The performance of the calibration is evaluated by looking at the final uncertainty on the model results taking into account the variances and correlation between the

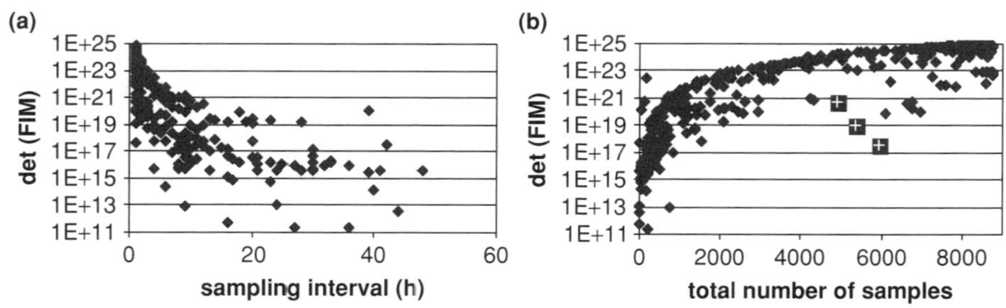

Figure 5.2.5 The inverse of det(**FIM**) as a function of the sampling interval (a) and the total number of samples (b). (Points marked by crosses in boxes are further investigated)

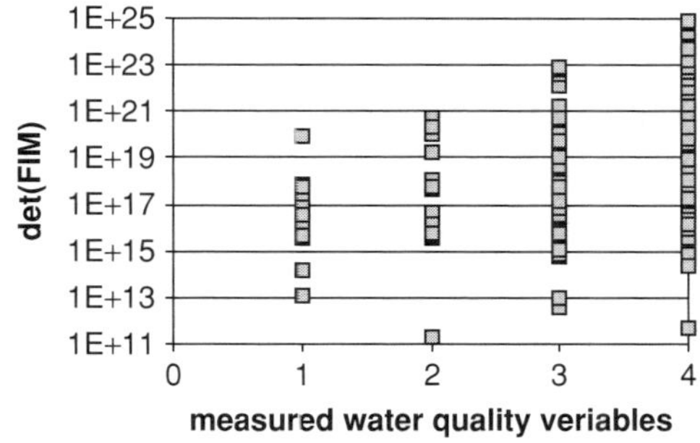

Figure 5.2.8 det(**FIM**) as a function of the measured water quality variables: 1, DO; 2, DO + NO_3; 3, DO + NO_3 + BOD; 4, DO + NO_3 + BOD + NH_4

those results. Next to the expression of uncertainty bounds on the results, uncertainty studies have mainly been used to provide insight in the parameter uncertainty. However, UA can also be a means to prioritise uncertainties and focus research efforts on the most problematic points of a model. As such, it can help to prepare future measurement campaigns and to guide policy decisions. Here we show an application of how UA can be used to point towards the most important inputs. The practical case study is again the river Dender in Flanders, Belgium modelled in ESWAT.

5.2.6.2 Methodology

To reduce the overall uncertainty on the model results for a certain variable the following steps are proposed:

(1) Identify which sources contribute mainly to the overall uncertainty on the model results.

(2) Estimate or calculate the uncertainty related to those main contributors.

(3) Propagate the uncertainty of all different kinds one by one through the model.

(4) Analyse these results to set up a future monitoring campaign.

(5) Perform the measurements.

(6) Recalibrate the model with new inputs.

(7) Repeat steps 3–6 until satisfying results are obtained.

For every step of this process different techniques exist that can be chosen according to the experience of the modeller.

5.2.6.3 Results and Discussion

Here an evaluation is done of the uncertainty on model results for nitrate in the river water.

Identification of the main uncertainty contributors

We evaluated the sensitivity of the model on the following result: the time that NO_3 is higher than 3 mg/l at Denderbelle, near the mouth of the river in 1994.

The technique is a global sensitivity analysis based on regression with Latin Hypercube Monte Carlo sampling (Vandenberghe *et al.*, 2002). For each of the sub problems the parameters or data that contribute significantly to the output (5 % level) are then taken together in one overall sensitivity analysis to compare the contribution of the different outputs. The column with the standard regression coefficient (SRC) as a result of that analysis is indicated in Table 5.2.3 with 'combined parameter input'. The SRC has the following meaning:

$$\mathrm{SRC}_i = \frac{\Delta y / S_y}{\Delta x_i / S_{x_i}}$$

Table 5.2.3 Results of the sensitivity analysis for the model output 'hours NO_3 >3 mg/l' at Denderbelle, 1994. Pa16, amount of fertilisation on pasture in subbasin 16; Fa4, amount of fertilisation on farming land in subbasin 4; gropa, growth date of pasture; plfa, Plant date on farming land; Co5, amount of fertilisation on corn in subbasin 5; Co15, amount of fertilisation on corn in subbasin 15; Pa12, amount of fertilisation on pasture in subbasin 12; Co11, amount of fertilisation on corn in subbasin 11; Ai5, O_2 uptake per unit of NH_3 oxidation; Rk5, denitrification rate; Rk2, oxygen reaeration rate; Ai6, O_2 uptake per unit of HNO_2 oxidation; Bc2, rate of NO_2 to NO_3; Rk3, rate of loss of BOD due to settling; Ai4, O_2 uptake per unit of algae respiration; Rs5, organic phosphorous settling rate

Diffuse pollution input	SRC	Point pollution input	SRC	Parameter	SRC	Combined Parameter-input	SRC
Pa16	−0.30	BOD point 6	−0.61	Ai5	−0.7	Ai5	−0.51
Fa4	0.23	NO3 point 7	0.42	Rk5	−0.34	Ai6	−0.50
gropa	−0.18	BOD point 5	−0.38	Rk2	0.32	Rk5	−0.40
plfa	0.17	BOD point 8	−0.24	Ai6	−0.21	Bc2	0.38
Co5	−0.17	NH3 point 1	0.23	Bc2	−0.2	Ai4	−0.31
Co15	−0.16	BOD point 3	−0.23	Rk3	0.17	Rk2	0.12
Pa12	0.16	BOD point 7	−0.22	Ai4	0.12	plfa	−0.08
Co11	0.15	BOD point 1	−0.14	Rs5	−0.09	BOD point 6	−0.07
		NO3 point 5	0.11		0.07	BOD point 1	−0.07
		BOD point 4	−0.09			Pa16	0.07
		NH3 point 2	0.09				
		BOD point 2	−0.08				
		NH3 point 3	0.06				

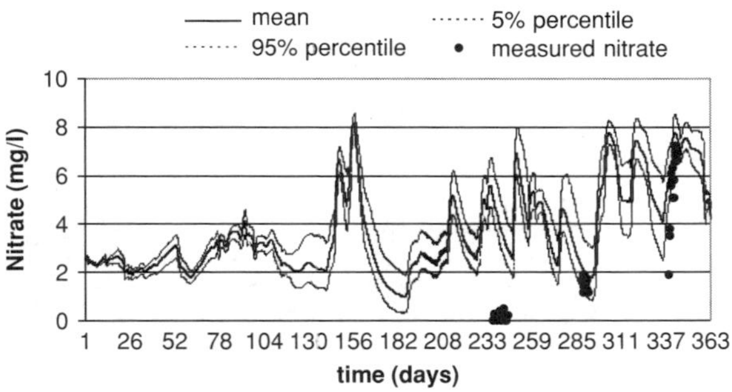

Figure 5.2.9 Simulation of nitrate with confidence intervals related to parameter uncertainty at Denderbelle, 1994

with $\Delta y/\Delta x_i$ being the change in output due to a change in an input factor and S_y, S_{x_i} the standard deviation of, respectively, the output and the input. The input standard deviation S_{x_i} is specified by the user. The technique of ranking the parameters according to their sensitivity based on regression analysis is further explained in Chapter 2.3.

For the parameters, the sampling for the sensitivity analysis was based on own experience and literature ranges.

For both the point and diffuse pollution input, the same uncertainties were taken to calculate the uncertainty on the time series results, as the sampling range used for the sensitivity analysis because we obtained no new information between the sensitivity analysis and the UA. Parameter uncertainty, diffuse pollution uncertainty and point pollution uncertainty were considered separately. Figures 5.2.9 and 5.2.10 show the time series of nitrate in the river water at Denderbelle, situated near the mouth, with the 5 % and 95 % uncertainty bounds with, respectively, uncertainty on diffuse input and point pollution input. Figure 5.2.11 shows the uncertainty bounds for nitrate at the same location due to parameter uncertainty.

We can now link the inputs to external circumstances. When considering the rain and flow rate (Figure 5.2.12), we can see that diffuse pollution inputs are important during periods with high rainfall and high flows. During dry weather flows, the

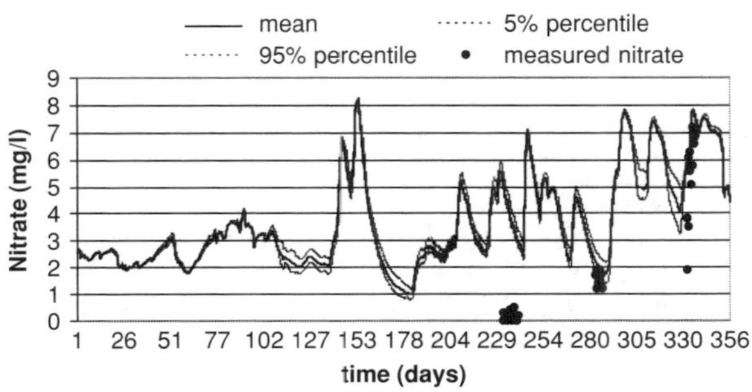

Figure 5.2.10 Simulation of nitrate with confidence intervals related to point pollution input uncertainty at Denderbelle, 1994

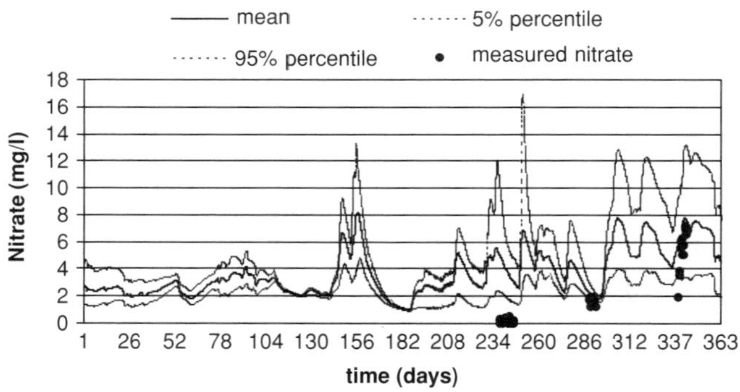

Figure 5.2.11 Simulation of nitrate with confidence intervals related to diffuse pollution input uncertainty at Denderbelle, 1994

input uncertainty of the loads is also propagated. Hence, this UA shows that we can obtain a better calibration for the diffuse pollution part of the model with data that are taken during wet periods with high flows, because the model output nitrate is more sensitive towards inputs of diffuse pollution in those periods. If one focuses on calibrating the in-stream behaviour and point pollution then measurements during dry periods are needed, as the model is in such conditions not sensitive towards input of diffuse pollution.

This case study shows that too often a model is calibrated with only one comprehensive measurement campaign. This is mostly not the most efficient way. When, for example, only measurements during dry periods are made, the model cannot be well calibrated for the diffuse pollution part. So it is better to perform two separate smaller measurement campaigns with the first one 'exploring', while the second campaign is guided by previous analysis of the model results. The combination of the two monitoring campaigns can guarantee that at least some measurements are performed at 'the right moment', making the calibration process easier and more reliable.

The limit of the OED methodology is that it highly depends on the completeness of the processes and its implementation in the model. The river water quality component

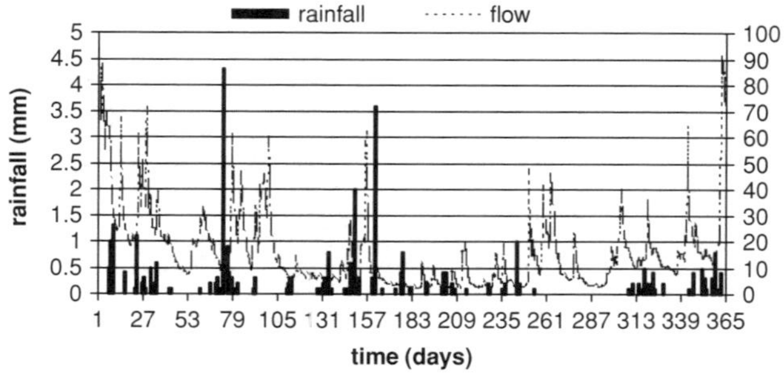

Figure 5.2.12 Rainfall/flow rate at Denderbelle, 1994

in ESWAT gives a one-dimensional description for the water column that does not provide any information on potential stratification in depth that might occur in the channel (Vandenberghe *et al.*, 2005b). Also the limited representation of the riverbed in QUAL2E does not reflect properly the riverbed activities that are known to be important (Vandenberhge *et al.*, 2002). The model selection and set-up is thus a crucial basic step that determines to a large extent the success of an OED analysis.

5.2.7 CONCLUSIONS AND PERSPECTIVES

Monitoring is carried out for the evaluation of groundwater and surface water quality. Each instance responsible for data gathering about the state of the water quality has its own methods of monitoring. This means a network of places where a selected amount of water quality variables with a certain frequency are measured. Till now, most of these monitoring networks have evolved historically: measurement places are situated at bridges, previous or still existing outlet pipes of the sewer system, industrial load outlets, borders of regions and nearby confluences of rivers. The frequency of measuring is regular, e.g. once per month and the water quality variables are the ones considered most important for evaluation purposes.

There exist however techniques of experimental design, combined with modelling of the groundwater or surface water, that give guidelines for monitoring campaigns. These monitoring campaigns can serve next to evaluation of the water quality state also as new data for model input. The models will be able to make more accurate predictions and the water quality can as such be better evaluated on locations or on moments where no measurements are done or the future state of the water body can be predicted. The monitoring will not be once per month any more, but only some periods of the year will be monitored more intensively. The places will change to places where the data gives most information about the water system and not all variables will need to be measured as frequently to attain the water quality state with good reliability, because modelling results can replace part of the data. The application of OED for guiding monitoring campaigns will make the monitoring more effective with less cost. Since OED is dependent on the knowledge and information that is incorporated in the model, special care should be given to the model set-up from the beginning. In addition, an OED analysis should not automatically overrule any other additional information on the system (such as expert knowledge).

REFERENCES

Ambrose, R. B. and Martin, J. L. (1993) *The water quality analysis simulation program, Wasp5, part a: model documentation.* US EPA, Athens, GA.

Arnold, J.G., Williams, J.R., Srivnivasan, R. and King, K.W. (1996) *SWAT Manual.* USDA, Agricultural Research Service and Blackland Research Center, Texas.

Bard, Y. (1974) *Nonlinear Parameter Estimation.* Academic Press, New York.

Brown, L. C. and Barnwell, T. O. (1987) *The Enhanced Stream Water Quality Models, QUAL2E and QUAL2E-UNCAS: Documentation and User Manual.* US EPA, Athens, GA.

Cai, X., McKinney, D.C. and Lasdon, L.S. (2003) *J. Water Res. Plann. Manag.*, **129**(1), 4–17.

Casman, E. A., Naiman, D. Q. and Chamberlain, C. E. (1988) *Water Resour. Res.*, **24**(3), 409–415.

Catania, F., Massabò, M., Minciardi, R., Paladino, O. and Robba, M. (2004) In: *Proc. iEMSs 2004 Int. Congress: Complexity and Integrated Resources Management*, Osnabrueck, Germany, June 2004.

Deksissa, T. and Vanrolleghem, P.A. (2004) In: *Environmental Biotechnology: Advancement in Water and Wastewater Application in the Tropics*, Z. Ujang and M. Henze (eds), Water and Environment Management Series (WEMS). IWA Publishing, London, pp. 241–248.

De Pauw, D. and Vanrolleghem, P. A. (2004) Optimal experimental design for model calibration: general procedure. In: *Proc. 4th PhD Symposium*, Ghent, Belgium.

Dochain, D. and Vanrolleghem, P. A. (2001) *Dynamical Modelling and Estimation in Wastewater Treatment Processes.* IWA Publishing, London.

Doherty, J. (2000) *Model Independent Parameter Estimation (PEST).* Watermark Computing, Brisbare, Australia.

Duan, Q., Gupta, V.K. and Sorooshian, S. (1992) *Water Resour. Res.*, **28**(4), 1015–1031.

European Commission. (2000) Water Framework Directive. Council Directive 2000/6/EG.

European Commission. (2004) Common Implementation Strategy for the Water Framework Directive (2000/60/EC) Guidance Document No. 11, Planning Processes. European Communities. Available at http://forum.europa.eu.int/Public/irc/env/wfd/library?l=/framework_directive/guidance_documents/

Faller, D., Klingmüller, U. and Timmer, J. (2003) *Sim.* **79**(12), 717–725.

Fedorov, V. V. and Leonov, S. L. (2001) *Drug Inform. J.* **35** (4), 1373–1383.

Godfrey, K.R. and Distefano, J.J. (1985). Identifiability of model parameters. In: *Identification and System Parameter Estimation.* Pergamon Press, Oxford.

Goodwin, G. C. and Payne, R. L. (1977) *Dynamic System Identification. Experiment Design and Data Analysis*, Academic Press, New York.

Hemmis NV. West. Kortrijk, Belgium.

Ko, Y., Kim, Y., Park, D. and Yun, I. (2004) *Rob. Comp-int. Man.*, **20**(2), 121–125.

Masliev I., Somlyody, L. and Koncsos, L. (1995). *On Reconciliation of Traditional Water Quality Models and Activated Sludge Models.* Working paper. Laxenburg, Austria.

Nahor, H. B., Scheerlinck, N., Verniest, R., De Baerdemaeker, J. and Nicolai, B. M. (2001). *J. Food Eng.*, **48**(2), 109–119.

Petersen, B. (2000) *Calibration, identifiability and optimal experimental design of activated sludge models.* PhD thesis, Faculty of Agricultural and Applied Biological Sciences, Ghent University, Ghent Belgium.

Reichert, P. (1994) *Water. Sci. Technol.*, **30**(2), 21–30.

Reichert P. and Vanrollegem P.A. (2001) *Water Sci. Technol.*, **43**(7), 329–338.

Shanahan, P., Henze, M., Koncsos, L., Rauch, W., Reichert, P., Somlyody, L. and Vanrolleghem, P. A. (1998) *Water Sci. Technol.*, **38** (11), 245–252.

Vandenberghe, V., van Griensven, A. and Bauwens, W. (2002) *Water Sci. Technol.*, **46**(3), 1–7.

Vandenberghe,V.,van Griensven, A. and Bauwens, W. (2005a) *Water Sci. Technol.*, **51**(3–4), 347–354.

Vandenberghe, V., Goethals, P., van Griensven, A., Meirlaen, J., De Pauw, N., Vanrolleghem, P.A. and Bauwens, W. (2005b) *Environ. Monit. Assess.*, **108**(1–3), 85–98.

Vandenberghe, V., van Griensven, A. Bauwens, W. and Vanrolleghem, P. (2006) *Water Sci. Technol.*, **53**(10), 277–284.

van Griensven, A. and Bauwens, W. (2001). Integral modelling of catchments. *Water Sci. Technol.*, **43**(7), 321–328.

Vanrolleghem, P. A., Van Daele, M. and Dochain, D. (1995) *Water Res.*, **29**, 2561–2570.

Wald, A. (1974) *Sequential Analysis*. John Wiley & Sons, Ltd, New York.

Walter, E. and Pronzato, L. (1999. *Identification of Parametric Models from Experimental Data*. Springer Verlag, Heidelberg.

Zelic, B., Vasic-Racki, D., Wandrey, C. and Takors, R. (2004) *Bioproc. Biosys. Eng.* **26**(4), 249–258.

5.3
Discharges in Sensitive Receiving Waters

Giuliano Ziglio, Marco Vian and **Claudia Lasagna**

5.3.1 INTRODUCTION

In urban areas two types of discharge systems are common. The separate sewer system collects municipal wastewaters from households and industrial and commercial wastes which is compatible with sewer materials and the biological process implemented in the wastewater treatment plant (WWTP). The surface urban runoff is collected separately and via a storm sewer is directly or indirectly (after some pretreatment) discharged into receiving waters.

Wastewater Quality Monitoring and Treatment Edited by P. Quevauviller, O. Thomas and A. van der Beken

In the combined system, municipal wastewaters, wastes from compatible pollution sources and urban runoff are collected in a single pipe. In dry conditions, only municipal and industrial or commercial wastes are flowing in the sewer. In wet conditions (rain, melting snow) also part of the surface runoff is discharged in the sewer. When the maximum capacity of the collecting system and of the WWTP is reached, the excess flows are discharged as the so-called combined sewer overflow (CSO), with or without any pollution control.

In reality, numerous combinations of the two systems frequently exist. The WWTP in wet conditions can receive from none to the whole of the runoff, with the consequent exceeding of treatment capacity.

5.3.2 KEY VARIABLES

The most critical pollutants (variables) that can enter sensitive receiving waters, such as lakes, rivers and coastal zones, from WWTP discharges are summarized and classified at different levels of priority in Table 5.3.1. Criteria for defining priority can be general or water body specific and site specific, according to climatic, environmental and social conditions. For more detailed information see Lijklema *et al.* (Lijklema *et al.*, 1993) and House *et al.* (House *et al.*, 1993).

However, in the context of discharge in sensitive waters, the concentration of these variables is far less important than the corresponding mass loading over different time periods.

Therefore, wastewater discharges have to be described by measuring the concentration (of the pollutants) in wastewater samples together with the flow rate condition at the moment of sampling (being mass loading flow rate times concentration).

5.3.3 CONTAMINANT MASS LOADING: IMPACT AND MANAGEMENT

The potential impacts of WWTP discharges on the receiving waters can be related to the continuous loading or the transient high loadings due to loss of treatment efficiency or to exceeding plant capacity.

The receiving water responses are determined by a series of physical, chemical and biological processes (Lijklema *et al.*, 1993).

These impacts can be verified by implementing monitoring strategies of key state variables or can be predicted and simulated with the use of deterministic and stochastic mathematical models, in order to develop appropriate management strategies and practices.

Among others, Lijklema *et al.* (Lijklema *et al.*, 1993) and House *et al.* (House *et al.*, 1993) reviewed and summarized potential ecological acute and chronic impacts that can be related to wastewaters from WWTP and to urban runoff waters

Table 5.3.1 Variables to be quantified in WWTP discharge and general level of priority

	Unit (as example)	Comments
High priority variables		
Flow rate	m^3/h	
Total phosphorus	kgP/day	Suspended and dissolved organic and inorganic P. Suspended organic P can be indirectly calculated by the content of volatile suspended solids multiplied by a standard factor of 0.005–0.010 or a calculated specific one (see also dissolved IP)
$N\text{-}NH_3$	kg N/day	In combination with temperature and pH
$N\text{-}NO_3$	kg N/day	
BOD_5	kg O_2/day	Low priority in coastal zones
Readily biodegradable COD	kg O_2/day	Low priority in coastal zones
Slowly biodegradable COD	kg O_2/day	Low priority in coastal zones
Escherichia coli	Log (/day) or N/100 ml	Might have intermediate or low priority in lakes Concentrations are used as input for hydraulic models computing initial dilution of the effluent concentration
Faecal Streptococci	Log (/day) or N/100 ml	Might have intermediate or low priority in lakes Concentrations are used as input for hydraulic models computing initial dilution of the effluent concentration
Pathogens (enteric viruses, protozoas)	Log (/day) or N/100 ml	Might have intermediate or low priority in lakes Concentrations are used as input for hydraulic models computing initial dilution of the effluent concentration
Intermediate priority variables		
Dissolved inorganic phosphorus (DIP)	kg P/day	DIP concentration is variable depending on process applied and relative treatment performance. DIP is easier to be analysed with conventional methods and on-line sensors Might have high priority
Total organic nitrogen	kg N/day	Dissolved and suspended N. Suspended organic N can be indirectly calculated by the content of volatile suspended solids multiplying with a standard factor of 0.01–0.06 or a calculated specific one
Total Kjeldahl N	kg N/day	Sum of total organic N and $N\text{-}NH_4$
Toxic metal (Zn, Cd, Pb, etc.)	g/day	Might have high priority in WWTP treating combined wastes (industrial activities and urban runoff)
Low priority variables		
Total suspended phosphorus	kg P/day	Might have moderate priority in eutrophic lakes
chloride	kg Cl/day	Might have an intermediate or high priority in cold climate (use of deicing chemicals)
Toxic organic compounds (i.e. PAH, AOX, EDCH)	mg/day	Might have intermediate or high priority in WWTP treating combined wastes (industrial activities and urban runoff)

AOX, adsorbable organic halogens; PAH, polycyclic aromatic hydrocarbon; EDC, endocrine disruptors compounds.

(CSO) as regards energy dynamics, food-web structure and complexity, biodiversity, genetic diversity, ecosystem development and migration, and sensitivity. In addition, important consequences could arise for public health, in relation to bathing and recreational activities and food (shellfish) and drinking water consumption. Important variables are human pathogens, and organic and inorganic micropollutants dissolved or accumulated in the food chain. Furthermore, the typical impact on algal growth stimulated by N and P nutrients can also produce indirect risk to humans, due to the (possible) presence of microtoxins related to the metabolism of several specific algae.

All these impacts are related to site specific characteristics such as the size of the discharge, the frequency of WWPT transient conditions, the size and nature of the sensitive receiving water body and the prevalent uses of the water.

5.3.4 MASS LOADING CALCULATION

5.3.4.1 Strategies

In mass loading calculation two 'time dimensions' have to be defined and considered: the duration of sample collection and the total time of measurement activity (period of record).

Sampling time may be on the order of only a few minutes, if on-line sensors are used, while in general it does last not more than a day.

The period of the measurement activity depends on the different management strategies that have to be fulfilled with the collected data (Somlyódy, 1995; Jeppsson and Hellström, 2002). Table 5.3.2 shows a nonexhaustive list of the potential use of data/information gathered.

Loads can be averaged on a seasonal basis or per year, such as for the nutrient balance in a lake, or in trend monitoring of toxic micropollutants.

In other situations, daily or specific single event duration might be more appropriate, as in the case of oxygen depletion or public health effects related to bathing activities.

Table 5.3.2 Example of potential use of mass loading data in management of lake and coastal zone water quality

Statistical analysis of plant discharges of key parameters
Development of sustained pick mass loadings of WWTP to be expected or allowed
Trend evaluation and analysis
Sources control and land use: master plant emission inventory and reduction strategies, source of nutrient accounting, etc.
Definition of sensitive areas
Modelling at different spatial and temporal scales water quality impact (Research or Decision Making Institutions)
Integration of models with advanced informatics tools (GIS, Expert Systems, etc.)

5.3.4.2 Source of Uncertainties

WWTP effluent concentrations and loadings of the parameters/contaminants usually selected have a diurnal, weekly or seasonal variation (van der Graaf, 1995; Metcalf and Eddy, 2002).

Seasonal variations are typically related to temperature (i.e. physical and biological kinetics of constant reactions). In the case of combined sewer the contribution of melting snow might also assume an important role.

Weekly variations usually depend on the variability of loads and heterogeneity of industrial discharges reaching the WWTP or fluctuations related to transient tourist populations.

Diurnal variation mainly depends on plant size. In addition, a day by day variation can be observed on weekend days.

In order to evaluate diurnal load fluctuations of some of the most significant variables, data collected from a detailed study carried out at the WWTP of Genova (Italy) can be used (Bazzurro *et al.*, 1999). The Darsena 220 000 equivalent inhabitants (EI) WWTP is fed by a combined sewer system, collecting civil, commercial and industrial wastes. The designed capacity of the sewer system is currently reached. As a consequence, the flow entering the plant is quite constant over time. The excess flow is discharged directly into the sea.

The plant is a conventional activated sludge system fed with liquid oxygen, having an anaerobic section for sludge treatment and biogas production. No final disinfection is applied, in accordance with the Liguria Region regulations for WWTPs discharging into the sea with a submarine outfall. Twenty-four instantaneous samples were collected automatically in wet and dry weather conditions (i.e. when urban runoff waters were entering the sewer system and not) with the aim of evaluating the daily fluctuation of the most significant parameters: biological oxygen demand in 5 days (BOD_5), chemical oxygen demand (COD), total suspended solids, total phosphorus and total inorganic nitrogen (TIN), which were analysed following the conventional reference procedures. Furthermore, flow was recorded at each sampling time.

Figure 5.3.1 shows diurnal variation of TIN, total phosphorus and BOD_5 loads of the Darsena WWTP, calculated from 24 instantaneous samples.

Figure 5.3.1(a) refers to dry weather conditions, while Figure 5.3.1(b) describes diurnal evolution in wet weather conditions. Differences can be appreciated between variables (e.g. BOD_5 versus nitrogen and phosphorus) and between the same variables in different weather conditions (e.g. BOD_5 and nitrogen). Frequency of sampling is consequently of paramount importance for an unbiased estimation of daily loads, as will be further discussed in this chapter.

In the case of a long-term period of record, an average value has to be calculated. This means that the frequency of sampling/analysis will influence the accuracy of the final data. In other words, what are the errors of calculated mass loading resulting from different sampling frequency and what level of accuracy can we accept considering sustainable costs?

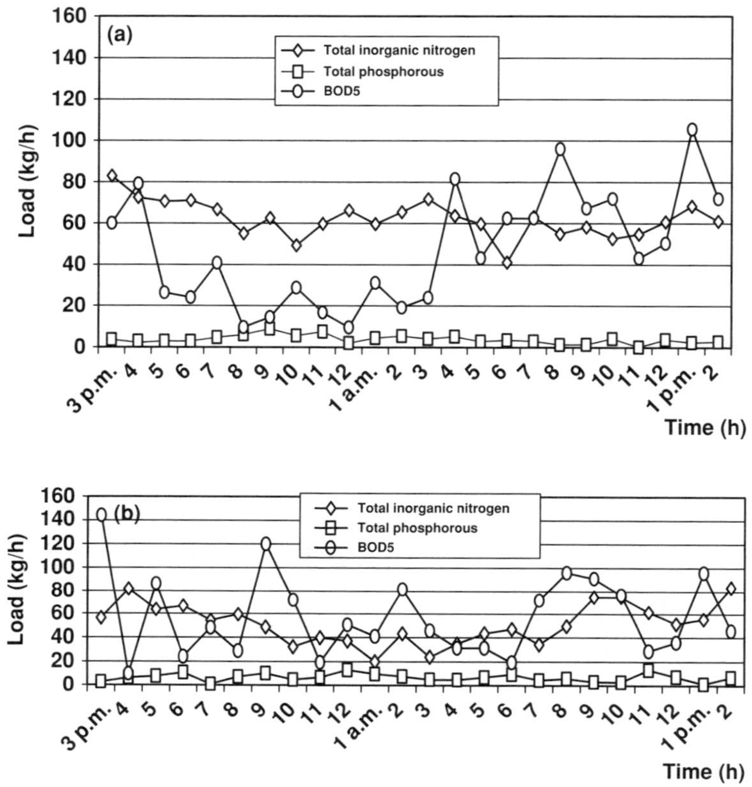

Figure 5.3.1 Evolution of hourly load of some typical variables during dry (a) and wet (b) weather conditions (see text)

Clement and Buzás (Clement and Buzás, 1999) evaluated errors that can be made in calculating annual nutrient (N, P) loads from monthly to weekly sampling frequency, in a study of the Danube and Zola rivers. The standard deviation/mean ratio distribution (assumed Gaussian) has been used as statistical methodology, following Cochran (Cochran, 1962), with a contextual use of a sophisticated Monte Carlo procedure for error evaluation.

In this study, independent sets of data for concentration C (366) and flow rate Q (52) were used. As the sampling frequency of the two variables was different, a further approximation was introduced to overcome this condition, assuming that sample size for C and Q were the same. A correcting factor was developed, on the basis of the experience that load and flow rate are generally correlated (Richards, 1989). This study refers specifically to hydraulic sections of the Danube and Zola rivers and so does not allow us to transfer final results and data interpretation to a WWTP context. However, valid general conclusions can be drawn from the study:

- The errors in mass loading estimation increase mainly with reducing sample size (i.e. the number of single measurements during the period of record).

- The error of load estimates for monthly sampling compared with weekly sampling is 20–30 % for the Danube, while for the Zola it is higher, due to the faster dynamic in the smaller water shed.
- The error in the estimates with Monte Carlo simulation does not differ substantially from the analytical statistical results.

5.3.4.3 Calculation for Long-term Periods

Because sampling frequency is generally reduced, the load of a specific sampling day is extrapolated to the entire period of days up to the following sampling day. A theoretical interpolation could moreover be made from the measured single day loads. Table 5.3.3 summarizes the main characteristics of different procedures that can be used for calculating mass loadings over long-term periods (from months to one year).

The unit loading factor (mass of specific pollutant discharged per person per day) is the least accurate, but may be a rapid, easy and no (or low) cost method of estimation that could be used as a guide (e.g. in calculating annual loads of nutrients).

The uncertainties of estimation lie in the use of standard (national) and fixed factors, regarding pro capita daily discharges into the sewer system and removal efficiencies of influent loads during treatment. A more precise estimation could be obtained utilizing local or site specific factors, including a better definition of the contributing population (inhabitants, equivalent and transient population) if such data were available at low cost.

Another procedure involves the utilization of dependent or independent data (as regards sampling time). To calculate the annual average of mass loading, the daily average load is required. The frequency of measurement of each daily load depends on the technique of sampling (concentration) and measuring flow rate which is used Considering the level and nature of fluctuation of both variables, the accuracy of the annual load depends on the number of daily data available, their distribution throughout the period and the methodology used for collecting waste samples: one or more grab samples, composite samples or continuous automatic sampling.

The more accurate (and generally used) procedure is based on the collection of a composite sample, representing a 24 h waste discharge. The composite sample is made up of waste portions collected at regular intervals (1 h or less). These fractions are put together proportionally to the (averaged) rate of flow at the time each sub-sample was drawn. The final sample (composite sample) gives the daily weighted average concentration of the parameter considered, which multiplied by the corresponding 24 h cumulative flow gives the corresponding daily load. Alternatively, each subsample can be analysed separately. The concentrations are multiplied by the corresponding flow rates, summing all the products. Therefore, for both procedures, the annual load is obtained on the basis of the daily loads available.

The composite sample can be obtained manually or using a refrigerated continuous sampler, which automatically collects individual subsamples or the final

Table 5.3.3 Characteristics of different methodologies for the calculation of mass loading on long-term periods [L(kg/period)]

Procedure	Calculation	Comments	Resources
Unit loading factor (LF)	$L = LF(\text{kg/day}) \times \eta \times$ population $\times$ period (day) where η = efficiency of treatment	Available national LF and literature η for BOD_5, COD, TSS and different fractions of N and P (see text)	Simple No (or low) procedure cost
Average loads [Calculated by time dependent or independent measure of daily average concentration (C_{av}) and flow rate (Q_{av})]	$L = C_{av}(g/m^3) \times Q_{av}(\text{m}^3/\text{day}) \times$ period(day)	Accuracy of calculation depends on the quality of daily average data, sampling strategies and time correlation between C and Q (see text)	Personnel for sampling and analysis Conventional laboratory instrumentation
24 h composite sample (flow rate weighted average) [Calculated by combining average weighted concentration (C_w) with corresponding 24 h cumulative flow rate (Q_c) (see text)]	$L = \left[\sum_{i=1}^{n} C_w(g/m^3) \times Q_c(\text{m}^3/\text{day})\right] \times \text{n} \times \text{CIS}$ where n = number of samples collected and CIS = constant interval of sampling (day) $= \frac{\text{period of record}}{n}$ This formula refers to sampling of regular frequency e.g. one composite sample any 7 days	Calculation gives different estimates in relation to the frequency and the strategy of sampling	Personnel for sampling and analysis Automated sampler Conventional laboratory instrumentation
On-line sensor [Calculated by combining continuously measured concentration (C_{OL}) with corresponding time flow rate (Q_{OL}). C_{OL} and Q_{OL} are frequently averaged to 1 h (C_H and Q_H)	$L = \sum_{i=1}^{p} C_H(\text{g/m}^3) \times Q_H(\text{m}^3/\text{day})$	This technique gives the best accurate evaluation of the annual mass load (see text)	Completely automated system Skills in on-line sensor use, data acquisition and handling The system is generally installed for real time advanced process control and monitoring purposes Sensor for inorganic dissolved P and N forms are available in the market

composite one, if a flow rate sensor electronically controls the amount of each waste sample.

A totally automatic system, based on an on-line sensor for monitoring concentration of important variables and flow rate, can calculate mass load for any time period with the most accuracy. However, notwithstanding the reliability of the new instrumentation available, the use of on-line sensors for real time control of WWTP processes is not common in Europe (Jeppsson *et al.*, 2002).

This procedure is the most complex, requiring three levels of information-related activities (Frey and Lynggaard-Jensen, 2002):

- level 1: data acquisition (calibration, preventive and emergency maintenance, communication interface for data collection supervisory control);
- level 2: data handling (wide range of possible actions, including data aggregation, synthesis and 'fusion', quality assessment);
- level 3: operation and management (planning, decision and cost / benefit analysis).

5.3.5 COMPARISON OF DIFFERENT STRATEGIES FOR NUTRIENT ANNUAL LOADING CALCULATION: AN EXAMPLE USING REAL DATA

The WWTP of Levico (Trento, Italy) was built for a capacity of 100 000 EI, and currently treats 50 000 EI. The fluctuating tourist population is estimated to be in the order of 10 000–15 000 EI/year.

When the combined sewer system is exceeded, the excess flow is allowed to escape in the form of CSO into the receiving river Brenta, a sensitive receptor, in accordance with the regulations of the Autonomous Province of Trento.

After a conventional pretreatment, the waste is split into two lines. The first line undergoes a pre-denitrification, followed by a conventional biological treatment (activated sludge) with ammonia nitrification; the second line undergoes only oxidative biological treatment.

The effluents coming from the respective secondary settlers are recombined, and subsequently filtered and (possibly) disinfected with chlorine before reaching the surface water receptor.

Flow rate, inorganic dissolved nitrogen (N-NH_4, N-NO_3) and dissolved inorganic phosphorus (IP) are continuously monitored using on-line sensors (SOIS, 2004).

Instantaneous waste samples are collected every hour and analysed following a European-patented batch-wide procedure (ECOFIELD), which utilizes conventional colorimetric principles. The overall precision is better than 5 %. The monitors installed are checked monthly with external reference procedures and recalibrated should a difference ±10 % be found. Single data can be validated performing a semi-automatic quality control (SOIS, 2004).

Figure 5.3.2(a–c) shows the relationship between hourly concentrations of the measured variables and the corresponding average hourly flows relating to the 2004 monitoring period.

About 80 % of the points for $N\text{-}NH_4$ and $N\text{-}NO_3$ are confined to a narrow domain of C and Q values, having similar relative size. About 85 % of the points for IP are confined to a domain having C values approximately 10 times the size of Q values.

Figure 5.3.3 (a and b) shows the relationship between hourly mass loading of TIN and IP and the corresponding average hourly flow rates. Loads of IP and TIN do not show any clear correlation with flow rates.

This database (more than 90 % of the total data for 2003) was utilized to compare the following different procedures, which can be easily simulated, for 'reconstructing' annual IP and TIN mass loads.

- Procedure 1 (reference): sum of each $C_h \times Q_h$ product calculated on an hourly basis.
- Procedure 2: sum of each $C_d \times Q_d$, where C is the daily arithmetical average and Q is the average 24 h total flow. This procedure simulates a continuous sampling not proportional to flow rate, combined with the cumulated value of the 24 h flow.
- Procedure 3: this procedure simulates 52 composite 24 h samples (weekly frequency).
- Procedure 4: this procedure simulates 26 composite 24 h samples (biweekly frequency).
- Procedure 5: this procedure simulates 12 composite 24 h samples (monthly frequency, on the first day of each month).
- Procedure 6: this procedure simulates four (seasonal) intensive composite sampling campaigns (in January, April, July and November, each lasting 4 consecutive weeks).
- Procedure 7 (a and b): these two procedures simulate a systematic sampling on each day of the year; for each day a continuous sample is collected in the interval 9–10 a.m. (procedure a) or in the interval 2–3 p.m. (procedure b). For both procedures, the single hour concentration is multiplied by the 24 h cumulative flow of the corresponding day; the products are summed throughout the year.

Procedure 2 simulates the common practice of using an automatic sampler operating under no flow control, using 24 h cumulative flow instead of instantaneous flow measurement.

Procedures 3–6 utilize different sampling strategies, based on the 24 h integrated sample and the 24 h cumulative flow.

Procedure 7 (a and b) represents extreme strategies, based on a very short time of sampling (manually or automatically for 1 h/day) for a period of 1 year. For

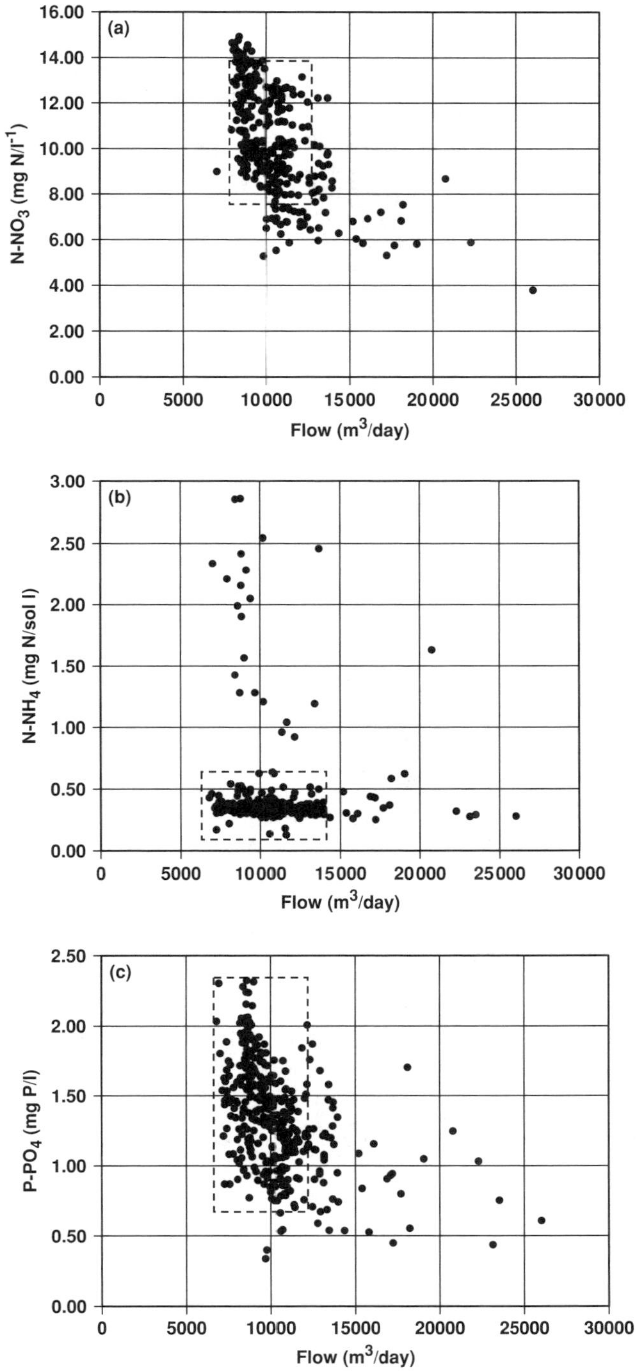

Figure 5.3.2 Correlation between flow rate and concentration of N-NO_3 (a), N-NH_4 (b) and P-PO_4 (c) (on-line monitoring, 2004). Single point combines hourly measure of variables

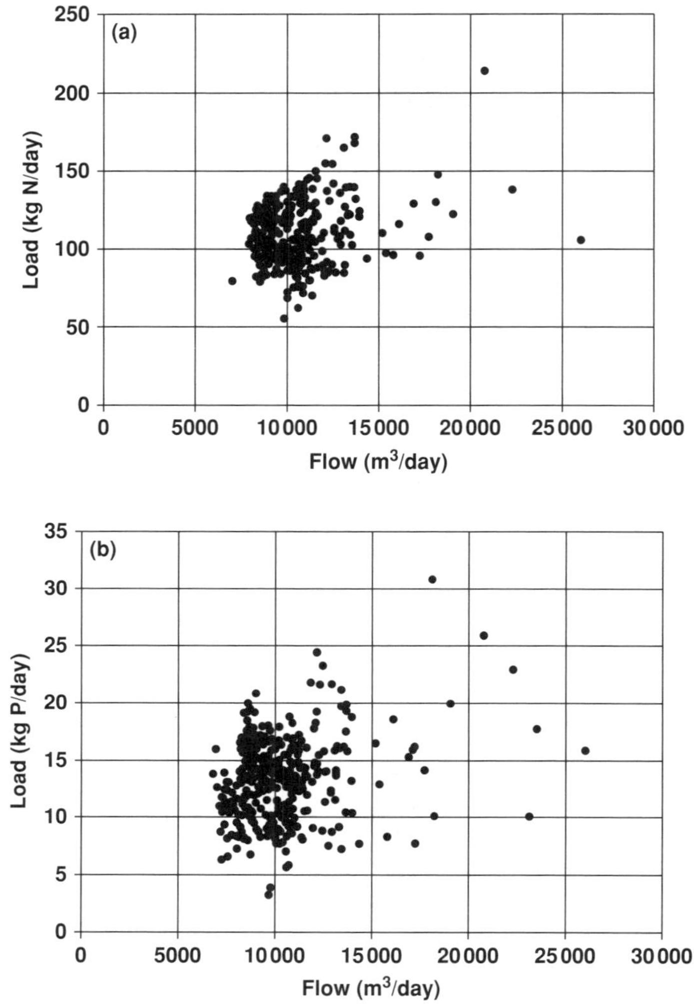

Figure 5.3.3 Correlation between load of TIN (a) and IP (b) and flow rate (on line monitoring, 2004). Single point combines hourly measure of variables

calculating daily mass loading, the single hour concentration is used to represent the 24 h average.

In Figure 5.3.4 (a and b) the profiles of daily mass loadings of TIN and IP, respectively, are represented. Daily loads are calculated by summing respective hourly loads.

Tables 5.3.4 and 5.3.5 summarize the results obtained following the different simulated strategies.

The set of data used have an intrinsic uncertainty, related to sensor accuracy, calibration procedures, maintenance operations, validation, etc., during the period of record (2004).

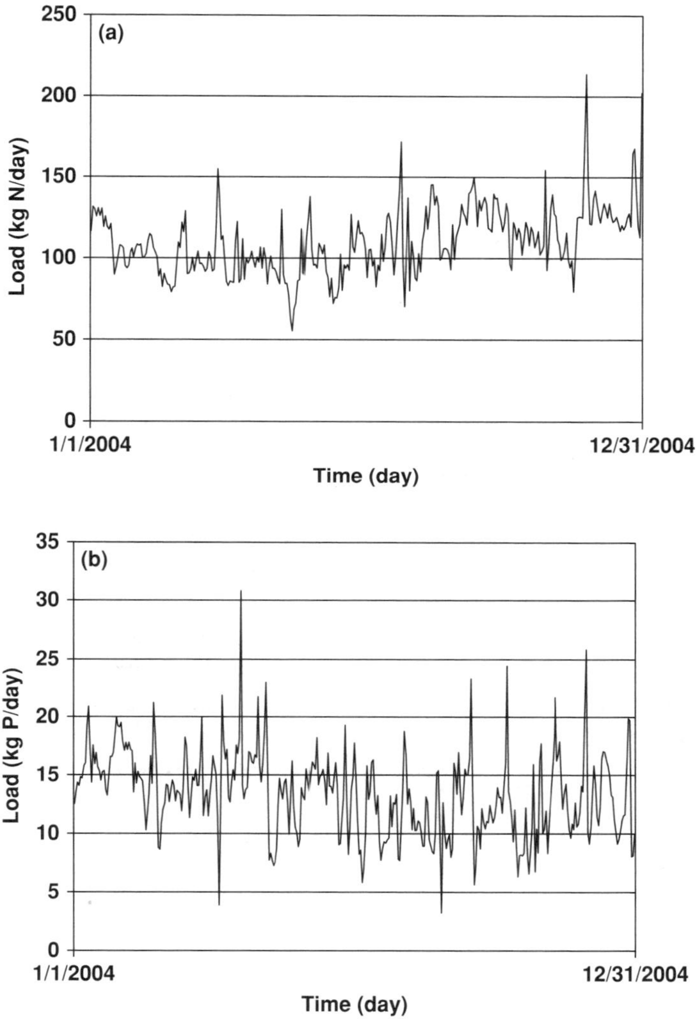

Figure 5.3.4 Profile of daily loads of TIN (a) and IP (b) during the year 2004 (on-line monitoring)

Therefore, the accuracy of each 'reconstructed' year load estimate, based on a different combination and use of these data, is only affected by the dimension (representativeness) of the sample extracted from the population, assuming that the error associated with analytical and collecting procedures is unchanged.

The importance of the frequency of sampling can be inferred from procedure 2. Notwithstanding the fact that all the daily average concentrations are calculated from (simulated) samples continuously collected rather than proportionally with flow rate, the estimated loads maintain a high level of accuracy.

Comparing procedures 3–6 with the reference procedure 1, it can be concluded that the nature and extent of N and P concentrations (see Figures 5.3.1 and 5.3.2)

Table 5.3.4 Comparison of yearly mass loading of TIN among different strategies of data collection

Code	Description of the procedure (see text)	Mass loading (kg/year)	Differences among the reconstructed loads and the 'true' load (kg/year)	(%)
1	On-line hour (reference)	40 067	—	—
2	Daily average, not Q weighted	40 185	118	0.30
3	52 integrated 24 h (weekly sampling)	40 316	249	0.62
4	26 integrated 24 h (biweekly sampling)	39 889	−177	−0.44
5	12 integrated 24 h (monthly sampling)	41 751	1684	4.20
6	4 weeks × 4 months (seasonal sampling)	40 171	104	0.26
7a	1 h × year (9 a.m.)	37 821	−2746	−6.85
7b	1 h × year (2 p.m.)	42 708	2641	6.59

have a major role, when the number of samples is progressively reduced, in order to contain the costs involved.

Procedure 7 (a and b) simulates a strategy of collecting samples throughout 1 year, based on minimization of the costs (personnel and technology) involved. The error that may be introduced (evident over- and underestimations of the annual loads) is the consequence of the lack of representativeness inherent in the entire procedure.

The main conclusions that can be drawn from this exercise are as follows:

- The differences between the 'true' annual load and averaged estimated loads, derived from other strategies of data collection, are strongly related to the global variability and intercorrelation of available data.

Table 5.3.5 Comparison of yearly mass loading of IP among different strategies of data collection

Code	Description of the Procedure (see text)	Mass loading (kg/year)	Differences among the reconstructed loads and the 'true' load (kg/year)	(%)
1	On-line hour (reference)	4903	—	—
2	Daily average, not Q weighted	4892	−11	−0.22
3	52 integrated 24 h (weekly sampling)	4811	−91	−1.86
4	26 integrated 24 h (biweekly sampling)	4610	−292	−5.96
5	12 integrated 24 h (monthly sampling)	5263	360	7.35
6	4 weeks × 4 months (seasonal sampling	5071	168	3.43
7a	1 h × year (9 a.m.)	5136	234	4.76
7b	1 h × year (2 p.m.)	4531	−371	−7.57

- The magnitude of 'error' of annual loads estimates is dominated by the frequency of sample collection.
- Variables showing a narrower pattern of dispersion (such as N in comparison with P data) do not substantially modify the (absolute) error of the estimate, when flow weighted sample frequency is reduced from yearly to biweekly and seasonal (16 samples).

5.3.6 CONCLUSIONS AND PERSPECTIVES

In the context of discharges in sensitive water bodies, the concentration of the variables involved is less important than the corresponding mass loading. As a consequence, wastewater discharges must be described by measuring the concentration of the pollutant in the sample, together with the flow rate condition at the moment of sampling.

In mass loading calculation, two time dimensions have to be considered: the duration of sample collection and the period of record. Sampling time may be on the order of only a few minutes and it lasts no more than a day. Loads can be generally averaged on a seasonal basis or per year but, in other situations, daily or specific single event duration may be more appropriate.

Sampling frequency being generally reduced due to economic limitations, the load calculated on a specific day is in some way extrapolated to the entire period up to the following sampling day.

The frequency of measurement of daily loads during the period of record depends on the technique of sampling and measuring flow rate used.

An accurate and generally used procedure is based on the collection of a composite 24 h sample, made up of waste portions collected at regular intervals put together proportionally to the flow rate. The final sample (composite sample) gives the daily weighted average concentration of the parameter under consideration, which, multiplied by the corresponding cumulative flow rate, gives the corresponding daily load. Annual or seasonal mass loads are obtained from the daily load available.

The composite sample can be obtained manually or using a continuous sampler that automatically collects individual subsamples or the final composite one. A totally automatic system, based on an on-line sensor for monitoring concentration and flow rate, can calculate pollutant mass load for any time period with the highest accuracy.

The accuracy of the annual load depends on the number of daily data available, their distribution throughout the period and the methodology used for collecting waste samples.

Generally, a weekly frequency is considered adequate to obtain an accurate estimate. The covariance dynamic of measured variables (parameter concentration and flow rate) can allow reduction, and possibly significant reduction, of the sampling frequency.

Unfortunately, this possibility, which depends on the type and the pattern of discharge of the pollution sources, cannot be predicated on theory. Potential indicators of stability of the whole system may be the capacity of the WWTP (increasing) and the presence of a separate urban collecting sewer.

In accordance with the recent European Water Framework Directive, the emission limit value (ELV) approach to tackling water pollution will become widely used in the near future. The ELV approach focuses on the maximum permitted quantity of a specific pollutant that may be discharged from a selected source, particularly in sensitive areas. ELV deals with wastewater treatment performance, mass load and pollutant balance (with particular focus on nitrogen and phosphorus) and also with strategy evaluation for restoring compromised water resources.

Conventional technology available for calculating mass loading can fulfil these tasks. Furthermore, the new instrumentation developed for control and automation in WWTPs could be utilized with the advantage of improving performance control and reducing energy consumption during treatment.

ACKNOWLEDGEMENTS

The authors thank the WWTP Management Service of the Autonomous Province of Trento (SOIS-PAT) for allowing access to and use of its database, containing all the analytical data acquired during the WWTP effluent quality control program.

REFERENCES

Bazzurro, N., Gallea, A. and Lasagna, C. (1999) Integrated Planning and Management of Urban Drainage, Waste Water Treatment and Receiving Water System: the Experience of AMGA. In: *Proceedings of the 8th International Conference on Storm Urban Drainage,* Jolisse, I. B. and Ball, J.E. (Eds). International Association for Hydraulic Research: Sydney, Australia, pp. 340–348.

Clement, A. and Buzás, K. (1999) Use of Ambient Water Quality Data to Refine Emission Estimates in the Danube Basin. *Water Science and Technology*, **40**, 10, 35–42.

Cochran, W.G. (1962) *Sampling Techniques.* John Wiley & Sons, Ltd, New York.

Frey, M. and Lynggaard-Jensen, A. (2002) Data Handling and Validation. In: *Online Monitoring for Drinking Water Utilities*, Hargesheimer, E., Conio, O. and Popovicova, J. (Eds). Awwa Research Foundation CRS PROACQUA, Denver, USA, pp. 289–312.

House, M.A., Ellis, J.B., Herricks, E.E., Hvitved-Jacobsen, T., Seager, J., Lijklema, L., Aalderink, H. and Clifforde, I.T. (1993) Urban Drainage – Impact on Receiving Water Quality, *Water Science and Technology*, **27**, 12, 117–158.

Jeppsson, U., Alex, J., Pons, M.N., Spanjers, H. and Vanrolleghem, P.A. (2002) Status and Future Trends of ICA in Wastewater Treatment: a European Perspective. *Water Science and Technology*, **45** (4–5), 485–494.

Jeppsson, U. and Hellström, D. (2002) Systems Analysis for Environmental Assessment of Urban Water and Wastewater Systems. *Water Science and Technology*, **46**, (6–7), 121–126.

Lijklema, L., Tyson, J.M. and Lesouef, A. (1993) Interaction between Sewers in Treatment Plant and Receiving Waters in Urban Areas: a Summary of the INTERURBA '92 Workshop Conclusion. *Water Science and Technology*, **27**, 12, 1–29.

Metcalf and Eddy (2003) *Wastewater Engineering Treatment and Reuse*, 4th Edn. McGraw Hill, Boston, pp. 170–197.

Richards, R.P. (1989) Determination of Sample Frequency for Pollutant Load Estimation Using Flow Estimation Only. In: *Proceedings of the International Symposium on the Design of Water Quality Information Systems*, Fort Collins, USA, pp. 136–144.

SOIS (2004) Database owned by the Service managing all the WWTP of the Autonomous Province of Trento (Italy).

Somlyódy, L. (1995) Water Quality Management: Can We Improve Integration to Face Future Problems? *Water Science and Technology*, **31**, 8, 249–259.

van der Graaf, J.H.J.M. (1993) Interactions of Sewerage and Wastewater Treatment: Practical Example in the Netherlands. *Water Science and Technology*, **27**(5–6), 1–9.

5.4
Water Reuse

Davide Bixio, Thomas Wintgens, Aldo Ravazzini, Chris Thoeye, Haim Cikurel, Avi Aharoni, Jaap De Koning and **Thomas Melin**

Wastewater Quality Monitoring and Treatment Edited by P. Quevauviller, O. Thomas and A. van der Beken

5.4.1 INTRODUCTION

This chapter reviews the status and advancements of monitoring practices in water reclamation and reuse, with particular reference to the European Union. The account is based on field knowledge of the authors, a conventional literature review and an enquiry covering 40 projects in the European Union, Israel and Australia. In the review, only reuse schemes larger than 2.5 million m^3 a year for secondary treatment and 0.5 million m^3 a year for tertiary or quaternary treatment were considered. This review has been conducted in the framework of the AQUAREC project on 'Integrated Concepts for Reuse of Upgraded Wastewater'. The AQUAREC project is co-funded by the European Commission within the 5th Framework Programme as well as in Australia by the Commonwealth Department of Education Science and Training and is coordinated by RWTH Aachen University (Germany). The general objective of the AQUAREC project is to provide knowledge for a rational strategy for municipal wastewater reclamation and reuse as a major component of sustainable water management practices.

In Europe the last two decades have witnessed growing water stress, both in terms of water scarcity and quality deterioration, which has prompted many municipalities to a more efficient use of the water resources, including a more widespread acceptance of water reuse practices (Bixio *et al.*, 2006). In 2004, 750 million m^3 of treated municipal wastewater was reclaimed, and within the next 20 years this could increase to 1540–4000 million m^3 a year (Hochstrat *et al.*, 2006).

Water reuse projects need to be effectively managed to ensure an appropriate use and acceptance of this practice, and monitoring and control plays a paramount role in this respect. A sub-optimally monitored scheme may result in adverse health, environmental or financial outcomes that may quickly dampen any enthusiasm for water reuse, hindering its development in the region.

Main determinants contributing to the setting of an appropriate monitoring and control strategy are:

- The requested water quality of the final product (Section 5.4.2).
- The treatment technology for water reclamation that is involved (Section 5.4.3).
- The inherent constraints associated with analytical procedures (sensitivity and response time), sensor availability and their cost (Section 5.4.4).

5.4.2 THE REQUIRED RECLAIMED WATER QUALITY

Contrary to drinking water production, water quality requirements are set up on the basis of the intended use. There can be as many uses as those of conventional water supply. No consensus however exists yet at Community level neither on their classification, nor on the water quality indicators (and their absolute value). The Italian legislation, for instance, foresees 4 classes of use and 53 parametes while the

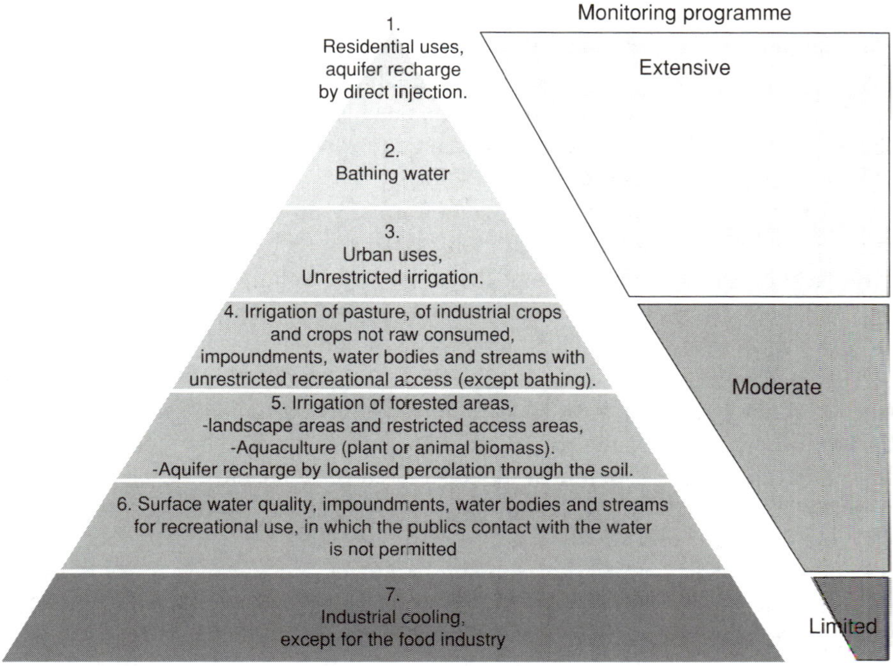

Figure 5.4.1 Hierarchy of water reuse classification (derived and adapted from Salgot *et al.*, 2006) and the related monitoring programme requirements

Spanish legislation that is being drafted, foresees 5 classes but a more limited set of parameters.

Primary reasons for the disagreement between experts and authorities in different countries are that the implementation of the precautionary principle is a regional/local matter (Anderson *et al.*, 2001) and that scientific information is not always conclusive (Toze, 2006).

Figure 5.4.1 shows the water use classification recommended by the AQUAREC project for harmonizing the various Member States' approaches. The water quality hierarchy is represented by a pyramid, with increasing water quality requirements when stepping up the pyramid.

Because of the lack of uniformity in regulatory water quality requirements, management practices of quality control obviously vary from region to region and in some regions even from project to project. Nevertheless, the surveyed projects indicate that there are also several common traits.

5.4.2.1 Agricultural Irrigation

In the last decades, there has been a number of different water quality guidelines related to the use of reclaimed water in irrigation. Reference international guidelines are the World Health Organization health guidelines for the use of wastewater for agriculture and aquaculture (World Health Organization, 1989).[1] These guidelines are directed mainly toward developing countries to encourage a safer use of wastewater than the common practice of using it untreated, and thus set a high level of tolerance for microbiological contamination compared with the guidelines of most of the European countries (Bixio *et al.*, 2006).

Regulatory water quality requirements normally do not include considerations about the chemical elements which plants need for their growth but rather focus on the health aspects of agricultural irrigation with reclaimed water. On the other hand, the monitoring of the chemical and physical properties of reclaimed water is a crucial aspect in any investigated monitoring programme.

Major concerns are inorganic salts, such as sodium chloride and a host of trace elements including heavy metals. Their follow-up is particularly necessary in dry climates, where much of the irrigation water evaporates and the concentration of salts in the drainage can be much higher than in the water itself, posing potential threats to the crop yield (Metcalf and Eddy, 2003) and groundwater quality (Bouwer, 2000).

Salinity is already a major environmental problem in many parts of the world and the use of inadequately treated reclaimed water can exacerbate this problem. Major determinants are:

- the total dissolved solids (TDS) concentration in soil water, as the salts that contribute to a salinity problem are water soluble;
- in case of saline water, the sodium content;
- the presence of specific toxic ions, the most commonly found in reclaimed water being boron, sodium and chloride (Bixio and Wintgens, 2006).

In the follow-up of salinity hazards, the electrical conductivity (EC) is used as a surrogate measure of TDS concentration (Pettygrove and Asano, 1985) and saline waters, while the sodium adsorption ratio $\{SAR = Na / [(\sqrt{}Ca + Mg) / 2]\}$ and the ratio between calcium and magnesium is used to predict problems related to the high sodium concentrations – notably, those related to the water infiltration rate (Ayers and Westcot, 1985).

Trace elements tend to accumulate in plants and soil causing phytotoxicity in plants or a health hazard for humans and animals. As a rule, irrigation water supplies do not need to be checked for trace elements unless toxicity is suspected. Suggested maximum trace element concentrations for irrigation waters are given in Ayers and Westcot (Ayers and Westcot, 1985).

Concerning microbial pathogens, within the European Union one widely accepted public health criterion for agricultural irrigation with wastewater is to restrict the use

[1] These are being revised (Kamizoulis, 2006).

of secondary treated sewage to crops that are cooked before being consumed and allow only water that has been reclaimed through advanced treatment and advanced disinfection to be applied to crops normally eaten raw (respectively, classes 4 and 3 in Figure 5.4.1).

Additional hazards that need to be monitored are related to the presence of *excess nutrients* (excessive nitrogen supply in the later part of the growing period may cause excessive vegetative growth, delayed or uneven maturity, or reduced crop quality), *abnormal pH* (normal range being between 6.5 and 8.4) and the possible formation of *scale deposits* (although there is no toxicity involved, the deposit reduces the marketability of the products, and in the case of fruits like apples and pears, requires an expensive treatment before marketing).

5.4.2.2 Urban and Residential Applications

There is a large variety in the type of urban water reuse schemes. A differentiation may be possible by location and type of use (toilet flushing, garden watering, parks, commercial and public complexes), type of source water (municipal wastewater, greywater, harvested rainwater) and type of treatment (centralized versus decentralized).

Urban and residential water reuse is most strongly impaired by hygienic parameters, but nutrient level [biological oxygen demand (BOD), chemical oxygen demand (COD), N, P], heavy metals, hardness, amount of suspended solids, salt content, odour and colour may also be critical for safe use and public acceptance. Representative quality criteria are illustrated in Table 5.4.1.

A commonly applied risk management action in centralized urban water reuse schemes is the introduction of a multi-barrier concept that leads to a lower likelihood of failure and generally to larger margins between performance and minimum requirements (Bixio and Wintgens, 2006).

Table 5.4.1 Quality criteria for Class A water in the Victoria Environmental Protection Agency guidelines (derived from Victoria Environmental Protection Agency, 2003)

Parameter	Requirement
BOD5	<10 mg/l
Suspended solids	<5 mg/l
Turbidity	<2 NTU
pH	6–9 (90th percentile range)
Residual Cl_2	<1 mg/l at point of supply)
E. coli	<10 org/100 ml
Helminths	<1 viable helminth egg/ l
Protozoa	<1 protozoa/50 l
Viruses	1 enteric virus/50 l

5.4.2.3 Industrial Uses

As for agricultural irrigation, industry-related regulatory water quality requirements tend to be oriented towards occupation and customer safety and less towards the particular industrial process water needs. Therefore in the regulated uses a distinction is commonly made between closed and open water circuits, for occupational safety reasons, and between specific classes of industrial processes, for customer safety reasons (e.g. food processing).

General water quality concerns relating to industrial uses include (adapted from Rommelmann *et al.*, 2003):

(1) Foaming due to residual organics.

(2) Potential corrosion in copper-based alloys from sulfates and ammonia.

(3) Interference with free chlorine disinfection due to ammonia.

(4) Stimulation of microbial growth (and consequently slime formation) due to ammonia, nitrates, phosphorus and residual organics.

(5) Deposition/clogging from suspended solids.

(6) Scale formation from calcium, magnesium, iron, silica and phosphorus.

(7) Staining from iron and manganese.

The actual choice of the monitoring and control action depends to a great extent on the type of industrial processes and on the water treatment equipment that is available at the industrial site.

Desirable water quality varies from industry to industry. Minimum water quality requirements of the most common industrial applications can be summarized by a water-quality pyramid where the water quality gradually increases when climbing up the pyramid (Figure 5.4.2).

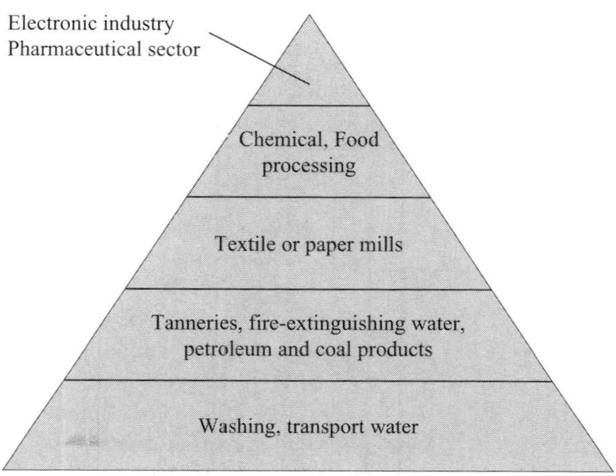

Figure 5.4.2 Hierarchy of water reuse classification for industrial process applications (Bixio and Wintgens, 2006)

Within each of the categories above, there exists a large heterogeneity of water (quantity and) quality requirements. The desirable water quality is site-specific, as most industries want reclaimed water quality to be equivalent to the water supply they have become accustomed to. In general, desired water qualities include neutral pH, moderately soft, low turbidity, low suspended solids, and low silica; dissolved solids and chloride content in reclaimed water are not critical (Water Pollution Control Federation, 1989).

5.4.2.4 Artificial Groundwater Recharge

In several guidelines, groundwater recharge is classified as a use of reclaimed water when in reality it can form a valuable part of the treatment, storage and distribution train prior to end uses such as indirect potable reuse, irrigation, industrial supplies and ecosystem support such as protection against saline intrusion and land subsidence or support of groundwater dependent ecosystems (Pyne, 1995; US Environmental Protection Agency, 2004).

Groundwater recharge is playing an increasing role in indirect water reuse, and this role is expected to grow as research continues to give increased certainty on the processes and rates of attenuation of contaminants, notably pathogens and some trace organics, within the subsurface (Dillon, 2005).

The Environmental and Water Resources Institute of the American Society of Civil Engineers Standards Committee developed standard methods for investigating establishing and operating recharge enhancement projects (EWRI/ASCE, 2001) in an attempt to reduce risks of failures in future projects. However, because of the large number of variables at any site; social, hydrogeological, hydrological, water quality, it is not possible to provide a recipe on the investigation requirements at any site.

Table 5.4.2 provides a list of tests that are proposed in the research project Reclaim Water on water reclamation technologies for safe artificial groundwater

Table 5.4.2 Protocol for routine analysis in water reuse and aquifer recharge case studies (www.reclaim-water.org)

Basic wastewater analysis	Microbiological analysis	Trace elements analysis	Salinity related analysis
Suspended solids, BOD (total and soluble), COD (total and soluble), dissolved organic carbon, ammonia, nitrite, nitrate, total N, phosphorus, alkalinity, pH, turbidity, detergents, mineral oils, phenols	Total bacteria count, faecal coliforms, *E.coli*, faecal *streptococci*, *clostridium* spores, bacteriophages	Boron (inorganic tracer), cadmium, chromium, cobalt, copper, fluoride, iron, lead, manganese, molybdenum, nickel, selenium, barium, cyanide, zinc	Chloride, electrical conductivity, sodium, potassium, calcium and magnesium, carbonate/ bicarbonate

recharge (Specific Targeted Research Project in the 6th Framework Programme of the European Union) for monitoring of the included treatment and reuse schemes used to confirm the safe use of reclaimed water at different sites within the European Union, Israel, Australia and China.

5.4.3 THE WATER RECLAMATION TECHNOLOGY INVOLVED

Monitoring is not restricted to reclaimed water quality parameters but includes parameters that characterize the state of the technical system, from incoming wastewater to distribution, and assures the water reclamation system's overall integrity.

A vast array of tertiary treatment technologies is available that has proven to meet a variety of water quality standards. Figure 5.4.3 shows the number of water reuse schemes per field of application and the level of treatment – secondary (with

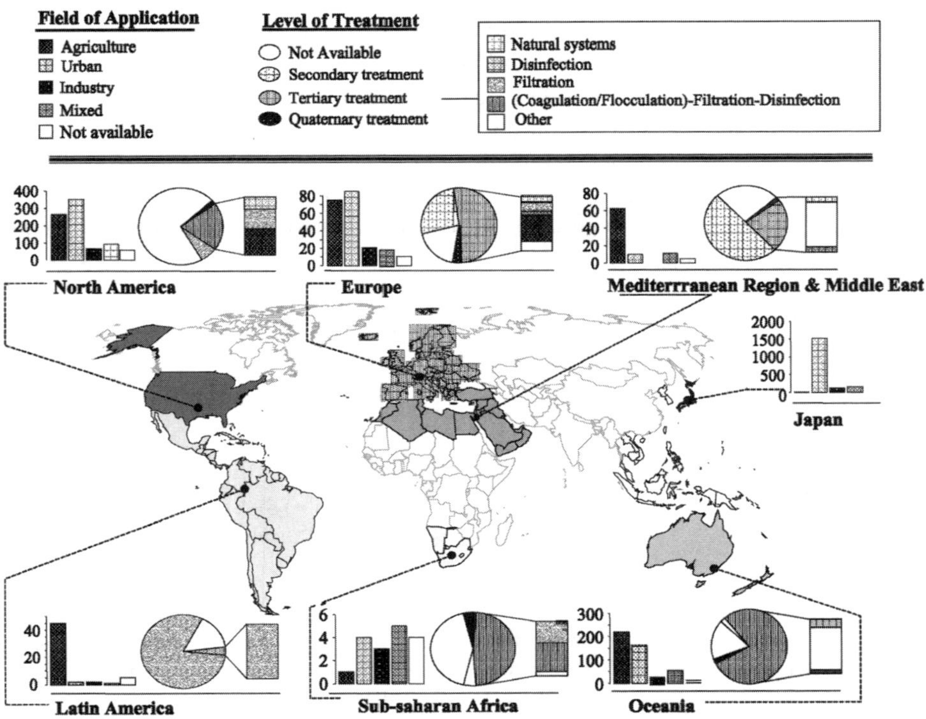

Figure 5.4.3 Water reuse schemes per field of application (bar charts) and level of treatment (pie charts with attached bar for main tertiary treatment processes) in different regions of the world (Bixio *et al.*, 2005)

or without nutrient removal), tertiary (with attached bar for main tertiary treatment processes) or quaternary (multi-barrier systems including a double membrane unit) – attached to them in different regions of the world (Bixio *et al.*, 2005).

For each treatment process, there are typically well-defined standard Quality Assurance/Quality Control (QA/QC) practices to assure that the system is functioning as designed (Bixio and Wintgens, 2006). While their description goes beyond the scope of this document, it is worth mentioning that on-line, real-time water quality monitoring is extensively used for this purpose and that the degree of satisfaction of the respondents of the Aquarec questionnaire with on-line sensors is very high.

Such indirect indicators that detect, on-line, possible integrity defects, are particularly desired for those parameters that cannot be monitored directly on-line or in rapid repetition, e.g. hygienic parameters.

Moreover, some technologies can introduce additional water quality concerns. For instance, by-products of disinfection processes (DBP) may yet prove to be among the greatest chemical concerns in reclaimed water. In recent years attention has been given to the detection of the potent carcinogen, nitrosodimethylamine (NDMA) in chlorinated sewages intended for reuse. Besides, we have seen that chloride levels can be of concern also for certain reuse application, for instance in irrigation because high levels are toxic to many plants. Dechlorination is sometimes required.

The type of distribution system also contributes to the setting of the appropriate monitoring strategy. For instance, long distribution systems may need a disinfection residual, or contamination in the distribution system or cross-connections may negate any of the benefits of the water reclamation step. Many respondents of the Aquarec questionnaire gave particular importance to the control of reticulation systems especially where reclaimed water is reticulated in close proximity to potable supplies.

Several types of contractual agreements were identified to limit the liability of the water supplier on the one hand, and to ensure that operation and maintenance is practised in a safe and responsible manner, on the other. The experience in Australia, at Sydney Olympic Park (Listowski, 2004) and at Rouse Hill in Sydney (De Rooy and Engelbrecht, 2003) is to manage the system with comprehensive inspection during construction and with follow-up inspections and certification. In Israel, in the case of a pipe leakage the farmer is immediately notified and the water utility will have to repair the leakage in a time delay of up to 24 h. Water supply interruptions of 24 h can be sufficient for some crops to deteriorate.

Despite the fact that the Hazard Analysis and Critical Control Points (HACCP) concept is increasingly used to direct efforts in process control and monitoring to guarantee hygienically safe reclaimed water, very few surveyed projects have used this approach to set up their monitoring programme. Two examples are the multipurpose NEWater scheme in Singapore (Leong Yin Hou, personal communication) and the artificial aquifer recharge project in Wulpen, Belgium (Dewettinck *et al.*, 2001).

5.4.4 INHERENT CONSTRAINTS OF ANALYTICAL PROCEDURES AND SENSORS

Several water quality indicators are difficult – and costly – to analyse and the time to conduct the analysis is too long (e.g. several microbial parameters as well as chemical compounds such as copper, iron and manganese) to avert potential problems.

Because of the preventive measures that are generally in place (including a multi-barrier protection of the intended reclaimed water use), within the European Union monitoring of delivered water quality is simply verification that the preventative measures are effective, and often variables that can be monitored instantaneously can give a higher level of confidence in safety of supply and at less cost than analysing for an expanding number of chemicals.

The monitoring of operating parameters continuously is a standard component of all water reclamation schemes investigated. Minimum instrumentation consisted of alarms at critical treatment units to alert an operator of a malfunction. Sensors that are available in almost all the water reclamation schemes in order to identify and halt the use of unacceptable reclaimed water quality are conductance meters and turbidometers (high levels of turbidity can protect micro-organisms from the effects of disinfection, stimulate the growth of bacteria, and exert a higher chlorine demand for disinfection). The disinfectant residual is also measured continuously and often the signal is used in automatic control systems. Note that the consequences of barrier failure were not considered by the operators to be of significance, either because there were multiple barriers in the schemes or because the end-user can tolerate some water quality that is below normal standards.

The water quality parameters most commonly monitored in wastewater treatment plant effluent and water reuse schemes and their frequency for different types of end-uses are given in Table 5.4.3 (reference costs are based on Western European values).

Note that the frequency may be dependent on the size of the installation and that in some legislation defined frequencies may be reduced by demonstration of appropriate effluent quality.

5.4.5 TRENDS

5.4.5.1 A More Varied List of Reclaimed Water Quality Indicators

The state of science is not yet reflected in water reuse guidelines or regulation and there is an open debate about the relevance of several additional water quality indicators for water reuse applications.

In the first place, the debate is on the microbiological indicators. Besides the traditionally established indicator organisms for the potential occurrence of pathogenic organisms such as faecal coliforms or *Escherichia coli*, a range of emerging pathogens has been discussed and investigated as well as new detection methods such as

Table 5.4.3 Use/cost per type of reuse application

Parameter	Example/indicators	Frequency per type of reuse				Costs
		Private, urban and irrigation	Environmental and aquaculture	Indirect aquifer recharge	Industrial cooling	
Physico-chemical	pH, EC, turbidity, TSS, colour	+++	+++	+++	+++	€
	COD (TOC, DOC), DO	+++	+++	+++	+++	€/€€
Organic sum parameters	BOD	++	++	+++	++	€€/€€€
	AOX	+++	+	++	□	€€/€€€
Nutrients	Total-N, NH_4+-N, Kjeldhal N	+++	+++	+++	+++	€/€€
	Total P, dissolved phosphates	+++	+++	+++	+++	€/€€
Residual chlorine	Free and total chlorine (if chlorination)	+++	+++	++	+++	€
	Chloride					€€
Physico-chemical	Sodium absorption ratio [SAR = f(Na,Ca,Mg)], UV 254	++	□	□	□	€
Nonmetallic ions	NO_2^-, NO_3^-, SO_4^{2-}	++	++	+++	++	€/€€
(Heavy) metals	As, Cd, Cr(III,VI), Hg, Pb, Cu, Zn, B	+	+	+	□	€€/€€€
Organic micropollutants	Surfactants, mineral oil	+	+	+	□	€€/€€€
(Heavy) metals	Al, Ba, Be, Co, Fe, Li, Mn, Mo, Ni, Se, Sn, Th, V,	□	□	+	□	€€€
Organic micropollutants	Aldehyde, phenols	□	□	□	□	€€€
Pesticides	Diuron; 2,4-D	□	□	+	□	€€€
Complex-forming substances	EDTA	□	□	+	□	€€€€
Chloride solvents	If AOX > limit, e.g. TCE	□	□	□	□	€€€€
Aromatic organic solvents	Benzene	□	□	□	□	€€€€
PAHs	Benzo(*a*)pyrene	□	□	□	□	€€€€
Pharmaceuticals	Carbamazepine, X-ray contrast media, Sulfamethoxazole	□	□	□	□	€€€€€
Endocrine disrupters	E-Screen	□	□	□	□	€€€€€
Disinfection (by-)products	NDMA	□	□	+	□	€€€€€

Frequency: +++ = daily/weekly; ++ = 1, 2 per month; + = monthly – twice per year; □ = yearly or less.
Cost per analysis: €€€€€= very high, >200 €; €€€€= high, 50 – 200 €; €€€= medium, 20 – 50 €; €€= low, 5 – 20 €; €= very low, <5 €.
AOX, absorbable organic halide; 2, 4-D, 2, 4-dichloropheroxyacetic acid; EDTA, ethylenediaminetetraacetic acid; TCE, trichloroethylene.

fluorescence *in-situ* hybridization (FISH) and polymerase chain reaction (PCR). Some pathogens are already included in guidelines and legislation. Florida for instance recognizes that *Giardia* spp. and *Cryptosporidium* spp. are pathogens of increasing importance to water reclamation and now requires monitoring for these pathogens. Another increasing concern is caused by the still growing multiple antibiotic resistances of human pathogens (Martínez and Baquero, 2002).

Second, an increasingly documented class of organic trace contaminants in wastewater is that of the 'endocrine disrupting chemicals'. Much attention has been devoted to natural and synthetic steroidal hormones, which are shown to induce biological effects on some organisms at part per trillion concentrations. Some steroidal hormones are poorly removed in conventional wastewater treatment processes (Purdom *et al.*, 1994). Other chemicals exhibiting similar effects at higher concentrations that are known to be present in sewages include some plasticizers, pesticides and degradation products of some detergents (Körner *et al.*, 2000).

Third, there is also a broad range of pharmaceutically active compounds which have been detected in municipal wastewaters in many parts of the world (Heberer, 2002). At this point there are no indications for limitations to water reuse caused by these compounds, although their effect is largely unknown.

There are currently no analytical standard procedures for a range of emerging pollutants such as endocrine disrupters and pharmaceuticals.

There has been an extensive effort to investigate emerging organic trace contaminants in wastewater streams and assess the removal capacity of both conventional and advanced wastewater treatment options (Poseidon, 2004). It is very obvious that sophisticated analytical techniques are required to measure most of the trace organic contaminants in realistic concentrations as found in wastewater treatment plant effluents (Kuch and Ballschmiter, 2000).

Apart from chemical analysis approaches to monitor emerging trace pollutants there have been many approaches to establish and apply biological effect assays, which do not look at the identification of single compounds but of cumulative effects of a water sample on a test cell line (*in vitro* assay) or on complex organisms (*in vivo* tests) (Segner *et al.*, 2003). Besides acute toxicity, chronic and sublethal aspects such as genotoxicity and mutagenicity are of increasing importance as well as endocrine disruption and immunotoxicity (Pérez *et al.*, 2003). Bioassays can prove to be very effective in the monitoring of both reclaimed water quality and water reuse impact (Valat *et al.*, 2004).

As part of the ongoing investigation of the behaviour of emerging contaminants in wastewater reclamation and reuse the Reclaim Water project applies a comprehensive analytical package for the monitoring of a number of water reuse case studies (www.reclaim-water.org). Within the project, standard analytical parameters (see Table 5.4.2) as well as emerging microbial and chemical contaminants are considered in combination with biological effect assays (Table 5.4.4).

In some advanced wastewater reclamation schemes, particularly for indirect potable reuse, already very comprehensive monitoring campaigns are carried out. The example of the Singaporean NeWater Project given in Table 5.4.5 shows the

Table 5.4.4 Intensified analysis program for water reuse sites which is being carried out in the RECLAIM WATER project (www.reclaim-water.org)

Specific microbial contamination and antibiotic resistance gene	Organic contaminants
Bacteria *Salmonella* spp., *Campylobacter jejuni*, *Yersinia enterocolitica*, *Francisella tularensis*, *Helicobacter pilori*, *Mycobacterium avium*	Antibiotics e.g. sulfamethoxazole, tetracycline
Protozoa *Cryptosporidium* spp., *Giardia* spp.	Organic tracer compounds and endocrine disruptors e.g. iodated contrast media, adsorbable organic iodine, carbamazepine, benzotriazole, diclofenac
Viruses Enterovirus (Poliovirus, Echovirus, Coxsackievirus), Hepatitis A virus, Rotavirus	Dissolved organic matter characterization e.g. natural organic matter and effluent organic matter fractions
Helminth eggs Ascaris lumbricoides, Trichurus trichiura, Anclylostoma duodenale	Disinfection byproducts e.g. *N*-nitrosamines, trihalomethanes
Antibiotic resistance gene	

Table 5.4.5 Number and nature of monitored parameters per sampling location at the NeWater reclamation facility, Singapore (redrawn from Public Utilities Board, 2002)

		Sample location						
Water quality parameter		Plant feedwater	MF filtrate	RO permeate	UV effluent	NeWater	PUB raw water	PUB drinking water
	Physical	9	3	3	2	8	8	7
Inorganic	Disinfection byproducts	6	1	2	1	6	6	6
	Other	39	2	32		39	38	39
Organic	Disinfection byproducts	22		22		22	22	22
	Other	42				41	41	37
Pesticides/herbicides		50				50	50	50
Radionuclides		6				6	6	6
Wastewater signature compounds		4				4	4	4
Synthetic and natural hormones		3	3	3		3	3	3
Microbiological		10	9	7		10	9	3
Totals		191	18	69	3	189	187	177

intensive monitoring efforts that have been made in the early phase of the project to confirm that there is no concern with respect to residual levels of trace contaminants. Currently, for the routine operation an extensive sampling and monitoring programme, which includes 278 parameters, is in place (Leong Yin Hou, personal communication).

In summary, it can be noted that there is a clear trend for more comprehensive monitoring which is based on more powerful analytical techniques both in the microbiological and chemical sector.

A key recommendation for operators is to make use of specialized laboratories for a baseline assessment of emerging parameters without the necessity to include them in routine analysis at this stage.

It is a research task to assess both the relevance and most efficient removal techniques for a wider range of emerging contaminants. Instead of widely enlarged parameter catalogues an opportunity would be to establish advanced treatment standards which provide some safeguard for the mitigation of emerging contaminants below effect levels. The monitoring effort can then focus on the integrity control of those barriers for classes of compounds.

5.4.5.2 Finding Key Parameters for the Aggregate Evaluation of the Water Quality

The definition of easy-to-monitor and 'fit-for-all-purposes' parameters is certainly highly desirable from an operator's point of view. Nevertheless the different constraints given by regulations and case-by-case permits cannot be overcome with respect to the demanded water quality parameters and sampling frequencies. Water quality control might be appropriate on shorter intervals to improve process reliability, end-user satisfaction and simply to avoid technical problems in the longer run. A presupposition for a key parameter and the respective way to monitor it (measurement point in the process, frequency, sample processing, reporting, feedback possibilities) is that the parameter should be suitable as an early warning indicator which shows that the process performance is about to fluctuate either on the basis of input variations or technical malfunctions. Hence, it is required to set up a system of control points which are able to observe input, functionality of the treatment process as well as product quality.

5.4.5.3 The Timescale of the Measurement

While analyses are becoming more complex, market pressures are also dictating that the results are available more quickly.

There are a number of new emerging instruments and techniques, such as particle counters and gene probe technologies, which hold considerable promise as monitoring tools. It is expected that these technologies may provide reliable and inexpensive methods to conduct parameter testing. Future research and lower instrument costs will be required before such technologies become widely available.

Promising results are also coming from the latest development of early warning systems for microbiological contamination. Two examples are the development of protocols for the examination of *E. coli* that allow results to be obtained in the range of measurement 10–50 CFU (colony forming units)/100 ml within 1 h (Morikawa *et al.*, 2006) and in the range 50–400 CFU/100 ml within 10 h (Braathen *et al.*, 2005)

5.4.6 CASE EXAMPLES

The next section will provide three examples representing applications requiring different extents of monitoring in the context of water reuse:

- limited (industrial cooling): class 7 (out of 7) of Figure 5.4.1;
- moderately extensive (unrestricted irrigation): class 3 of Figure 5.4.1;
- full-fledged (aquifer recharge for indirect drinking water supply): class 1 of Figure 5.4.1.

5.4.6.1 Monitoring Strategy to Deliver Cooling Water Make-up in Tienen, Belgium

The Tienen (Belgium) water reuse scheme reclaims 2 000 000 m^3 per year in order to reduce the extraction of natural groundwater for close-circuit cooling water make-up at a nearby chemical company. The requirements for this application are stable quality and quantity, low turbidity, conductivity, suspended solids and temperature.

The wastewater treatment plant (WWTP) consists of a low loaded activated sludge system (oxidation ditch) with enhanced biological phosphorus removal and complies with the European Urban Wastewater Treatment Directive for Sensitive Areas (91/271/EC) (European Union, 1991). To prevent microbiological growth in the ducts, the effluent is disinfected with ozone, to obtain an ozone concentration of 3–4 mg/l at the entry of the cooling water make-up storage facility. Since the start-up of the use of reclaimed water in 2003, no problems have occurred regarding regrowth in the cooling circuit, with the reclaimed water having a total cell count of 100–200 CFU/ml.

To detect and retain reclaimed water of unacceptable quality, conductivity and turbidity are measured on-line. Should the conductivity exceed 1700 μS/cm or the turbidity be higher than 3 NTU (nephelometric turbidity units), the reclaimed water would be diverted to the surface water body and the system be provided with an alternative water supply (groundwater). Temperature is also measured continuously. Seasonal evolution of temperature implies that the reclaimed water cannot be used during summer months (Thoeye *et al.*, 2006).

5.4.6.2 Monitoring Strategy for Agricultural Irrigation in the Dan Region, Israel

The Dan Region Reclamation Project treats around 140 million m^3 per year (or 340 000 m^3/day) from the Greater Tel-Aviv. The effluent of the conventional activated sludge plant (WWTP) is conveyed to a tertiary treatment composed of a soil aquifer treatment (SAT), from where it is recovered and distributed throughout a 100 km pipeline and seasonal reservoirs. The project should ensure the distribution of effluents suitable for unrestricted irrigation quality, i.e. enabling the irrigation of all kinds of crops.

The monitoring programme includes analyses of physical, chemical and biological parameters, at various sampling points (Figure 5.4.4) and time intervals (Table 5.4.6).

Besides sampling in different points, there is 24 h monitoring of the WWTP and SAT operation in a manned control room in the WWTP and also visual inspections are performed along the distribution line up to the end-user.

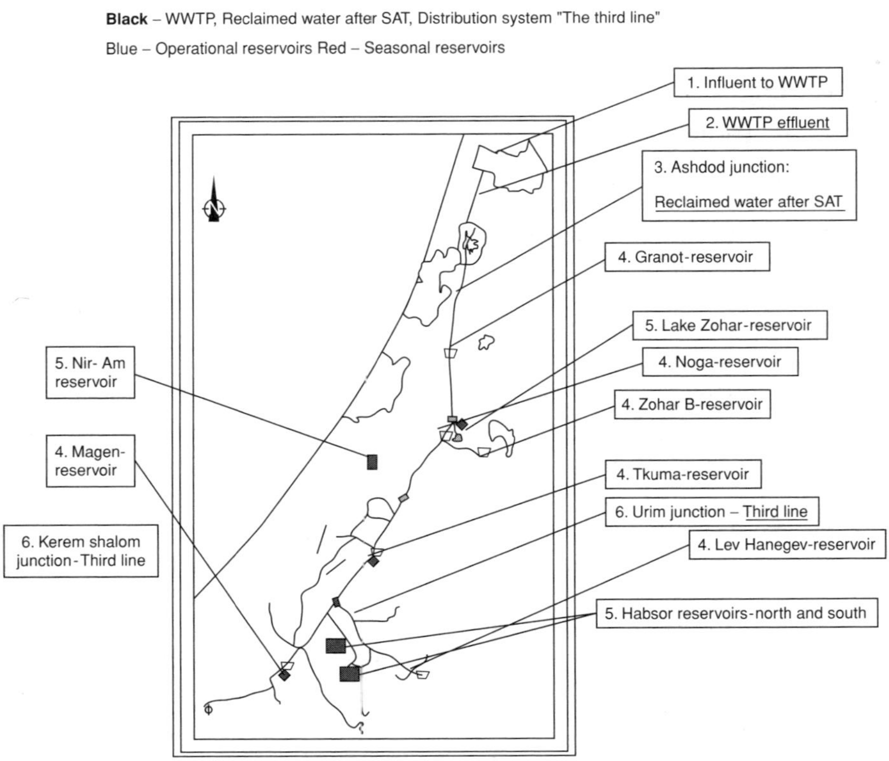

Figure 5.4.4 Scheme of Dan Region project with the critical control points for monitoring (Cikurel and Aharoni, 2004)

Table 5.4.6 Monitoring practice at the Dan Region reclamation scheme

Measuring point (frequency)	Indicator
On-line	Effluents: turbidity Distribution system: chlorine Pumping stations: sand detectors Infiltration fields (each basin): flow, water level and condition of the valve (opened/closed)
WWTP effluent (frequency/week)	COD_f, TSS, NH_4, turbidity, pH (daily) NO_2 (6), COD, TKN_f, PO_4 (5), BOD_f (4) , ISS(4), TKN(4), BOD, NO_3, UV_{254} (3) DS (2) , IDS (2) , Cl (2), TP, TP_f, conductivity and alkalinity (2) Hardness , Ca, detergents , fats and oils, DO (1)
WWTP effluent (monthly)	Trace elements, phenol
Observation/production wells (twice a year)	Cl, EC, detergent, UV_{254}, COD, DOC, turbidity, pH, alkalinity, NKJ, NH_4, NO_2, NO_3, hardness , Ca, B, Na, K, SO_4 , TP, PO_4, temperature, Fe, Mn, As, Cr, Cd, Ag, Ba, Zn, Cn , Se, Pb, Cu, Hg, Mo, Ni, Co, F, Sr, Li, Al, Sn, Be, Va, Phenol, oil and grease, Total bacteria, *E. coli*, faecal coliforms, faecal *streptococcus*, colour
Distribution system (monthly)	Mn, Al, Fe, DO, pH and turbidity
Intermediate reservoirs (monthly/twice a month in the hot season)	Bacteriology, Ervinia, chlorophyll, alga, TSS, VSS, UV, DO, pH, temperature, turbidity

DO, dissolved oxygen; TSS, total suspended solids; vss, volatile suspended solids.

Along the distribution line there is different equipment to monitor the clogging capacity in the line. At the exit of the seasonal reservoirs along the distribution system there are wire filters (80–120 mesh) to prevent algae and other clogging matter from the reservoirs to reach the pipe-lines. The degree of clogging (mainly by algae) is also measured by means of a by-pass system which also consists of a wire filter. The degree of clogging is indicated by the number of backwashes performed in a given time interval.

The monitoring practice implies:

1. The measurement of the infiltration velocity is conducted in an automatic way.
2. Daily automatic measurement and recording of the water level in the basin, integrated with daily inspection performed on-site by an operator.
3. Optimized cleaning routine, based on the infiltration velocity and assisted by the decision of the field operator.

4. A preventive cleaning programme, applied every 15–30 days, depending on the clogging rate.
5. Ploughing (by using disc or plough), chosen according to the field situation. Sand removal is seldom practiced.

5.4.6.3 Monitoring Strategy to Close the Water Cycle at the Flemish Coast, Belgium

At the Belgian coast, the Intermunicipal Water Company of the Veurne region (IWVA) is responsible for the potable water production and supply in the area; the source for the potable water production was the sandy aquifer in the dunes near the coast. Tourist activities in the area resulted in a largely varying water demand and saline intrusion, which was threatening the sustainability of the potable water production. In 2002 the IWVA started artificial recharge of an unconfined aquifer in its dune water catchment St André. Wastewater effluent was used as the source for the production of infiltration water. This plant, with a production capacity of 2 500 000 m^3/year, combined membrane filtration techniques to achieve the stringent standards set for the quality of the infiltration water. The whole project was developed to create a sustainable groundwater management; the natural groundwater extraction was reduced from 3 700 000 m^3/year to 2 700 000 m^3/year. By 2010 another 500 000 m^3/year will be saved (Van Houtte and Verbauwhede, 2005). Saline intrusion was gradually prevented and the sustainability of the water production was increased. The treatment scheme is presented in Figure 5.4.5.

The regulatory water quality requirements for the water reclamation scheme before infiltration are: temperature, pH, conductivity, chloride, sulfate, magnesium, sodium, total hardness, nitrate, nitrite, ammonia, aluminium, iron, manganese, copper, zinc, phosphorous, fluoride, cyanide, chrome, mercury, nickel, lead, antimony, selenium and trihalomethanes.

The operational monitoring practice to prevent the infiltration of substandard water is summarized in Table 5.4.7.

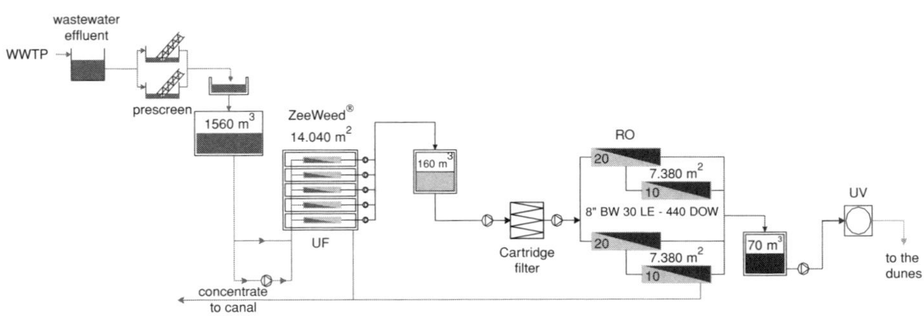

Figure 5.4.5 Tertiary treatment scheme of St André aquifer recharge project in Wulpen (IWVA)

Table 5.4.7 Operational monitoring practice at Wulpen reclamation scheme: double membrane system; Van Houtte, 2004

Frequency	Indicator
On-line	Flow, conductivity, pH and temperature on all waters, turbidity on MF filtrate, pressures and energy consumption of processes
Daily	SDI, conductivity, pH on MF and RO filtrate
Weekly	Bacteriology, salt and nutrient content
Monthly	Greater set of parameters (to comply with the norms)

5.4.7 CONCLUSIONS

Each water reclamation plant is unique, with its own requirements for an integrated monitoring and control instrumentation system. However, some common traits can be defined, especially considering the monitoring of the hygienic quality.

Because of the multiple barriers that are generally in place to prevent unsuitable reclaimed water reaching the customer, the regulatory monitoring of delivered water quality is often simply a verification that the preventative measures are effective, and often variables that can be monitored instantaneously can give a higher level of confidence in safety of supply and at less cost than analysing for an expanding number of chemicals.

Two opposite trends are in place in the monitoring of water reuse schemes: on the one hand, the need to follow up a more varied list of reclaimed water indicators and on the other, the development of easy-to-monitor and fit-for-all-purposes sensors that can provide an overall indication of the suitability of the reclaimed water for a range of intended uses.

ACKNOWLEDGEMENTS

This work was partly funded by the European Commission under the AQUAREC Project on Integrated Concepts for Reuse of Upgraded Wastewater (EVK1-CT-2002-00130) in the 5th Framework Programme. The authors would like to express their gratitude to Paul Hendrickx, Manu Van Houtte and the PUB Singapore whose responses in the questionnaire are used in this document.

REFERENCES

Anderson, J., Adin, A., Crook, J., Davis, C., Hultquist, R., Jimenze-Cisneros, Kennedy, W., Sheikh, B. and van der Merwe, B. (2001) *Water Sci. Technol.*, **43**(10), 1–8.

Ayers, R.S. and Wescot, D.W. (1985) *Food and Agriculture Organization Irrigation and Drainage Paper 29*. Water Quality for Irrigation, Rome, Italy.

Bixio, D., De Heyder, B., Cikurel, H., Muston, M., Miska, V., Joksimovic, D., Schäfer, A.I., Ravazzini, A., Aharoni, A., Savic, D. and Thoeye, C. (2005) *Water Sci. Technol.: Water Supply*, **5**(1), 77–85.

Bixio, D., De Koning, J., Joksimovic, D., Savic, D., Wintgens, T., Melin, T. and Thoeye, C. (2006) *Desalination*, **187**, 89–101.

Bixio, D. and Wintgens, T. (eds) (2006) *Manual on Management of Water Reuse Schemes*. European Publishing Services, Brussels, Belgium, (in press).

Bouwer, H., 2000. *J. Environ. Hlth*, **63**(3), 17–20.

Braathen, H., Ranneklev, S., Rydbert, H. and Sagstad, H. (2005) *Water Wastewater Int.*, **20**(2), 35.

Cikurel, H. and Aharoni, A. (2004) *AQUAREC Deliverable D10 Management Review Report*, Bixio, D. and Thoeye, C. (eds). Aartselaar, Belgium.

De Rooy, E. and Engelbrecht, E. (2003) *Proceedings Water Recycling Australia*, Brisbane, September 2003. Australian Water Association, Sydney, Australia [CD-ROM].

Dewettinck, T., Van Houtte, E., Geenens, D., Van Hege, K. and Verstraete, W. (2001) *Water Sci. Technol.*, **43**(12), 31–38.

Dillon, P.J. (2005) *Hydrogeol. J.*, **13**(1), 313–316.

European Union. (1991) *Council Directive Concerning Urban Wastewater Treatment*. 91/271/EC of 21 May 1991, OJ L135/40 of 30 May 1991.

EWRI/ASCE. (2001) *Standard Guidelines for Artificial Recharge of Groundwater*. Environmental and Water Resources Institute, American Society of Civil Engineers, EWRI/ASCE 34–01, Reston, VA.

Heberer, T. (2002) *Toxicol. Lett.*, **131**(1–2), 5–17.

Hochstrat, R., Wintgens, T., Melin, T. and Jeffrey, P.J. (2006) *Desalination*, **188**, 1–8.

Kamizoulis, G. (2006) *Proceedings Integrated Concepts for Reuse of Upgraded Wastewater*; Barcelona, Spain, 1–3 February 2006 [CD-ROM].

Körner, W., Bolz, U., Süßmuth, W., Hiller, G., Schuller, W., Hanf, V. and Hagenmaier, H. (2000) *Chemosphere*, **40**, 1131–1142.

Kuch, H. M. and Ballschmiter, K. (2000) *Fresenius J Anal. Chem.*, **366**(4), 392–395.

Listowski, A. (2004) *Proceedings Aquarec Workshop*, Thessaloniki, Greece, 11–12 March 2004 [CD-ROM].

Martínez, J.L. and Baquero, F. (2002) *Clin. Microbiol. Rev.*, **15**(4), 647–679.

Metcalf and Eddy. (2003) *Wastewater Engineering, Treatment and Reuse*, 4th Edn. McGrow Hill, Boston.

Morikawa, A., Hirashiki, I. and Furukawa, S. (2006). *Water Sci. Technol.*, **53**(4–5), 523–532.

Pérez, S., Reifferscheid, G. and Barcelò, D. (2003) *Environ. Toxicol. Chem.*, **22**(11), 2576–2584.

Pettygrove, G.S. and Asano, T. (eds) (1985) *Irrigation with Reclaimed Municipal Wastewater – A Guidance Manual*. Lewis Publishers, Inc., Chelsea, MI.

Poseidon. (2004) Assessment of Technologies for the Removal of Pharmaceuticals and Personal Care Products in Sewage and Drinking Water Facilities to Improve the Indirect Potable Water Reuse. www.poseidon-eu.org.

Public Utilities Board. (2002) Singapore water reclamation study expert panel review and findings: http://www.pub.gov.sg/NEWater_files/download/review.PDF

Purdom, C. E., Hardiman, P. A., Bye, V. J., Eno, N. C., Tyler, C. R. and Sumpter, J. P. (1994) *Chem. Ecol.*, **8**, 275–285.

Pyne, R.D.G. (1995) *Groundwater Recharge and Wells: a Guide to Aquifer Storage and Recovery*. CRC Press, Florida.

Rommelmann, D. W., Duranceau, S. J., Stahl, M. W., Kamnikar, C. and Ganzales, R. M. (2003) Industrial water quality requirements for Reclaimed water. AwwaRF Report no. 90979F. Awwa Research Foundation, Denver, CO.

Salgot, M., Huertas, E., Weber, S., Dott, W. and Hollender, J. (2006) *Desalination*, **187**, 29–40.

Segner, H., Navas, J. M., Schäfers, C. and Wenzel, A. (2003) *Ecotoxicol. Environ. Saf.*, **54**(3), 315–322.

Thoeye, C., Wintgens, T., Van Houtte, E., Bixio D., De Gueldre, G. and Van De Steene B. (2006) *Wastewater Reclamation and Reuse in Belgium*. Wastewater management in regions under water scarcity: the state-of the-art in the beginning of XXIst Century. Springer Publishing (in press).

Toze, S. (2006) *Desalination*, **187**, 41–51.

US Environmental Protection Agency. (2004) *Guidelines for Water Reuse*. EPA/625/R-04/108, September 2004.

Valat, C., Champiat, D., Degorce-Dumas, J. R. and Thomas, O. (2004) *Water Sci. Technol.*, **49**(1), 131–138.

Van Houtte, E. and Verbauwhede, J. (2005) *Proceedings 5th International Symposium on Management of Aquifer Recharge*, Berlin, Germany [CD-ROM].

Victoria Environmental Protection Agency (2003) http://epanote2.epa.vic.gov.au/EPA/Publications.nsf/2f1c2625731746aa4a256ce90001cbb5/64c2a15969d75e$FILE/464.2.pdf.

Water Pollution Control Federation (1989) *Water Reuse Manual of Practice*, 2nd Edn. Water Pollution Control Federation, Alexandria, USA.

World Health Organization. (1989) *Health Guidelines for the Use of Wastewater for Agriculture and Aquaculture*. Technical Report Series 778. World Health Organization, Geneva, Switzerland.

Van Houtte, E. (2004) AQUAREC WP6 Questionnaire on management practices IWVA Wulpen.

6.1
Collecting and Merging Data from Widespread and Disparate Sources

Michael J. Scott

Wastewater Quality Monitoring and Treatment Edited by P. Quevauviller, O. Thomas and A. van der Beken

6.1.1 INTRODUCTION

There are continuously growing commercial and regulatory pressures to improve the control of water and wastewater and hence the impact of industry and the populace on the environment. The problem is that point pollution is fairly well regulated and controlled. The attention is now on diffuse pollution and consideration on the health of complete structures, such as river basins and catchments. This poses several problems since disparate groups within and without companies and regulators have to work together if progress is to be made. Inevitably this means that data from widespread and disparate sources need to be compared and merged.

At the time of writing the problems of multidepartment/company/regulator co-operation and collecting and using disparate data, often sparse and lacking comparability is in its infancy; but with rapid progress being made. This chapter therefore raises some of the issues, indicates where information can be obtained and suggests some of the future actions.

6.1.2 DATA AND INFORMATION NEEDS

The EU Framework Directives, in particular for this chapter the Water Framework Directive (WFD), are recognising that it is not economically possible to make all the industrial and environmental measurements that the precautionary principle suggests. The solution is to use language that requires such goals as sustainability, lack of toxic effect and river basin management. The problem for industry is that because measurement has been a very small part of the capital budget it has been neglected and is low on management priority lists.

For individual measurements it is easy to get data but getting credible and useful data is difficult and convincing others with that data is even more difficult.

To meet future requirements information will come from a wide variety of sources as discussed below.

Those that are making the measurements and those that are using the resulting data need:

- higher confidence in the data
  - better reliability
  - better sampling
  - more complete audit trails;
- lower skill requirements;
- lower cost of ownership;
- better integration of disparate data.

Even at an individual level there are urgent needs for cooperative developments but the serious challenge is to make better use of the data already available and to develop ways of merging temporally and spatially sparse data to improve the understanding of complete mechanisms such as river basins.

In general the water and waste treatment and environmental industry is not limited by on-line measurement technology and it is likely that evolution of currently available techniques and products will be the dominant feature of market development. There are some opportunities for new measurements and for much lower detection limits. Examples of the former are the continued search for an on-line method for biological, chemical or total oxygen demand, the need to find an on-line method for the Mogden formula used for assessing the loading on a waste treatment works and the need for the on-line or at-line measurement of ammonia. Various optical spectral analysis techniques have been trialled over the last decade but each has foundered on the site and application specific nature of the measurements required.

However, rapid advances in data analysis and the sharp reduction in the cost of data processing might well provide commercial opportunities. Examples of the latter are the more widespread measurement of turbidity and the need to check the efficacy and the by-products resulting from disinfection techniques. Each of these might offer opportunities for optical technology.

In general the EU Directives do not specify actual measurements to be made; nor the acceptable levels of particular pollutants. This is likely to continue. Where chemicals are specified as in the WFD, the list is indicative only and other chemicals might be added later; but more likely will be controlled at the national or regional level.

(Annex X can be obtained at http://www.wfduk.org/about_wfd by going to the bottom of the page, and selecting daughter directive – priority substances.) The WFD sets out a methodology for determining environment quality standards (EQSs) for priority substances (including priority hazardous substances) and other specific pollutants. The UK, and other Member States, have existing EQSs for many of these of these substances; however, work is required to ensure their derivation is compliant with the WFD.

There will be an increasing need to measure specific chemicals at much lower detection levels, as required by the regulators. Although the WFD only gives the above list of measurands, without detection levels or ranges, which is mirrored in the UK Environment Agency's position, it is clear that many more measurements will arise in the future.

The problem is that there are, or will be, many thousands of chemicals that should be measured; at least on a precautionary principle. The resort to the 'precautionary' principle, when hard data toxicity or harm data are not available for specific chemicals, makes it almost impossible to predict which chemicals will fall under future regulations. Industry and society will not afford such a large number of measurements but this presents commercial opportunities for instruments that can be rapidly configured to new demands.

The WFD approaches the problem of the potentially large number of chemicals to be measured by calling for the monitoring of changes in, for example, catchments and uses descriptive phrases which are not yet defined. There is a great deal of research work in hand to explore the possibilities of the use of surrogates or broad band measurement that are strongly indicative of change in a particular body of water, or catchment. Techniques already used include the colour change of vegetation, chlorophyll levels and plankton blooms viewed by remote surveillance, and bird counts. This might well be where biosensors finally provide commercially valuable tools.

6.1.3 THE MEASUREMENT CHOICE

Data may come from a wide variety of measurement methods and the data quality is dependent on several *people skills* related issues:

- measurement selection;
- sampling considerations;
- installation design;
- installation, calibration and maintenance.

There is a bewildering choice of measurement technologies and formats by which the actual measurement device is presented to the sample. Note that for the majority of measurements only a small part of the fluid or gas under investigation is seen by the actual sensing part of the measurement technology. Figure 6.1.1 gives a partial

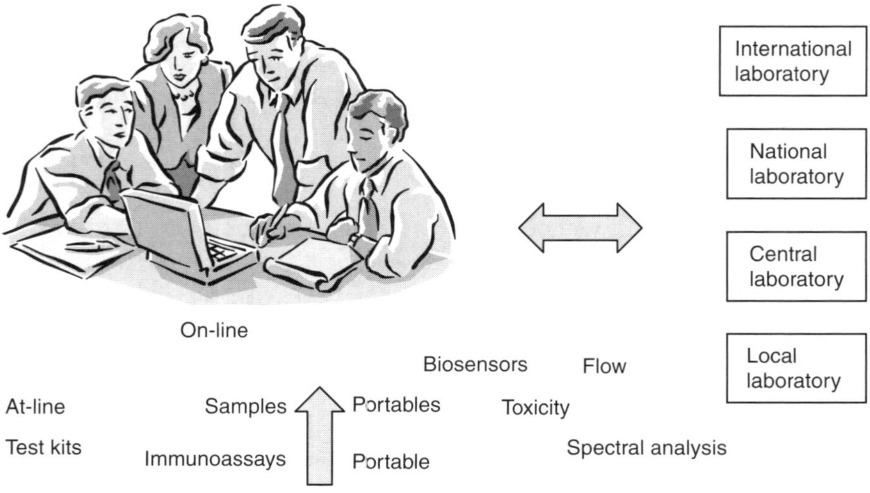

Figure 6.1.1 The measurement choice

idea of the bewildering number of terms that anyone wanting to gather data may face.

Examples of possible sources and formats are listed at the bottom of the diagram and the hierarchy of methods of indicating data quality are shown on the right of the diagram in ascending order of rigour and hence time and cost.

In the new climate, the comparability of data is increasingly important so the ascending list of reference sources, on the right of the diagram, must be considered, using risk analysis, to determine how far up the chain it is prudent to go.

The vast majority of measurements are application sensitive and site specific and this even applies to flow metering; particularly open channel flow measurement. So the selection and placement of the best sensor, for the specific business need, is a complex multidiscipline, multidepartment matter.

The data, particularly for environmental issues, are sparse and derived from disparate sources which must be merged if the user is to have a comfortable feel for the quality of the underlying data, and some idea where improvements might be made. So yet another discipline is brought into play.

6.1.4 FRAGMENTATION

The structure of the innovation, supply and demand mechanisms for the water, waste treatment and environmental markets is highly fragmented and the market is of modest size. Communication between the many players is challenging and very few people have an overall broad range of contacts. Although some large supplier companies are involved they are almost always compartmentalised and act in an SME (small to medium enterprise) manner. The overall market is characterised by very small suppliers and small groups of users dealing with niche measurement needs.

There are a very large number of instrument manufacturers and supply companies and together with the very large number of measurands and instrument formats this provides a bewildering fragmentation to those who are often only infrequently required to select and use sensors. More technologies and more formats are coming to market. For example, biosensors will finally impact the market in the next 5–10 years and there is far more to come with the miniaturisation of mass spectrometers, chromatographs and optical spectra. To gain some appreciation of the range of measurands one has only to look at the Products section of the Institute of Measurement and Control's Year Book or the Omega web site (www.omega.com). Further study shows that even a simple sensor such as a thermocouple or resistance thermometer has a number of basic types and a myriad of configurations; depending on the intended use. Pressure measurement has even more variations and suppliers often confine themselves to one range of applications. So, for example a supplier of pressure measurement to the chemical industry would probably not supply to the automobile industry or aerospace industry and might not supply to the pharmaceutical industry.

There are a large number of 'Supply companies' who trade in a particular geographical area or industrial sector. They may have their own core instruments, to which they have added branded or factored instruments from others; or they may have reached an agreement with one or more of the large companies to sell in a defined geographical area or sector. These companies are almost always highly application skilled and will source any very low volume instruments that their regular customers sometimes need.

To provide a benchmark it has been speculated that the market for on-line industrial turbidity instruments might be 20k to 30k units per year and yet there are at least ten front line suppliers and twenty to thirty that claim to be suppliers. The recent absorbing into conglomerates of one or two of the specialist suppliers does little to reduce the competitiveness and fragmentation. Close inspection of this market shows that niche expertise is the pacing issue for commercial survival and often results in market leadership in that niche. Such niche leading companies might have total sales of £1–5m.

6.1.5 DATA QUALITY AND COMPARABILITY

Once the management issues have been addressed there is the need to consider the already mentioned issue of establishing the user's confidence in the data. It is no longer enough to present data. Regulators are already requiring that the quality of the data at the point of use, and in the archive, is stated and justified.

There is a serious lack of up to date standards, methodologies and training. Some Blue Book methods are incomplete, or not widely tested and accepted and, for some measurands, Blue Book or equivalent methods do not exist. This is a particular problem for recently identified pollutants. Some standards have been shown to be based on laboratory work which does not translate to on-line or field-based measurements; an example being the British Standard for open channel flow.

Updating standards and codes of practice is time consuming and normally takes a number of years to get to the approval stage; a further few years is usually necessary for them to be adopted as the norm. Companies are increasingly reluctant to allow engineers time off to sit on standards and other policy setting committees which not only means that timescales are extended but the work may not have a balanced input.

MCERTs, which is an Environment Agency initiative, is an example of a quality control procedure for the whole data to data use chain and is indicative of the future for environmental measurements (www.MCERTS.net). MCERTs has taken nearly a decade to become well established for Stack emissions and is now beginning to be the required code of practice for open channel flow; the initial focus for aqueous measurements. Work on MCERTS for continuous on-line has been in progress for several years and a consultation document was issued in the middle of 2005 for on-line continuous water quality measurements for:

- temperature;
- pH;

- conductivity;
- dissolved oxygen;
- turbidity;
- ammonia;
- nitrate;
- nitrite;
- orthophosphate;
- chlorophyll a.

A consultation document was released at the end of 2005 for the following portable water quality measurements:

- temperature;
- pH;
- conductivity;
- dissolved oxygen;
- turbidity;
- ammonia;
- nitrate;
- nitrite;
- orthophosphate;
- Chlorophyll a.

This particular consultation document comes from the consolidation of extensive work on earlier work for a workable format for ensuring high quality chemical measurements individual measurands. It is thus likely to be the basis of an Environment Agency mandatory approach to administering their self monitoring policy.

MCERTs is only an input for part of the uncertainty chain. IPPC (Integrated Pollution Prevention and Control) and PPC (Pollution Prevention and Control) do not now rely on 'end of pipe' measurements and it is to be expected that audit trails and uncertainty chains will increasingly be demanded for key process variables.

Calibration, maintenance, training and consumables management practices and controls form an integral part of understanding the quality of measurement data.

Obviously there is a lot of judgement and reliance on skills involved and it becomes clear that comparability of data is a challenge.

Pre-normative work (Standardised testing of on-line sensors/analysers, Anders Lynggaard-Jensen, ISO Working Group) under an EU project in the late 1990s was reported at the ISA-Tech Conference at Interkama in October 1999:

> *In order to support the protection of the environment there is an urgent need for improvement of comparability, reliability and quality of measurements obtained from in-situ on-line sensors/analysers used to determine the composition and flow of effluents from wastewater treatment plants and industries. Users cannot check and compare different products before purchase without individually completing costly experimental trials and so the number of potential users is restricted. Furthermore, there are no standardised procedures to provide users and regulators with acceptable and robust validation of compliance data.*
>
> *As a result, the widespread adoption of systems is limited and the benefits of this in terms of more efficient and effective pollution control are lost. Moreover, the instrumentation industry does not have any standard to develop systems to, and so on the one hand is unable to properly define its development costs and on the other hand has no means of independently demonstrating that its products are 'fit for purpose' in the European market.*
>
> *The paper described a pre-normative research work and the subsequent work in a working group under ISO concerning standardisation of in-situ on-line measurements of water quality determinands. The work includes development of a draft test protocol for validation of the performance of in-situ on-line sensors/analysers to guarantee comparability of results, the practical testing of the draft test protocol to access its applicability to provide a final test protocol and turning the final test protocol into an agreed standard/guideline.*

The EU thus recognised that there was no comparability of data between Member States. A current EU 6th Framework Project to deal with this problem is SWIFT (Screening Methods for Water Data Information in Support of the Implementation of the Water Framework Directive; www.swift-wfd.com).

Establishing cost effective and useful measurements requires significant inputs from a wide range of disciplines:

- establish application need;
- design application and data flow;
- procurement;
- installation;
- calibration;
- maintenance;
- audit trail.

An UKWIR Report in 2000, 'On-line instrumentation Standards and Practices Ref No. 00/PC/03/1' concluded, among other things, that measurement was below senior management's horizons in the user industry and recommended the generation of multidepartment understanding and cooperation. Dealing with widespread sparse data will be even more of a challenge and will require significant changes in management structures skill sets and cooperation in the users.

During the UKWIR project it was discovered that at least one UK water company switched off two archives because there was no reasonable way to normalise the data with other data and the future format.

The quality control of the transmission and archiving of measurement data is a further part of data management if comparability of data is to be achieved.

6.1.6 DISPARATE AND SPARSE DATA

Up to this point the generation of data has been from man-made devices (test kits, portable instruments, samples to laboratories, etc.). These data generators may be widely spread (sparse) in, for example, a sewer network or water distribution scheme or for the environmental control of a river catchment. As we have already seen the methods of generating data are many and various and time lines may be difficult to establish. Data derived from test kits may be occasional (sparse) and since the operators may vary or change substantially over time it may be difficult to estimate uncertainties for that data.

However, for large processes, particularly environmental, there are other inputs that will affect the merging of data. Diurnal changes and weather conditions are known to affect the generation of data but are frequently ignored. Other disturbing events may occur and may not have been recorded.

SCADA data are part of the information being analysed as part of a EU 'MicroRisk' project whose primary aim is to undertake quantitative microbial risk assessment on 12 complete water supply systems (source/catchment, treatment, distribution, consumption, infection) – essentially Monte Carlo style simulations of the doses of pathogens that consumers receive in model full scale systems. Obtaining high quality data is turning out to be much more of a challenge than was expected.

A considerable amount of such data and analogous high resolution data sets (e.g. microbiological data) collected daily for several years is available. However, systematic inspection of SCADA data and associated diary data from a particular water treatment plant has proved much slower and more problematic than expected.

Several issues have arisen.

(1) SCADA data management with high resolution data sets quickly exceeds the capacity of MS Access to manage and distilling the raw data takes a long time. Also, there do not seem to be many standard/statistical algorithms for distinguishing baseline from event data in long time series. It seems that eyeballing is the most effective technique.

(2) SCADA data are often provided as stand alone without clear links to what is actually happening in a water treatment plant.

(3) It is often not clear whether departures from targets represent problems or maintenance or poor instruments.

(4) Measurement facilities are often not well maintained and/or the quality control regime for them is not well documented.

The performance of water treatment plants is often not recorded electronically and sifting through paper records makes the task of quantifying and understanding hazardous events very time consuming and difficult. The use of SCADA data is obvious but it is important to recognise its limitations in the role of risk assessment. After all the data are mainly for real time process control therefore their interpretation for hazard analysis requires care. Although not yet established with the regulators and courts, there is a growing awareness that on-line data can be legally robust.

Samples analysed in a laboratory are the established source of regulatory or other control data; despite the known potential problems with the relevance of the sample taken, transport to the point of analysis and the quality of the determination itself. Establishing a sound audit trail is still a challenge and, surprisingly, it is not usual for a measurement uncertainty chain, based on the full sample to data presentation trail, to be available. For many chemical measurements Blue Book methods are not widely accepted, not available, or more than one method is available.

Underlying this all is the ‘know your system’ philosophy being promoted in the new ‘Water Safety Plan’ approach for managing water supply risks.

The above discussion has concentrated on a process for which it is reasonable to expect a reasonable amount of data; although some of it may be from disparate sources. In this EU project the instrument data and the diaries need to be matched.

If we broaden the information needed to company wide or environmental considerations it can rapidly be seen that for economic reasons the data that must be merged may be sparse and disparate.

Ways will need to be developed to instil confidence for infrequent users of the data such as senior management, pressure groups and regulators. The data need to be presented with quality statements, which is a challenge, but there is also a urgent need to find graphical methods to provide confidence that the complete ‘system’ is in balance. Senior management, pressure groups and regulators do not have the time, and possibly the expertise, to deal with all the data. Concepts like ‘mass balance’ and ‘ecological stability’ can be used to provide a framework against which the detail might be considered. Most factory HMIs (human machine interfaces) allow a selection of a number of measurements to be overlaid (Figure 6.1.2). Some of the measurements are continuous and some are infrequent samples analysed in the laboratory. The traces would be selected so that if the particular combination is ‘stable’ then it is a reasonable assumption that the other measurements are within acceptable bounds. It is a relatively simple matter to train a wide range of people to understand what is normal and what needs closer inspection. The UK Environment Agency has already declared a policy of self monitoring so it is to the advantage of ‘polluters’ to demonstrate and communicate indicative information with the detail

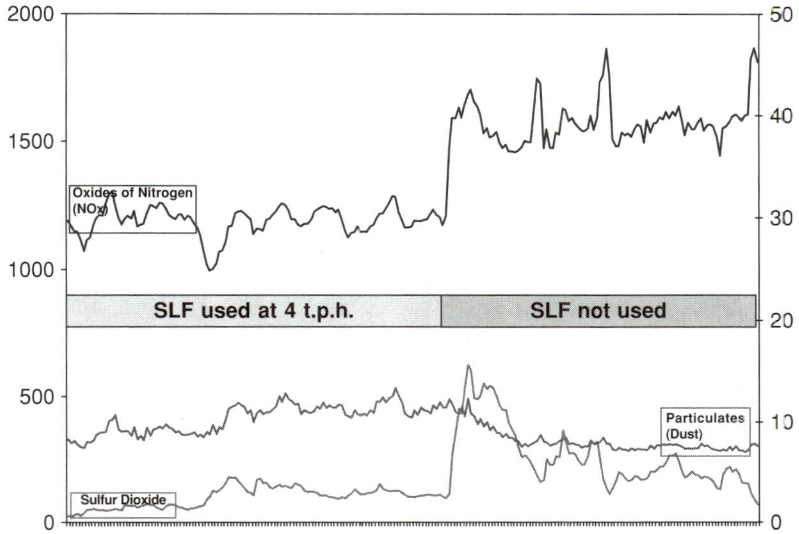

Figure 6.1.2 Tracking operational performance with overlaid measurement traces

readily available if necessary. The timescale can be rapidly changed so that both incidents and long term trends can be inspected or studied.

The WFD implies a massive need for data, which is unlikely to be met. Many believe that modelling is the only viable approach and the selection and quality of indicative data will be the pacing issue. It will be challenging to ensure that the users of the resulting information have a comfortable feel for the quality of the underlying data and some idea where improvements in measurement and in modelling might be made and to what effect.

Elsewhere it is argued that information will increasingly come from the merging of widespread and disparate data in the form of models. In this case measurements may be calculated and inferred and it is probable that some of the input data may have a high uncertainty but adequate for the needs; with consequent cost savings.

6.1.7 RESEARCH AND DEVELOPMENT SUPPORT

The very large instrumentation companies have provided support for long-term enabling technologies; some of which only became 'open' after the market had been established, i.e. HART protocol. Some long-term support, such as the Foxboro/Invensys work towards 'self diagnostic' sensors at Oxford University has yet to gain commercial rewards but spin-off work on the way seems to have justified the effort. A common framework for self diagnostic sensors might well be an important donation to the worldwide sensors industry.

It is not unknown for a major user to spend a considerable amount of money over many years and apparently have nothing material to show for it. Often this happens because of people mobility but more often because of unclear objectives. However, closer inspection often uncovers benefits that are real, but hard to quantify; ideas and prejudices challenged, broader perspectives considered, etc.

The interoperability of telemetry systems is being taken seriously by the sensor and telemetry community and a cross industry activity (WITS; www.ukwits.org) is already making progress on common methodology. A methodology for communication from the management network out to the sites will be undergoing interoperability trials in 2006.

A cooperation exercise called WASP (www.wasp-protocol.com) is starting to make headway at the next level down so as to bring some sort of order to the connection of the WITS level to the sensors and actuators. Generally the WITS level is driven by information technology people whereas the WASP level is driven by instrumentation people and the two cultures need to develop a common understanding of the issues.

Of course all is not sweetness and light, as the decade or more of inter company and country rivalry in the Fieldbus wars showed. However, some enlightenment is showing and real progress is being made on the various modes of serial communication. The breakthrough was probably the broader acceptance that a single serial communication (fieldbus) was never going to happen and inter-operability was a better objective. The uptake of Ethernet and the internet was the final nail in the coffin of formal international standards having any precedence. Standards setting with timescales of a decade or more cannot be relevant when viewed against the very high rate of change of serial communication technologies.

6.1.8 USERS

The users, whether industrial concerns, regulators or governments, are usually compartmentalised by discipline or market sector, whereas measurement is usually a multidepartment, multidiscipline activity. The users have little or no wish to become engaged with the sensor technologies. The days of the separately identifiable instrument department or senior engineer are long gone in the majority of industries. There are often pockets of identifiable expertise but these will disappear with time since industrial on-line measurement is rarely an identifiable higher education topic. On-line measurement is much less than 1 % of capital and does not figure in senior management planning unless a commercially viable project plan is placed before the project selection committee.

Measurement decisions may well end up at the end of a long line of subcontractors as part of the outsourcing approach to business. This generates problems for the instrument supplier since the decision making unit (DMU in marketing terminology) is often large and diffuse with quite a lot of mobility in its constituent parts. Who do you approach and convince with a new idea and who has the budget to evaluate the idea?

It is easy to see that there is a 'disconnect' between the suppliers and the users. Each has an honest and hard working approach to the needs and responsibilities but the business and cultural differences need to be addressed.

A further problem is that although there is a now grudging acceptance of the need for 'maintenance' and other OPEX issues the out-sourcing and multidepartment nature of the user makes budgeting, coordination and resource allocation close to impossible.

A normal asset life is taken as 10 years or more and this is challenging for the fast moving technologies that underpin industrial on-line sensors. For example electronic chip, embedded computers and serial communications are all moving with much shorter time constants than 10 years. Given that an industrial on-line sensor may take several years from inception to established commercial viability the planners, developers and marketing departments have an interesting challenge.

An idea mooted some years ago and progressed by some suppliers was to lease the 'measurement'. The supplier takes total responsibility for the supply, installation, calibration, maintenance, updating, etc. This approach has been used for decades in the North Sea oil and gas production industry.

It is increasingly difficult for automation suppliers to sell 'boxes'. The majority of suppliers have recognised that they have to sell 'solutions'. However for small value items it is difficult to achieve the critical mass where the overheads become a small part of the total cost. Of course the user still has to be satisfied that the measurement(s) has a business benefit and has to take ultimate responsibility for the data, which is a nontrivial challenge.

A simple solution for individual measuring instruments is for the supplier to design for 'fit and forget' and 'point and shoot'. In other industries, such as medical, auto and home goods, there seems to be no alternative and this may be the only approach for demands such as monitoring sewers for the onset of flooding or the monitoring of drinking water supply networks for unwanted substances or the monitoring of river catchments (Figure 6.1.3).

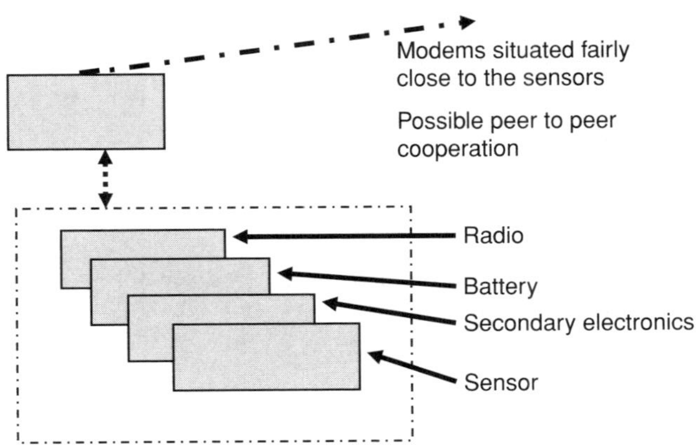

Figure 6.1.3 Structure of an autonomous instrument

Available technology already allows autonomous on-line instruments, as in Figure 6.1.3, with a life of 5–10 years and some are being trialled. We can expect that the Homeland Security and Battle Field industries will provide the fundamental research and useful components. The problem will be gaining adoption in process industries with long asset lives and well established methodologies and cultures.

In the past sensors and measurements have been specified without proper cost–benefit analysis with a high proportion of instruments being improperly installed and supported with proper OPEX budgets and defined ownership. A 2000 report for the water industry (UWKWIR Ref 00/PC/03/1) estimated that at much as 25 % of the installed instrument base could be removed with consequent cost advantages. It was suggested that a further advantage would be an increased confidence in the remaining measurements. Action was taken and there is a notable increase in attention for the measurements that directly affect performance.

It is generally accepted that all environmental measurements are driven by regulations or other socially driven requirements. The user is understandably reluctant to make measurements that might then be used against it or to make improvements in data availability or quality that will then be used to increase costs. Thus the market for new or improved measurement instruments or procedures will come from changes to regulatory or other controls; usually with long timescales and demonstration costs.

Both the users and the regulators become 'fixated' on a particular measurement practice or procedure when it has apparently 'worked' for a significant time. An example is the use of ISO 7027 as the standard for turbidity instruments. It has been well known for many years that polar plot of the scattered light gives much more information about the particles but ISO 7027 specifies a narrow beam and a specific wavelength. A further example is the dependence on residual chlorine as an indicator of 'safe' drinking water in the distribution system when it is known that cryptosporidium is unaffected by residual chlorine. Changing established standards or guidelines can take many years even when the enthusiastic volunteers have been found and motivated.

6.1.9 TECHNOLOGY

As already stated, in general the water and waste treatment and environmental industry is not limited by on-line measurement technology and it is likely that evolution of currently available techniques and products will be the dominant feature of market development.

A wide range of techniques and technologies are used for on-line measurement in the target industries including:

- various pressure techniques;
- various height measurements for open channel flow;

- magnetic and ultrasonic techniques for closed pipe flow;
- pH;
- conductivity;
- ion selective electrodes;
- chromatography;
- various light obscuration and scattering techniques for turbidity and particle analysis;
- various optical techniques to measure colour changes on paper strips or in cuvettes or other transparent containers;
- various uses of spectral changes in the UV/IR/NIR/VIS regions;
- biosensors;
- Clarke cell;
- Chemiluminescence.

There are four main technical drivers influencing developments for measurement technology: onboard intelligence; miniaturisation; serial communications, radio communications.

It is expected that separation techniques and spectral analysis will become available at the basic sensor level. Chromatographs, ion mobility spectrometers and mass spectrometers are already deployed in the field but still need specialist support. Data analysis is not yet robust enough for the casual user but, as shown below, the technology is available for this position to change.

It is expected that highly capable portable and hand-held instruments will have an increasing role, particularly when time, date, location data are automatically collected with each measurement. Such instruments will have simple operator-free downloading facilities with operator identity thus providing more secure audit trails than is generally available in this industry.

It is tempting to see an increased role for test kits but the need for secure audit trails needs to solved, if test kits are not to be limited in their use to investigative and process development roles.

Elsewhere the argument is made that there will be regulatory and social pressures for many more measurements to be made with lower and lower detection limits, but this poses a capital expenditure problem. However, there are technological drivers and enablers already being used and on the horizon which will permit autonomous sensors with much lower CAPEX and OPEX costs.

Sensor development and commercialisation has been a multidiscipline activity for decades but the increasing intensity of the need for a multidiscipline approach may well be a pacing issue. The higher education institutions do not provide multidiscipline courses and there is a notable reduction in networking and a reliance on internet searching becomes the norm. For example, blending microminiaturisation with the latest optical, biosensor, battery and radio technologies is not a trivial exercise.

6.1.9.1 Miniaturisation

Nanotechnology and microminiaturisation in its many and various forms has had huge publicity for a decade or more and is revolutionising sensors in other fields but the small scale of the water and waste treatment and environmental industries has hindered takeup. However the massive amounts of resource being deployed for Homeland Security in the USA and elsewhere will provide useful modules and enabling techniques. The pacing issue is the cost of the 'foundries' to 'machine' the structure; usually silicon but other materials, such as polymers, are beginning to be explored. We already know from the semiconductor industry that very large numbers of very low cost units are required to cover the overhead costs of a silicon foundry. At the miniature sensor level there are some promising polymer ideas that might help parts of the market needs.

An article in the November issue of Intech magazine (www.isa.org/intech) describes microChemLab, developed by National Nuclear Security Administration's Sandia National Laboratories. The suitcase-sized equipment uses separation techniques and detection in the gas phase with a surface wave detector. While it is not clear from the article what miniaturisation is involved the development direction is clear.

A brief article in the February issue of The Engineer announced a miniaturised ion mobility spectrometer that will make hand-held detectors for lung diseases and airport explosives security. While this is a gaseous phase measurement earlier phases of this work have been shown to be useful for land remediation investigations using a head space analysis approach.

Even a brief web search finds a very large amount of activity in the many forms on miniaturisation and many 'clubs'.

The idea of 'smart pebbles' – autonomous sensors that can be scattered over a battlefield and self network to discover 'hot spots' of some chemical or activity of interest – has been discussed for nearly a decade. Outside of the military secrecy curtain they do not yet exist but there is now talk of 'smart dust'.

6.1.9.2 Battery

Battery technology already permits sensors with an installed life of 5 years and some of the more ambitious companies are claiming a 10-year installed cost. This massively decreases the cost of installing sensors in areas where services are not already readily available.

6.1.9.3 Chemometrics

Chemometrics, modelling and other data manipulation techniques will have a major impact in the future. The collection and manipulation of widespread disparate and sparse, both temporally and spatially, data will allow the economic control of large wastewater collection, drinking water distribution systems and prescribed environmental areas. The derived inferential and surrogate measurements will also be useful for the larger processes. It is probable that the uncertainty demands for much of the collected and manipulated data will be low with consequent economic benefits.

Chemometrics and other data analysis techniques are developing rapidly and computing power costs and size also dropping rapidly. It is expected that sensing techniques, such as spectral analysis, will be enabled by the adoption of large in built libraries. For example the use of UV/visible optical techniques for such measurements as biological oxygen demand (BOD) have failed because of diurnal and other changes. The problem might use neural nets and site history to develop more robust information.

The use of libraries of spectral plots is well known in the laboratory when looking for a particular peak. Such libraries integral with the sensor are now affordable.

6.1.9.4 Spectrometry

On-line spectrometry has promised much for many years with convincing on-site demonstrations of correlations with a number of parameters, including chemical oxygen demand (COD) and total organic carbon (TOC). However, the downfall has been the site and time specific nature of the correlations and attendant calibrations. The water and waste treatment industry is notoriously subject to seasonal and diurnal changes and as the circumstances at the site changed so did the correlation. However, at least two companies seem to have solved the problems with one of them claiming 800 installed multi-analyte instruments over a 6-year period measuring parameters such as COD, BOD, TOC, dissolved organic carbon (DOC), NO_2, NO_3, benzene, toluene, xylene and many pesticides.

6.1.9.5 Data Collection

Collection from distributed sources has been a costly challenge but this is changing. The integration of legacy telemetry systems is expensive and sometimes not economically viable. The water industry has worked to come up with standards and the work to date may be seen at www.UKWIR.org.

The urgent need to track the contents of tank farms and silos has bred level instruments which communicate/power mininetworks of level devices that communicate directly with Ethernet networks. These instruments can send emailed level readings.

embedded if required in excel or access formats, at programmed intervals or on the initiation of an alarm.

Wireless systems are becoming very attractive alternatives to hard-wired systems. While innovative wireless enabled instruments proliferate, they tend to operate via different protocols, technologies and access points making mixed technology industrial schemes difficult to implement. The existence of legacy systems further complicates the position.

Wireless technology is already available as very small, low power, chip set for integration into sensors and deployment has started in the subject industries. Wireless sensor networks are easy to deploy at exceptionally low cost compared with hard-wired networks but there are issues regarding cost, reliability of service and mobility of the service providers. Worldwide, there is a wide range of standards and regulations for radio but, in general it is the low cost local, unregulated radio that offers most promise and may well be starting to achieve some critical mass. There are complications such as the availability of methods for onward transmission and connection to management systems and/or the Ethernet.

While such tools as Blue Tooth, Zigbee and Mesh networks are enjoying dramatic growth these are, in general, for short distance communications in controlled environments. Mesh networking, where modules provide radio pathways to a master point, as required, will reduce communication costs, particularly where support services are unavailable. This is the so-called smart pebbles or smart dust that is being proposed for battlefield monitoring.

Data rates are low or even very low but early trials with SMS were unpromising due to patchy coverage by service providers. Now that the mobile phone is ubiquitous these problems seem not to be a limiting factor. At least one international supplier of water meters is using SMS for data and meter health transmission with a modest extra cost. The extra is likely to trend to zero. The same manufacturer is not yet using the technology for water quality meters, probably because water quality meters are often in more remote locations. However, the massive increase in the use of mobile phones and other electronic gadgets suggests that SMS will become ubiquitous for low data rate transmission from environmental and similar data sources.

6.1.9.6 Test Kits and Portables

Test kits and portable instruments are a well established method of monitoring processes and environments. In some cases the data generated can challenge the total uncertainty chain of data derived from samples taken to a laboratory. However, merging any such data with on-line and 'sample to laboratory' data has been difficult since, for the most part, the data is operator skill dependent and records are hand written and subject to human errors. The new breed of personal data devices and mobile phones has allowed portable instruments to provide time, date and location stamp together with a reasonable idea of who took the data. This can all be sent immediately to an archive and multiple users or downloaded later back at base; in either case with a fairly strong audit trail.

Inevitably the question of the future application of *lab on a chip* technology arises. An article in the November issue of the magazine *Design Solutions* argues that:

> *Lab-on-Chip is a concept without a viable market entry point and although potential end users extol the possibilities that the technology will offer them, it appears that the very advantages that makes Lab-on-Chip so appealing – advanced technology for the masses and small throw away single use devices that can be used to move the laboratory processes out of the lab – is inhibiting its wide spread rollout.*

The article goes on to discuss the tens of thousands of devices that must be sold every month to make the technology commercially viable.

While it would be very useful to have multi-analyte portable instruments, this seems to be some way off. Elsewhere in this chapter the small size and fragmentation of the water and waste and environmental industries is discussed and the 'few per month' volumes mean that this industry will have to wait for the military and medical industry to provide adoptable modules and processes.

6.1.10 STRUCTURE OF THE MEASUREMENT INDUSTRY

As has already been noted the users face significant challenges and this is not helped by the small size and fragmentation of the instrument supply industry.

There are no reliable statistics available at the sensor level in the public domain. Even overall statistics are hard to find so it is important to gain a perspective for a particular sector by first considering the instrumentation industry as a whole. The numbers often quoted in magazines and articles are usually at too high a level of aggregation to be useful for the developer or supplier of specific measurement instruments or measurement packages in the subject industries. Furthermore, there are serious definition problems so that data from different sources cannot easily be compared. Individual specialists and many companies have high quality market data for their own specific use but this is often hard to track down and to obtain.

The magazine *Control* publishes a list of the world's top 50 instrumentation companies by sales (Figure 6.1.4). The data have been collected for many years by a highly respected group of individuals and are probably the most reliable sales figures available. However, it is an aggregate of all process instrumentation and includes process control as well as measurement for a very wide range of industries.

Even brief inspection shows that there are only seven or eight companies with more than \$1bn sales and then there is a very long tail of smaller companies. If the list were to be extended it would be seen that there are a very large number of worldwide companies with sub £10m sales. Usually it is the companies in this long tail that provide the innovation and the more successful are hoovered up by the large companies. Often, inspection of the large instrumentation companies shows that the more successful acquisitions continue to operate as self contained units.

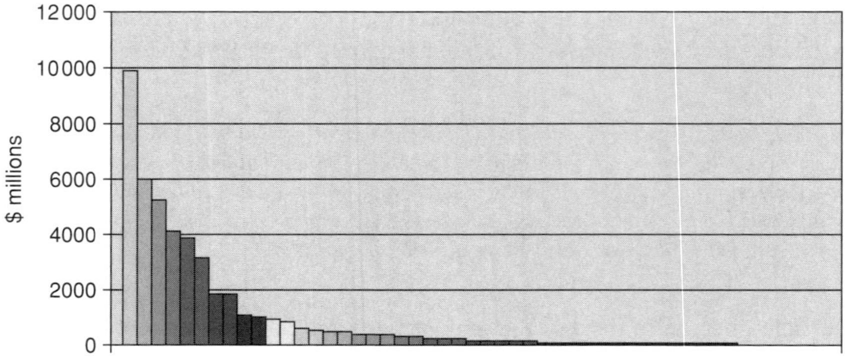

Figure 6.1.4 Top 50 instrumentation and process control companies worldwide

The water and waste treatment and environmental market is of modest size in global terms and highly fragmented at all levels (innovation, development, sales). Close inspection shows that often the apparently large companies are compartmentalised with highly protective and skilled fiefdoms.

A further complication, for the subject industries, is that there are surges in the sales of individual measurements caused by the introduction of a new regulation or other business drive. For example, there was a surge in the sales of nitrate analysers, with two or three major orders, to meet the Nitrate Vulnerable Zones regulations and a surge in sales of turbidity measurements to combat the threats of cryptosporidium.

At the APACT 05 Conference, Analytical Solutions for the Process Industries, Professor R. S. Benson FREng, an ABB Associate, gave his view that the process analytics market had the following characteristics:

- approximately $1800m/year;
- many suppliers;
- even the largest suppliers only around $210m;
- sampling systems, systems integration and maintenance approximately 50 % of the turnover;
- spread across a wide range of analysers;
- chromatography (gas and liquid);
- spectrometry (FTIR/NIR, mass, Raman);
- photometry (UV/NDIR/VIS);
- thermal conductivity (TC);
- paramagnetic oxygen (PMD);
- flame ionisation (FID);

- water analysis (pH, conductivity);
- physical properties;
- Innovation spend $3–5m/year in even the largest companies;
- average innovation spend <$200k/analyser year;
- custom applications/domain knowledge;
- analyzer network components;
- integrated analyser systems (shelters);
- aftermarket services.

Professor Benson thus concludes that not only is the market for process analytics small in global terms it is, itself, highly fragmented. Given Professor Benson's background with ICI it is probable that his perspective is skewed toward the oil and chemical industry but this does not reduce the usefulness of the view.

The approaching 2006 deadline for the generation of reportable data for the Urban Wastewater Directive (see Chapter 1.1) is particularly expected to provide a strong impetus for growth. In terms of product types, the market is dominated by pH, ORP and conductivity meters (on-line and portable/hand-held) (revenue share of 43.6 % in 1999), followed by BOD, COD and TOC meters (15.1 %) and ammonium, nitrate and phosphate analysers (13.1 %).

It is expected that the relative importance of the markets for pH, ORP and conductivity meters, and for DO meters in the total market will dwindle in future years. The market segments promising the most dynamic growth potential include BOD, COD and TOC analysers, nutrient analysers and multiparameter systems. In future years, a growing demand for automation is likely to encourage growth in the market for sophisticated on-line analysers.

Process and environmental measurement is a very minor part of capital expenditure on projects and has therefore not been seen as a priority for user's management. However, pressure from the Regulator, increases in the cost of power and the WFD are all bringing management's attention to the need for, and the quality of, data. The WFD has added clarity to the need to consider the overlap between process control considerations and environmental requirements. The WFD is also driving the need to provide relevant data, with a known and reported quality in a form understood by the user of that data at the point of use.

- End of pipe measurement is no longer good enough.
- It is the data and the data quality at the point of presentation, to the many and various users, that matters.
- Data must be turned into relevant information.

6.1.11 INFORMATION SOURCES

Even for a narrow field such as sensors for aqueous measurements there is an information overload with excellent information dispersed in a myriad of articles, conference, papers, books and manufacturers support documentation. The Further reading section of this chapter identifies some worthwhile reading but the list is representative only and falls dramatically short of being balanced or thorough.

The SWIG (Sensors for Water Industry Group) web site (www.swig.org.uk) has a list of past and future SWIG Workshop titles and the presentations from past SWIG Workshops going back a number of years. Even a quick viewing gives an idea of some of the hot measurement topics over the period covered.

Suggestions such as the US Environmental Protection Agency's Review on EWS (Emergency Warning Systems) have a very specific agenda and therefore Sections 5, 6 and 7 on sensor technologies, although useful and thought provoking, need to be used with caution.

The RUNES (Reconfigurable Ubiquitous Networked Embedded Systems) suggestion is because of the technological drive for automatic data collection and thus the need for sensors to be developed with appropriate structures and interfaces. RUNES is a European Commission funded project which started in September 2004 with the main objective to enable the creation of large-scale, heterogeneous networked embedded systems, which interoperate and adapt to their environments. Networked embedded systems merge a wide range of information technologies that spans hardware, systems software, networking and programming methodologies. Europe has leading positions in sectors where embedded technologies are central (avionics, automotive, space, consumer electronics, telecom devices, energy distribution). In these sectors there are certainly many future alternatives, however, the process of technology roadmapping helps narrow down the field of possible solutions to those more likely to be pursued. Most of the issues and technical requirements are not orthogonal and as such trade-offs depend on applications.

The RUNES technology roadmaps are a synthesis of the knowledge and views collected between October 2004 and April 2005 across Europe. The roadmaps reflect the following five major market classifications for networked embedded systems:

(1) building and home automation;

(2) medical care;

(3) disaster management and emergency response;

(4) automatic meter reading;

(5) industrial control and automation.

The RUNES technology roadmaps are designed to communicate the relevance and potential of networked embedded systems to a number of different stakeholders:

scientists, technologists, industry leaders, funding bodies, research students and the general public. This is done by portraying the structural relationships, among technology and applications and how organisational and social issues affect them. The focus of the roadmaps is approximately 10 years although timescales for technological progress are notoriously difficult to predict.

It is hoped that the RUNES technology roadmaps will identify the research breakthroughs necessary, which can be translated, into successful products and services. The technology road-mapping process encourages discussion and debate throughout the community about the requirements for success.

SENSCOPE is an Intersect Faraday Strategic Scoping Study on the Convergence of Enabling Technologies: Microsystems, Nanotechnology and Sensors under the DTI Basic Technologies for Industrial Applications (BTIA) Programme, Technology Transfer Project in July 2004. This Report and Roadmap has useful descriptions of sensor technologies and is recommended reading. (Copyright is held by SIRA Ltd and contributors.)

The UK Micro and Nanotechnology (MNT) Network has been established by the DTI and the 12 Regional Development Agencies and Devolved Administrations working together, to provide a market-oriented focus for the facilities, people and organisations engaged in Micro and Nanotechnologies in the UK. The Network is helping to lower entry barriers and drive the widespread market development and exploitation of these technologies – building a prosperous, world-class MNT sector in the UK.

The benefits of the Network include:

- Improving the access to a critical mass of world-class knowledge and facilities in the UK and overseas.
- Providing a catalyst to drive the specialist training and development of people to fuel growth in these emerging markets.
- Facilitating the integration of the complete supply-chain and better use of facilities, to take 'blue-skies research' through to high-volume and high-value-added manufacture by UK companies.
- Identifying the demand for, and working with stakeholders to provide the new facilities needed to build the UK MNT capability.
- Encouraging a coordinated approach to applied research programmes and business support.
- A focus for information and advice to provide the support for UK business to drive innovation and new products.

The general area of micro and nanotechnologies will certainly provide inputs to new generations of sensors and it is possible that there will be public fears to some of the longer term possibilities. Terms such as Smart pebbles and Smart dust have been used for some years in Research and development circles and anything which might prejudice the use of possibly very low cost autonomous, self networking sensors must be taken into consideration. In preparation for FP7 the EU is collecting ideas on nano(eco)toxicology and an EU published nanotechnology action plan is available (http://www.cordis.lu/nanotechnology/actionplan.htm). The objective is to provide a connection between all those working on nanotechnology so as to enhance cross-fertilisation of ideas and minimise duplication of effort.

The UK Water industry statistics are available via Water UK's WaterFacts which is available from www.water.org.uk WaterFacts lists all the water companies and their assets and so is an invaluable way of estimating the maximum possible number of any particular application. That, of course, does not mean that any particular water company would take such a simplistic approach to dealing with an issue.

6.1.12 CONCLUSIONS

There is an increasing need for data in the water and waste and environmental industries and the actual needs, if the precautionary principles are fully applied, are not economically and socially achievable. There is already a move towards merging what data are available and then 'plugging the holes'. However, there are serious challenges since available data are in many different forms and quality, often with low or no comparability. Merging such data is a non-trivial exercise. Not only is data generation multidiscipline but the quality is dependent on multidepartment, multicompany cooperation.

There is a need for some fascinating European level cooperation, training and generation of codes of practice. But before that there is a need for more discussion and debate so that the many participants can become familiar with many facets that need to be considered.

FURTHER READING

Taming contaminants, John Fawell. *Water & Waste Treatment* 2005, **48**(10), 15–19.

Flying by wireless – is it safe? Steve Russell, WRc. *Water & Waste Treatment* 2005, **48**(10), 29–30.

Less wire, more speed, John Hother, Proneta. *Water & Waste Treatment* 2005, **48**(10), 34–37.

Year Book. Institute of Measurement and Control.

EPA Review on EWS. Technologies and Techniques for Early Warning Systems to Monitor and Evaluate Drinking Water Quality: State-of-the-art Review. US EPA, December 2005 (http://www.epa.gov/nhsrc/pubs/reportEWS120105.pdf).

What is the future for Lab-on-chip?, Alec Reader, Innos. *Design solutions* 2005, **November**, 20.

Portable chemistry labs will test water toxins. *Intech* 2005, **November** (www.isa.org/intech).

Easy and cost efficient on-line monitoring for water treatment plants. *International Environmental Technology* 2005, **November/December**, 36–37 (www.iet-pub.com).

Senscope. Strategic Scoping Study on the Convergence of Enabling Technologies: Microsystems, Nanotechnology and Sensors, July 2004. SIRA Ltd for a DTI Technology Transfer Project.

Proceedings of the RUNES Industry Forum. Sixth Framework Programme on 'Information Society Technologies' (www.ist.runes.org/publications.html).

Performance Standards and Test Procedures for Portable Water Monitoring Equipment: Temperature; pH; Conductivity; Dissolved Oxygen; Turbidity; Ammonia; Nitrate; Nitrite; Orthophosphate; Chlorophyll a. Environment Agency, Version 1, January 2006.

Performance Standards and Test Procedures for Continuous Water Monitoring Equipment: Turbidity; pH; Ammonia; COD; TOC; Dissolved Oxygen; Total Phosphorous; Nitrates; Total Oxidised Nitrogen. Environment Agency, Draft V2, June 2005.

6.2
Training

Jean-Luc Cécile and **Evelyne Touraud**

6.2.1 INTRODUCTION

The treatment of wastewater, whether industrial or urban, and its impact on the natural environment is an important theme in training, both for courses leading to a qualification and for continuous professional training. However, most courses cover treatment processes rather than their monitoring. The most complete degree courses, which are described later, include this monitoring aspect, particularly by dealing with the usual methods which comply with the requirements of the various regulations: self-monitoring and self-control of facilities. The impact on the natural environment is reduced to an examination of the evolution of ecosystems. Approximate

Wastewater Quality Monitoring and Treatment Edited by P. Quevauviller, O. Thomas and A. van der Beken

coverage of water bodies is achieved through a systematic follow-up of surface water, underground water and seawater in the framework of national and international programmes. As a general rule, the presentation is structured around three topics:

- Water chemistry, in particular physical and chemical properties and forms of pollution.
- Existing regulations and their application.
- Parameters for water quality characterization and relevant methods to measure them: standard methods, alternative methods, bioindicators and biosensors.

These training programmes consist of a series of conferences and studies of applications. The didactic approach is based on practical cases. Participants' specific cases are also often used. Training courses include practical work, which is conducted in pilot units or in the field, on operational sites. They are intended for technical agents, supervisors and engineers responsible not only for the running of wastewater treatment installations and facilities but also for quality assurance, environmental and safety services. In addition, specific training is given to staff working in public administration (appointment to a new post or training leading to a qualification).

The length of the course varies from 1 to 5 days (in continuous professional training) or from 10 to 80 h for training leading to a qualification.

6.2.2 TYPE OF TRAINING AND TRAINING INSTITUTES

Considering training, there are two approaches: to offer students a course leading to a vocational qualification; or to provide additional training for active professionals. In the first case, student training can be offered at three levels, for operating technicians, supervising technicians and engineers. University programmes train students for positions in administration. In the second case, the same three levels exist, although they are not so clearly separated (nonhomogeneous staff). In this section, these types of training are discussed briefly, along with the main training providers.

In France, two main institutes, the International Office for Water, OIEau, based in Limoges (www.oieau.fr) and the Instrumentation and Control Institute for Industrial Processes, IRA, based in Arles (www.poleira.com) provide professional training in the water field.

The OIEau is intended for the managers of water and sewerage system services, manufacturers, designers, constructors and fitters of work and equipment, and administrative bodies in charge of aquatic resources, implementing regulations, technical assistance and control services. The IRA provides industrial personnel with training in the field of measurement and control, logics and industrial data processing and on-line analysis dedicated to water management. Both institutions provide technological and educational platforms that enable real-life practical training to be performed.

Engineering training in French '*Grandes Ecoles*' is based on a rare blend of Science, Engineering, Management and Human Sciences combined with project work for industry. The Ecole des Mines d'Alès, EMA (www.ema.fr) offers students a specialization in Environmental Engineering, including the industrial environmental regulation context, water analysis and monitoring, and wastewater treatment processes. The EMA provides professional training in environmental chemistry and water metrology. These courses are specifically intended for administrative bodies in charge of water quality control. The Ecole Nationale du Gènie de l'Eau et de l'Environnement in Strasbourg, ENGEES (www-engees.u-strasb.fr) offers specialized master and bachelor degrees in water management and treatment. Polytech Montpellier (www.polytech.univ-montp2.fr) trains general engineers in the water field through its Water Science and Technology Department.

Elsewhere in Europe, water monitoring is an important issue but is not specifically taught, except perhaps in the UK. The Community European Management School, CEMS (www.cems.org) delivers a course which runs over 3 days and is aimed specifically at those people working within the emissions monitoring field such as Environmental Managers, Stack Testers and Environment Agency Inspectors. The course provides a broad view of the subject including continuous flow monitoring and continuous methods for particles. Other issues covered include quality assurance for automated measurement systems and information on forthcoming CEMS legislation. The course consists of 60 % theory and 40 % practical work. Students work on a comprehensive range of *in-situ* analysers and systems for practical work, and networked PCs with flat screen monitors for all theoretical work.

In North America, we can cite the example of the American Water Works Association, AWWA (www.awwa.org). This association offers a training programme for continuous water monitoring. The seminar lasts 1 day and its objective is to respond to future users considering beginning or expanding an on-line monitoring system. The seminar schedule goes from the definition of on-line monitoring in the current water supply and treatment context to the organizational issues involved. It includes the principle components of on-line monitoring systems, evaluating the costs and benefits of on-line monitoring, general equipment selection guidelines, an overview of available on-line instruments and data handling issues.

Finally, another example can be given in New Zealand: the National Institute for Water and Atmospheric Research, NIWA (www.niwa.co.nz). The science campus focuses on various disciplines including aquaculture, climate, freshwater, coastal, marine, fisheries and atmospheric research. It also has a number of field offices primarily for the collection of environmental data. The institute offers various training courses in the field of water monitoring such as Optimizing Data Quality from Environmental Monitoring Stations, and Successful Use of Handheld and Continuous Water Quality Sensors/Loggers.

6.2.3 WATER CHEMISTRY

6.2.3.1 Objectives

The aim of these courses is to study the physical and chemical properties of water in order to improve its management. The implementation of various methods is covered: laboratory methods (standards), alternative methods (rapid and on-line). Finally, courses deal with the interpretation and exploitation of results.

6.2.3.2 Content

Courses consist of an introduction in which the natural water cycle is presented in order to understand the problems raised by its exploitation. Water resources are described. Because of their vulnerability, regulated management is required, combined with means of control. On the basis of the various uses described in the urban water cycle, including industrial activities, the causes and consequences of the degradation of water body quality are presented. Then, water chemistry is introduced through various concepts concerning atomic and molecular structure, covered at differing levels of detail, in order to provide a better comprehension of reactional mechanisms. The main characteristics of water are recapped. The nature of the substances found in water: gas and dissolved salts, ions, nonionic molecules, is addressed in detail. Reactions are grouped into families:

- Oxido–reduction reactions;
- acid–base reactions;
- complexation reactions;
- reactions combining oxido–reduction and acid–base;
- reactions combining acid–base and complexation;
- reactions combining oxido–reduction, acid–base and complexation.

Sometimes, the application of metrology/quality assurance to water quality analysis is examined, which is illustrated by the international course series on quality assurance for chemical analysis (QUACHA) organized as an *ad hoc* activity in various languages by the European Commission in the years 1998–2002, but the tendency is to minimize, even neglect this aspect which, nevertheless, is an essential aspect of the measurement approach. The main characteristics of a measurement are described: accuracy, reliability, uncertainty, range of measurements, validation and limits of quantification and response time. The physico-chemical measurements are listed: water level, flow measurement, pressure, basic and specific parameters. Sampling conditions are examined in detail. An important part is dedicated to the principles and methods applicable in laboratory. The aim is to present existing principles and standards rather than commonly used procedures. Measurements carried out using alternative methods, the so-called micromethods, and on-line measurements using sensors, probes or industrial analysers are very rarely covered. The presentation and exploitation of results are often neglected.

The IRA institute offers professional training modules, from basic to advanced level, dedicated to measurement and control and which cover the major points mentioned above. Several courses are available: basic water chemistry (4 days), self-monitoring of treatment processes (4 days), operating of water sensors and analysers (4 days) and industrial environment: water and wastewater (4.5 days). Similarly, in the framework of a course in environmental chemistry (2.5 days), the EMA covers water chemistry reactions and the fate of pollutants through real and practical applications.

OIEau, together with the National Centre for the Water Profession (CNFME), provides professional training courses focused on the management of water resources and public services, municipal treatments and networks (potable water and sewage) and the processing and decontamination of industrial wastewater.

Finally, some companies which provide advanced analytical systems and technical support for water quality testing also deliver specific training courses. For example, the Hach Technical Training Center (www.hach.com) offers several workshops (2–3 days) dedicated to water and wastewater training. Some of them concern environmental metrology with the testing techniques and systems available for quality measurement in applications including drinking water, wastewater, environmental water and industrial water. Other are dedicated to on-line monitoring, especially chlorine and turbidity process analysers and their maintenance (calibration, verification and troubleshooting techniques).

6.2.4 REGULATIONS AND THEIR APPLICATION

6.2.4.1 Objectives

The main aim is to study the existing regulatory context and the obligations upon project managers. Another objective is to understand what parameters are to be measured (flow, quality parameters). An important part is dedicated to the exploitation and interpretation of results. However, conformity criteria are rarely addressed.

6.2.4.2 Content

A very considerable part is dedicated to the presentation of the existing regulatory context applicable to the management of urban and industrial wastewaters. The problems linked to industrial wastewater running into collective urban networks (subject to authorization and agreements) as well as specific cases of industrial sites with potential risks for the environment are examined. The principle of self-monitoring and self-control of installations is often proposed. The methodology and procedures used to measure the quantity and the quality of effluents: flow, quality parameters: total suspended solids (TSS), chemical oxygen demand (COD), BOD_5, global nitrogen, total phosphorus, specific parameters for nonurban effluents are given in detail. As mentioned above, the presentation and interpretation of results are often not covered in much detail. The same applies to the conformity of urban or industrial water treatment processes.

6.2.5 PARAMETERS, METHODS AND PROCEDURES FOR WATER QUALITY CHARACTERIZATION

6.2.5.1 Objectives

The main objective is to study the metrology applicable to water measurements, in particular the rules governing implementation and the existing standards. In addition, courses examine how measurement equipment is operated and maintained. Finally, the basis for the organization of a 'Metrology and Measurements' Department is given.

6.2.5.2 Content

After giving background information on the chemistry of water, the main basis of the metrology applicable to water measurements is explained. A pragmatic or logical approach structured around four main points is used: Why perform measurements? What are the parameters to be measured? Where are the measurements to be made? How are the measurements to be made? Applicable standards are presented and the conditions for their use analysed. Courses are mainly concerned with laboratory methods and procedures, much less with alternative methods. Only rapid methods or micromethods are beginning to be taken into account. The equipment and the conditions of their use for measuring physical parameters, pressure, level and flow (pipes, open channels) are examined in detail. Various modules cover the measurement of basic physico-chemical parameters (nitrogen and phosphorus compounds, hydrocarbons, heavy metals, pesticides). The role of biosensors and biodetectors is rarely studied. The organization of a metrology laboratory for the operation and maintenance of sensors and analysers is made up of several parts:

- control and verification operations;
- calibration;
- quality assurance is sometimes carried out according to standard EN17025.

As an example, the parameters for water and wastewater monitoring are presented in the water metrology course (2.5 days) delivered by the Ecole des Mines d'Alès. Hands-on practice is performed on real samples, on site, using various test methods and portable field instruments. The exploitation and validation of results are studied.

6.2.6 CONCLUSION

As a conclusion, it can be said that wastewater monitoring courses are of short duration, as components of a larger whole. There is no occupation dedicated to wastewater monitoring, or to other types of water (drinking water, natural waters). There is clearly a pressing need for this work to be duly recognized, combining a traditional approach (specimen sampling, laboratory measurements) with alternative methods, in particular on-line methods. Some organizations are beginning to acknowledge this need and to offer specialized training.

In France, on-line analysis has been adopted as a part of industrial self-monitoring to monitor processes with a view to optimizing performance. Unlike atmospheric controls, however, alternative methods and on-line analysis for the purposes of effluent quality and environmental impact have not gained full acceptance. Thus, alternative methods remain to some extent outside the mainstream, and few organizations offer the training required.

Index

Note: Page numbers in **bold** refer to tables and those in *italics* refer to figures

Wastewater Quality Monitoring and Treatment Edited by P. Quevauviller, O. Thomas and A. van der Beken